Electricity from Sunlight

Electricity from Sunlight

Photovoltaic-Systems Integration
and Sustainability

Second Edition

Vasilis Fthenakis
Columbia University
USA

Paul A Lynn
formerly
Imperial College London
UK

Registered Offices
John Wiley & Sons, Inc., 111 River Street, Hoboken, NJ 07030, USA
John Wiley & Sons Ltd, The Atrium, Southern Gate, Chichester, West Sussex, PO19 8SQ, UK

Editorial Office
The Atrium, Southern Gate, Chichester, West Sussex, PO19 8SQ, UK

For details of our global editorial offices, customer services, and more information about Wiley products visit us at www.wiley.com.

Wiley also publishes its books in a variety of electronic formats and by print-on-demand. Some content that appears in standard print versions of this book may not be available in other formats.

Library of Congress Cataloging-in-Publication Data

Names: Fthenakis, Vasilis M., author. | Lynn, Paul A., author.
Title: Electricity from sunlight : photovoltaic-systems integration and
    sustainability / by Vasilis Fthenakis, Paul A Lynn.
Description: Second edition. | Hoboken, NJ : John Wiley & Sons, 2018. |
    Includes bibliographical references and index. |
Identifiers: LCCN 2017040584 (print) | LCCN 2017047711 (ebook) | ISBN
    9781118963777 (pdf) | ISBN 9781118963784 (epub) | ISBN 9781118963807 (cloth)
Subjects: LCSH: Photovoltaic power generation. | Solar cells. | Solar energy.
Classification: LCC TK1087 (ebook) | LCC TK1087 .F764 2018 (print) | DDC
    621.31/244–dc23
LC record available at https://lccn.loc.gov/2017040584

Cover Design: Wiley
Cover Images: (Top image) © skeijzer/Gettyimages;
(Left to right images) © Gyuszko/Gettyimages; © coddy/Gettyimages; © VioNet/Gettyimages;
(Bottom image) © Kativ/Gettyimages

Set in 10/12pt Warnock by SPi Global, Pondicherry, India
Printed and bound in Malaysia by Vivar Printing Sdn Bhd

10 9 8 7 6 5 4 3 2 1

Contents

About the Authors

Vasilis M. Fthenakis is the founder and director of the Center for Life Cycle Analysis (CLCA), Department of Earth and Environmental Engineering, Columbia University, New York, USA. He is also a senior scientist emeritus at Brookhaven National Laboratory (BNL) where he conducted research for 36 years and directed the National Photovoltaics (PV) Environmental Research Center and several international networks. Dr. Fthenakis is the coauthor and editor of four books and about 400 scientific publications on topics at the interface of energy life cycles and the environment. Currently, he is leading research on solar desalination, energy systems modeling, life-cycle analysis, and PV recycling.

Paul A. Lynn obtained his B.Sc. (Eng) and Ph.D. degrees from Imperial College London, UK. After several years in the electrical/electronics industry, he lectured at Imperial College and the University of Bristol, latterly as Reader in Electronic Engineering. In 1993 he became the founding managing editor of the prestigious Wiley journal *Progress in Photovoltaics* and held this position for 14 years. As a retired academic, Dr. Lynn's continued interest in renewable energy has led to a trilogy of Wiley books and, in his spare time, three solar-powered boats. He is the author of nine other books and numerous technical papers and articles.

Foreword

Just over 40 years ago, the idea that solar power could make the leap from powering satellites in space to powering the planet was the vision of only a few people brought together by the shock of the first oil embargo. Today, almost everyone sees solar panels on a daily basis. *Electricity from Sunlight: Photovoltaic-Systems Integration and Sustainability* describes the journey from the sun's use in earliest times to tomorrow's continuum of recyclable materials employed in producing energy from a manufactured good, rather than by consuming Mother Earth's resources. The exponential increase in applications powered by solar is shown to be directly tied to the predictable cost reduction experienced through economies of scale and continuous technology performance improvements. The book delivers insights that are both inspirational and quantitatively informative by thoroughly documenting many of the pathways for achieving scale that have been established upon foundations of science and experience proven over decades.

Electricity from Sunlight describes the quantification and consideration given to every dimension of the solar value chain: how feedstocks are prepared, how factories consistently operate, how manufacturing environmental health and safety is planned and audited, how life-cycle benefits are quantified, and how our planet wins at the same time business wins. We now have an existence proof. It is possible to have sustainable development that is economically sustainable.

Sunlight is freely distributed across our planet. The combination of information technology with low-cost solar technology has established the foundation for an irreversible growth connecting the dots between the supply of cost-competitive electricity and the need for power everywhere. The book is a roadmap to understanding the cornerstones upon which an industry has been framed so that ideas for totally optimizing that early vision can be explored and accelerated to meet real-world needs. Here in one place is the guidebook to a clean energy future.

Charlie Gay, Ph.D.
Director, Solar Energy Technology Office
U.S. Department of Energy
Washington, DC

Preface to the First Edition

Photovoltaics (PV), the 'carbon-free' technology that converts sunlight directly into electricity, has grown dramatically in recent years. Unique among the renewable energies in its interaction with the built environment, PV is becoming part of the daily experience of citizens in developed countries as millions of PV modules are installed on rooftops and building facades. People living in sunshine countries will increasingly live in solar homes or receive their electricity from large PV power plants. Many governments around the world are now keen to promote renewable electricity as an essential part of the 21st century's energy mix, and PV is set for an exciting future.

This book is designed for students and professionals looking for a concise, authoritative, and up-to-date introduction to PV and its practical applications. I hope that it will also appeal to the large, and growing, number of thoughtful people who are fascinated by the idea of using solar cells to generate electricity and wish to understand their scientific principles. The book covers some challenging concepts in physics and electronics, but the tone is deliberately lighter than that of most academic texts, and there is comparatively little mathematics. I have included many colour photographs, gathered from around the World, to illustrate PV's huge and diverse range of practical applications.

In more detail, Chapter 1 introduces PV's scientific and historical context, suggests something of the magic of this new technology, and summarises its current status. The treatment of silicon solar cells in Chapter 2 includes material in semiconductor physics and quantum theory, described by a few key equations and supported by plenty of discussion. The new types of thin-film cell that have entered the global PV market in recent years are also introduced. Chapter 3 covers the characteristics of PV modules and arrays, discusses potential problems of interconnection and shading, and outlines the various types of system that track the sun, with or without concentration. The two major categories of PV system, grid-connected and stand-alone, provide the material for Chapters 4 and 5 respectively, and Chapter 6 concludes the story with some of the most important economic and environmental issues surrounding PV's remarkable progress.

Photovoltaic technology seeks to work with nature rather than to dominate or conquer it, satisfying our growing desire to live in tune with Planet Earth. I trust that this book will inspire as well as inform, making its own small contribution to an energy future increasingly based on 'electricity from sunlight'.

Paul A. Lynn
Butcombe, Bristol, England
Spring 2010

Preface to the Second Edition

The eight years since *Electricity from Sunlight* first appeared have witnessed a remarkable development in the field of renewable energy—the explosive growth of photovoltaics (PV). Global installed capacity, which reached about 40 gigawatts (GW) in 2010, is approaching 400 GW. It is even possible to imagine 1000 GW by 2020—a thousand times greater than at the start of the new millennium.

PV's meteoric rise is due to a combination of factors: at the technical level, steady improvements in solar cells, modules, and systems; at the international political level, an ever-increasing awareness of the threats posed by global warming; and at the production level, a dramatic reduction in costs as PV exhibits the well-known "learning curve" of manufactured products experiencing exponential growth.

This new edition pays special attention to issues raised by PV's extraordinary progress, especially the integration of large amounts of solar electricity into existing grid networks and its sustainability in terms of markets, resources, and life-cycle impacts. Chapters 4, 5, and 7 have been revised and greatly expanded, and a brand new chapter on PV manufacturing has been inserted. The other chapters have all been updated.

We hope the new edition will act as an essential primer for entrants to the PV industry needing an up-to-date appreciation of the subject. It also offers a unique treatise on the sustainability of emerging transformative technologies, making it valuable to system analysts and energy policy strategists. Last but not least, we have included end-of-chapter questions and problems to support instructors and the ever-increasing number of college and university students taking courses in renewable energy and PV.

Vasilis Fthenakis
Columbia University and Brookhaven
National Laboratory, USA

Paul A Lynn
formerly Imperial College London, UK
Spring 2018

Acknowledgment to the First Edition

There is nothing like a good set of pictures to illustrate PV's extraordinary progress and I have enjoyed enlivening the text with colour photographs obtained from around the world. I hope that my readers will regard them as an important and inspirational aspect of the book. They come from widespread sources and I have received generous cooperation from people in many organisations and companies who have provided copyright permissions and, in several cases, suggested stunning alternatives to illustrate particular topics.

I am especially grateful to the two international organisations that have provided the lion's share of the photographs reproduced in this book:

1. The European Photovoltaic Industry Association (EPIA)

2. The International Energy Agency Photovoltaic Power Systems Programme (IEA PVPS)

3. Additional acknowledgements

I am also grateful to a further group of companies and organisations that have agreed to their photographs appearing in this book, and for help received in each case from the named individual:

Amonix Inc. (Nate Morefield)
3425 Fujita Street, Torrance, CA 90505, USA

Boeing Images (Mary E. Kane), USA
www.boeingimages.com

Dyesol Ltd (Viv Tulloch)
P.O. Box 6212, Queanbeyan, NSW 2620, Australia

Dylan Cross Photographer (Dylan Cross), USA
dylan@dylancross.com

First Solar Inc. (Brandon Michener)
Rue de la Science 41, 1040 Brussels, Belgium

Isle of Eigg Heritage Trust (Maggie Fyffe)
Isle of Eigg, Inverness-shire PH42 4RL, Scotland

Padcon GmbH (Peter Perzl)
Prinz-Ludwig-Strasse 5, 97264 Helmstadt, Germany

Steca Elektronik GmbH (Michael Voigtsberger)
Mammostrasse 1, 87700 Memmingen, Germany

Tamarack Lake Electric Boat Company (Montgomery Gisborne)
207 Bayshore Drive, Brechin, Ontario L0K 1B0, Canada

Wind and Sun Ltd (Steve Wade)
Leominster, Herefordshire HR6 0NR, England

The publishers acknowledge use of the above photographs, which are reproduced by permission of the copyright holders, and individually acknowledged where they appear in the text.

The use of three photographs from the NASA website, and several pictures from the Wikipedia website is also gratefully acknowledged.

The author of a comparatively short but wide-ranging book on PV – or any other technology – inevitably draws on many sources for information and inspiration. In my case several longer and more specialised books, valued companions in recent years, have strongly influenced my understanding of PV and I freely acknowledge the debt I owe their authors, often for clear explanations of difficult concepts that I have attempted to summarise. These books are included in the chapter reference lists, and you may notice that a few of them appear rather frequently. I have tried to give adequate and appropriate citations in the text.

My previous books on electrical and electronic subjects have been more in the nature of standard textbooks, illustrated with line drawings and a few black-and-white photographs. When the publishers agreed to my proposal for an introductory book on PV containing full-colour technical drawings and photographs, I realised that a whole new horizon was in prospect, and have enjoyed the challenge of trying to choose and use colour effectively. The photographs, many of them superb, have already been mentioned. It has also been a great pleasure to work closely with David Thompson, whose ability to transform my sometimes rough sketches into clear and attractive technical drawings has been something of an eye-opener.

For nearly 15 years my main involvement with PV was as Managing Editor of the Wiley international journal *Progress in Photovoltaics: Research & Applications*. Among the many editorial board members who gave valuable advice over that period, I should particularly like to mention Professor Martin Green of the University of New South Wales (UNSW), world-renowned for his research and development of silicon solar cells; and Professor Eduardo Lorenzo of the Polytechnic University of Madrid (UPM), whose encyclopaedic knowledge of PV systems and rural electrification was offered unstintingly. It was both a privilege and a pleasure to work with them for many years. And although any shortcomings in this book are certainly my own, any merits are at least partly due to them and other members of the board.

Finally I should like to thank the editorial team at Wiley UK for their enthusiasm and guidance during this project. They, and others, have eased into publication this account of an exciting new technology that magically, and quite literally, produces electricity from sunlight.

Paul A. Lynn

Acknowledgment to the Second Edition

Following Paul Lynn's lead in including color photographs in the first edition, illustrating PV's beauty and extraordinary promise, I have added plenty more in this new edition. For the new figures and valuable exchanges, I am grateful to the following individuals and organizations: Mahesh Morjaria, First Solar; Pierre Verlinden, Trina; William Shafarman, University of Delaware; Raed Bkayrat, First Solar; Jason Baxter, Drexel University; Jeff Britt, Global Solar; Steven Hegedus, University of Delaware; Christian Den Heijer, Hukseflux; Vahan Garboushian, Arzon Solar; Ronny Gløckner, Elkem Solar; Markus Gloeckler, First Solar; John Lushetsky, US-DOE; Vijay Modi, Columbia University; Craig Murphy, MEMC Electronic Materials; John Phufas, JFK Solar Enterprises; Thomas Shilling, Photon; Parikhit Sinha, First Solar; and Arnulf Jäger-Waldau, European Commission Joint Research Centre.

The organization and content of this second edition owe much to my teaching at Columbia University, and I am grateful to many undergraduate and graduate students whose interest in solar energy and the environment created this publishing opportunity. Especially helpful were three of my former doctorate students who are now pursuing auspicious careers in renewable energy: Marc Perez, currently with Clean Power Research; Rob van Haaren, currently with First Solar; and Thomas Nikolakakis, currently with the International Renewable Energy Agency. I also wish to acknowledge my former postdoctoral associates Annick Anctil, currently with Michigan State University; Jun-Ki Choi, currently with University of Dayton; Chul Hyung Kim, currently with Ford; Damon Turney, currently with City College of New York; and Wenming Wang, currently with First Solar. Their significant contributions are widely cited. Thanks are also due to my current graduate students Samet Ozturk and Zhuoran Zhang for their help with problem solving and elegant schematics.

I have also benefitted immensely by interacting during my tenure at Brookhaven National Laboratory with individuals leading the energy transition to solar in the United States, including Mike Ahearn, Charlie Gay, Larry Kazmerski, and Ken Zweibel with whom I shared the vision of the *Solar Grand Plan*.

This second edition is dedicated to my wife Christina Georgakopoulos Fthenakis, an accomplished skin research scientist, for her love and encouragement. We both hope that *Electricity from Sunlight: Photovoltaic-Systems Integration and Sustainability* succeeds in clearly defining a pragmatic solution to the risks that climate change and conflicts for energy and water resources present to our children and the generations that follow.

It has been a pleasure to work with Paul Lynn on this new edition, a transatlantic link appreciated by us both.

Vasilis M. Fthenakis

About the Companion Website

Don't forget to visit the companion website for this book:

www.wiley.com/go/fthenakis/electricityfromsunlight

There you will find valuable material designed to enhance your learning, including

1) Solution materials
2) Software program

Scan this QR code to visit the companion website.

1

Introduction

1.1 Energy and Sustainable Development

This book is written by a chemical engineer and an electronic engineer who believe that continuing to burn fossil fuels for energy is not sustainable and that a transition to renewable energy is feasible. Let us start the discussion by reflecting on what sustainability of certain development is all about. "Sustainable development" is characterized as the "development that meets the needs of the present without compromising the ability of future generations to meet their own needs." Now it becomes clear why continuation of using fossil fuels for energy is not sustainable. First, the combustion of fossil fuels results in the release of carbon dioxide (CO_2) and other pollutants (NO_x, SO_x, particulates, mercury, and other toxic metals) into the atmosphere. The increased atmospheric concentrations of these pollutants cause a series of environmental impacts, including global warming and respiratory health effects. Second, the rate at which we consume fossil fuels is much higher than the rate at which they are replenished so mankind cannot rely on this source of energy forever.

The main candidates for facing this dual challenge of carbon dioxide (CO_2) emissions and fossil fuel depletion are coal with carbon capture and sequestration (CCS), nuclear, and renewable sources of energy. However, safe and economic concepts for carbon sequestration have not been proven; nuclear suffers from high cost, radioactive waste management, fuel availability, and nuclear weapon proliferation issues; and renewables have been limited by resource limits, high cost, and intermittency problems. Biomass could be a substitute for fossil fuels, but enough land or water to both meet the demand for power and to feed the world's growing population is not available. Solar energy has huge potential—tens or hundreds of terawatts (TW) are practical, but it suffers from intermittency. Wind resources are less abundant and even more variable than solar, but in many regions they can complement the variability of solar resources.

The cost challenge for solar electricity is being resolved as recent drastic cost reductions in the production of photovoltaics (PV) paved the way for enabling solar technologies to become cost competitive with fossil fuel energy generation. Such cost competitiveness, called "cost grid parity," has already been accomplished for parts of the southwest United States, Chile, Spain, Italy, and other countries. PV rooftop systems and utility power plants are relatively easy to build and deployment grows fast. Nevertheless solar is still a minor contributor in electricity mixtures worldwide as

Electricity from Sunlight: Photovoltaic-Systems Integration and Sustainability, Second Edition.
Vasilis Fthenakis and Paul A Lynn.
© 2018 John Wiley & Sons Ltd. Published 2018 by John Wiley & Sons Ltd.
Companion website: www.wiley.com/go/fthenakis/electricityfromsunlight

inertia in energy policy is stalling the transformation urgently needed. To this end, our book aims in conveying the great potential of PV and helping accelerate their deployment in a world longing for sustainable development.

1.2   The Sun, Earth, and Renewable Energy

We are entering a new solar age. For the last few hundred years, humans have been using up fossil fuels that took around 400 million years to form and store underground. We must now put huge effort—technological and political—into energy systems that use the sun's energy more directly. It is one of the most inspiring challenges facing today's engineers and scientists and a worthwhile career path for the next generation. PV, the subject of this book, is one of the exciting new technologies that is already helping us toward a solar future.

Most politicians and policymakers agree that a massive redirection of energy policy is essential if planet Earth is to survive the 21st century in reasonable shape. The 21st Conference of the Parties (COP21) that brought 190 countries together in Paris in December 2015 agreed that consistent efforts are needed worldwide to keep the global temperature increase to below 2°C, or preferably below 1.5°C.

This is not simply a matter of fuel reserves. It has become clear that, even if those reserves were unlimited, we could not continue to burn them with impunity. The resulting carbon dioxide emissions and increased global warming would lead to a major environmental crisis if we do not curtail the CO_2 concentration in the atmosphere and do it soon, before we are locked into irreversible processes. So the danger is now seen as a double-edged sword: on the one side, fossil fuel depletion and, on the other, the increasing inability of the natural world to absorb emissions caused by burning what fuel remains.

Back in the 1970s there was very little public discussion about energy sources. In the industrialized world we had become used to the idea that electricity is generated in large centralized power stations, often out of sight as well as mind, and distributed to factories, offices, and homes by a grid system with far-reaching tentacles. Few people had any idea how the electricity they took for granted was produced, or that the burning of coal, oil, and gas was building up global environmental problems. Those who were aware tended to assume that the advent of nuclear power would prove a panacea; a few even claimed that nuclear electricity would be so cheap that it would not be worth metering! And university engineering courses paid scant attention to energy systems, giving their students what now seems a rather shortsighted set of priorities.

Yet even in those years, there were a few brave voices suggesting that all was not well. In his famous book *Small is Beautiful*,[1] first published in 1973, E.F. Schumacher poured scorn on the idea that the problems of production in the industrialized world had been solved. Modern society, he claimed, does not experience itself as part of nature, but as an outside force seeking to dominate and conquer it. And it is the illusion of unlimited powers deriving from the undoubted successes of much of modern technology that is the root cause of our present difficulties. In particular, we are failing to distinguish between the capital and income components of the Earth's resources. We use up capital, including oil and gas reserves, as if they were steady and sustainable income. But they are actually once-and-only capital. It is like selling the family silver and going on a binge.

Figure 1.1 Toward the new solar age, this rooftop PV installation at the Mont-Cenis Academy in Herne, Germany, is on the site of a former coal mine (*Source:* Reproduced with permission of IEA-PVPS).

Schumacher's message, once ignored or derided by the majority, is increasingly seen as the essence of sustainable development. For the good of planet Earth and future generations, we have started to distinguish between capital and income and to invest heavily in renewable technologies—including solar, wind, and wave power—that produce electrical energy free of carbon emissions. The message has been powerfully reinforced by former US Vice President Al Gore, whose inspirational lecture tours and video presentation *An Inconvenient Truth*[2] have been watched by many millions of people around the world. Most importantly, the vision was captured by industry leaders who made solar and wind systems affordable and continue to advance them.

Whereas the fossil fuels laid down by solar energy over hundreds of millions of years must surely be regarded as capital, the sun's radiation beamed at us day by day, year by year, and century by century is effectively free income to be used or ignored as we wish. This income is expected to flow for billions of years. Nothing is "wasted" or exhausted if we don't use it because it is there anyway. The challenge is to harness such renewable energy effectively, designing and creating efficient and hopefully inspiring machines to serve humankind without disabling the planet.

We should perhaps consider the meaning of renewable energy a little more carefully. It implies energy that is sustainable in the sense of being available in the long term without significantly depleting the Earth's capital resources, or causing environmental damage that cannot readily be repaired by nature itself. In his excellent book *A Solar Manifesto*,[3] the late German politician Hermann Scheer considered planet Earth in its totality as an energy conversion system. He notes how, in its early stages, human society

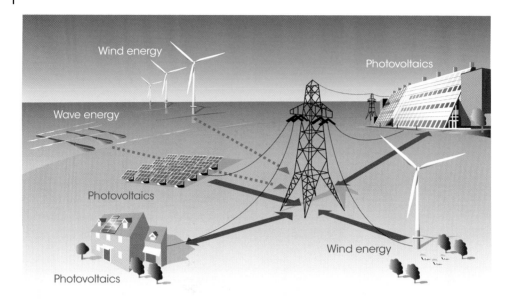

Figure 1.2 Three important renewable technologies: PV, wind, and wave.

was itself the most efficient energy converter, using food to produce muscle power and later enhancing this with simple mechanical tools. Subsequent stages—releasing relatively large amounts of energy by burning wood and focusing energy where it is needed by building sailing ships for transport and windmills for water pumping—were still essentially renewable activities in the previously mentioned sense.

What really changed things was the 19th-century development of the steam engine for factory production and steam navigation. Here, almost at a stroke, the heat energy locked in coal was converted into powerful and highly concentrated motion. The industrial society was born. And ever since we have continued burning coal, oil, and gas in ways that pay no attention to the natural rhythms of the Earth and its ability to absorb wastes and by-products, or to keep providing energy capital. Our approach has become the opposite of renewable and it is high time to change priorities.

Since the reduction of carbon emissions is a principal advantage of PV, wind, and wave technologies, we should recognize that this benefit is also proclaimed by supporters of nuclear power and carbon dioxide capture and carbon sequestration (CCS) technologies. But frankly they make strange bedfellows, in spite of sometimes being lumped together as "carbon-free." It is true that all offer electricity generation without substantial carbon emissions, but in almost every other respect they are poles apart. The renewables offer the prospect of widespread, both large- and small-scale electricity generations, but nuclear must, by its very nature, continue the practice of building huge centralized power stations. CCS could, in the best case, capture only a part of CO_2 emissions, but there are not safe sequestration options, and it would increase health, safety, and environmental impacts at the coal mining areas and in subsequent stages. Both fossil fuels and uranium are depletable resources. PV, wind, and wave need no fuel and produce no waste in operation; the nuclear industry is beset by problems of radioactive waste disposal. On the whole, renewable technologies pose no serious problems of safety or susceptibility to terrorist attack—advantages that nuclear power can hardly claim. And finally there is the

Figure 1.3 The promise of photovoltaics (*Source:* Reproduced with permission of European Photovoltaic Industry Association).

Figure 1.4 A rooftop residential system at Long Island, New York (Vasilis Fthenakis).

issue of nuclear proliferation and the difficulty of isolating civil nuclear power from nuclear weapons production. Taken together these factors amount to a profound divergence of technological expertise and political attitudes, even of philosophy.

It would however be unfair to pretend that renewable energy is the perfect answer. For a start such renewables as PV, wind, and wave are relatively diffused and intermittent.

Figure 1.5 A large PV utility system in South California (*Source:* Reproduced with permission of Desert Sunlight, First Solar/NEXTera Energy).

Often, they are rather unpredictable. And although the "fuel" is free and the waste products are minimal, up-front investment costs tend to be large. There are certainly major challenges to be faced and overcome as we move toward a solar future, and these are discussed comprehensively in this new edition.

Our story now moves on toward the exciting technology of PV, arguably the most elegant and direct way of generating renewable electricity. But before getting involved in the details of solar cells and systems, it is necessary to appreciate something of the nature of solar radiation—the gift of a steady flow of energy income that sustains life on the planet.

1.3 The Solar Resource

The sun sends an almost unimaginable amount of energy toward planet Earth—around one hundred thousand TW; 1 TW is 1 trillion watts. In electrical supply terms this is equivalent to the output of about 100 million modern fossil fuel or nuclear power stations. To state it another way, the sun provides in about an hour the present energy requirements of the entire human population for a whole year. It seems that all we need do to convert society "from carbon to solar" is to tap into a tiny proportion of this vast potential.

However some caution is needed. The majority of solar radiation falls on the world's oceans. Some is interrupted by clouds and a lot more arrives at inconvenient times or places. Yet, even when all this is taken into account, it is clear that the sun is an amazing source of energy, an enormous fusion reactor at a safe distance from the Earth.

Figure 1.6 Energy for ever: an installation in Austria (*Source:* Reproduced with permission of IEA-PVPS).

The opportunities for harnessing its energy, whether represented directly by sunlight or indirectly by wind, wave, hydropower, or biomass, seem limited only by our imagination, technological skill, and political determination.

The sun's power density (i.e., the power per unit area normal to its rays) just above the Earth's atmosphere is known as the *solar constant* and equals $1366\,W/m^2$. This is reduced by around 30% as it passes through the atmosphere, giving an *insolation* at the Earth's surface of about $1000\,W/m^2$ at sea level in the early afternoon on a clear day. This value is the accepted standard for "strong sunshine" and is widely used for testing and calibrating terrestrial PV cells and systems.

Another important quantity is the average power density received over the whole year, known as the *annual mean insolation*. A neat way of estimating it is to realize that, as seen from the sun, the Earth appears as a disk of radius R and area πR^2. But since the Earth is actually spherical with a total surface area $4\pi R^2$, the annual mean insolation just above the atmosphere must be $1366/4 = 342\,W/m^2$. However it is shared very unequally, being about $430\,W/m^2$ over the equator, but far less toward the polar regions that are angled well away from the sun. The distribution is illustrated in the upper half of Figure 1.7.

The lower half of the figure shows the reduction in insolation caused by the Earth's atmosphere. Absorption by gases and scattering by molecules and dust particles are partly responsible. Clouds are a major factor in some regions. We see that the daily (24-hour) average insolation at the Earth's surface is greatly affected by local climatic conditions, ranging from about $300\,W/m^2$ in the Sahara Desert and parts of the Pacific Ocean to less than $80\,W/m^2$ near the poles. Hourly irradiance levels on planes that track the sun's movement could reach and even exceed $1000\,W/m^2$ during clear summer days in sunny locations.

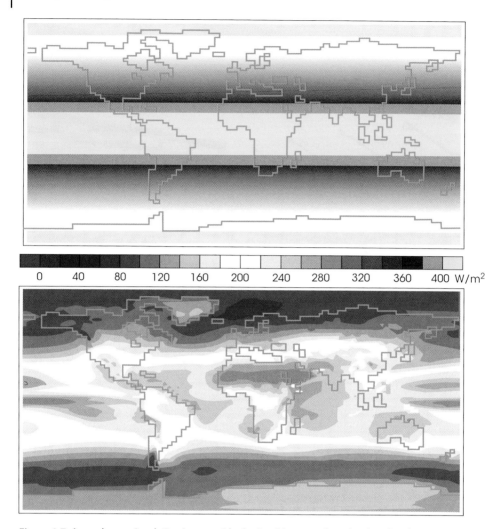

Figure 1.7 Annual mean insolation just outside the Earth's atmosphere (top) and at the Earth's surface (bottom) (*Source:* Adapted from Wikipedia).

If we know the average insolation at a particular location, it is simple to estimate the total energy received over the course of a year (1 year = 8760 hours). For example, London and Berlin, both with annual mean insolation of about 120 W/m^2, have annual energy totals of about $120 \times 8760/1000 = 1050$ kWh/m^2. Sydney's mean of about 200 W/m^2 is equivalent to 1750 kWh/m^2. These levels correspond to total (direct and diffuse) irradiation over a horizontal plane, called global horizontal irradiation (GHI). The highest GHI levels are measured in subtropical arid climates where many deserts lie. The high altitude Atacama Desert in North Chile receives a GHI as high 2530 kWh/m^2/year. However, higher irradiation levels can be obtained over the course of the year if the plane of incidence is tilted at an angle approximately equal to the latitude of a location, and even higher irradiation can be harvested if the plane tracks the movement of the sun. For example, global irradiation on a latitude tilt plane at the Atacama

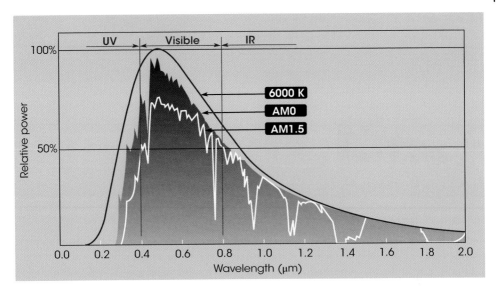

Figure 1.8 Spectral distributions of solar energy.

is 3160 kWh/m^2/year and at a one-axis sun-tracking plane is 3900–4000 kWh/m^2/year. Irradiation measurements and conversions on different planes are discussed in Chapter 3.

Such figures are useful to PV system designers who need to know the total available solar resource. However, we must remember that they are averaged over day and night, summer and winter, and are likely to vary considerably from year to year. It is also interesting to speculate how far global warming, with its interruptions to historical weather patterns, may affect them in the future.

So far we have not considered the sun's spectral distribution—that is, the range and intensity of the wavelengths in its emitted radiation. This is a very important matter because different types of solar cell respond differently to the various wavelengths in sunlight. It is well known that the sun's spectrum is similar to that of a perfect emitter, known as a *black body*, at a temperature of about 6000 K. The smooth curve in Figure 1.8 shows that such black-body radiation spreads over wavelengths between about 0.2 and 2.0 μm, with a peak around 0.5 μm. The range of wavelengths visible to the human eye is about 0.4 μm (violet) to 0.8 μm (red). Shorter wavelengths are classed as ultraviolet (UV) and the longer ones as infrared (IR). Note how much of the total spectrum lies in the IR region.

The figure shows two more curves, labeled AM0 and AM1.5, representing actual solar spectral distributions arriving at Earth. To explain these we need to consider the pathlength or *air mass* (*AM*) of sunlight through the atmosphere. AM0 refers to sunlight just outside the atmosphere (pathlength zero) and is therefore relevant to PV used on Earth satellites. In the case of terrestrial PV, the pathlength is the same as the thickness of the atmosphere (AM1) when the sun is directly overhead. But if it is not overhead, the path length increases according to an inverse cosine law. The widely used AM1.5 curve, shown in the figure, represents the early afternoon conditions when the solar beam to Earth is 48.2° from overhead and is generally accepted as a good average

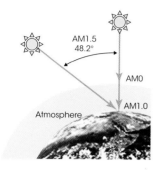

The Air Mass coefficient is equal to $1/\cos \theta_z$ where θ_z is the zenith angle, thus the angle between the vertical and the line to the sun.

For a thickness l_o of the atmosphere, the path length l through the atmosphere for solar radiation incident at angle θ_z is $l = l_o/\cos \theta_z$

Figure 1.9 Air mass (AM) definitions

for assessing PV cells and systems. The deep notches of the AM1.5 curve are due to absorption by oxygen, water vapor, and carbon dioxide. AM2 conditions would correspond to the sun being lower at the horizon (60° from overhead).

This is not quite the whole story because when solar cells are installed at or near ground level, they generally receive indirect as well as direct solar radiation. This is shown in Figure 1.10. The *diffuse* component represents light scattered by clouds and dust particles in the atmosphere; the *albedo* component represents light reflected from the ground or objects such as trees and buildings. The electrical output from the cells depends on the combined effect of all components—direct, diffuse, and albedo. In strong sunlight the direct component is normally the greatest. But if the cells are pointed away from the sun, or if there is a lot of cloud, the diffuse component may well dominate (clouds also cause blocking, or attenuation, of direct radiation). The albedo

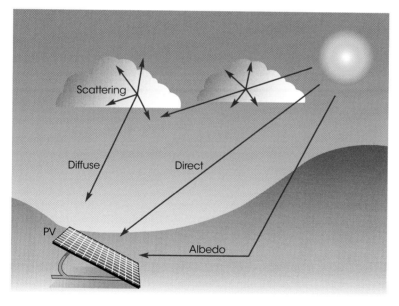

Figure 1.10 From sun to PV through the Earth's atmosphere.

contribution is often small but can be very significant in locations such as the Swiss Alps due to strong reflections from fallen snow.

We have now covered the main features of solar radiation as it affects terrestrial PV. We shall find this information useful when considering the mounting and orientation of PV cells and modules in Chapter 3.

1.4   The Magic of Photovoltaics

From time to time human ingenuity comes up with a new technology that seems to possess a certain magic. We can all think of examples from the past—the printing press, steam locomotion, radio communication, powered flight, and medical imaging—although our choices inevitably reflect personal tastes and priorities. In most cases such technologies were unimaginable to previous generations and caused amazement and even fear when they appeared. Quite often a technology that promises major social as well as commercial benefits turns out to have rather questionable applications. American aviation pioneers Wilbur and Orville Wright, whose first powered flights at Kitty Hawk in 1903 changed the world forever, initially believed that scouting aircraft would render wars obsolete by allowing each nation to see exactly what the others were doing. But by the end of World War 1, it had become clear that this view was overoptimistic and Orville instead declared that "the aeroplane has made war so terrible that I do not believe any country will again care to start one." Perhaps we are rather more realistic today and understand that technological advance almost always carries risk as well as social benefit. The magic is not without its downside.

Where does PV fit in the landscape of technological change? Half a century ago few people realized that sunlight could be converted directly into electricity. Even the early pioneers of PV could hardly have guessed that their researches would lead to a worldwide industry providing electricity to millions of people in developing countries. A generation ago it seemed unlikely that PV would branch out from its early success in powering space satellites and come down to Earth. More recently it would have taken courage to suggest that terrestrial PV would move into a multi-gigawatt era and start rivaling conventional methods of electricity generation in the developed world. From the technical point of view, it is certainly a remarkable story—and one that, on a historical timescale, is still in its early stages.

Just as importantly, it is difficult to see any major downside in PV's gentle technology. Few people find much to object to in the deployment of solar cells and modules. True, some worry that the aesthetics of existing homes, offices, and public buildings can be marred by having PV attached to them, although this is a matter of taste. There is also the question of land use: It seems that PV takes up a lot of space compared with a conventional power plant, but not if we take account of the whole life cycle of the fuel. For example, the coal life cycle in the United States takes more space than PV for the same amount of electricity generation, because coal surface mining takes a lot of land. Furthermore, PV does not damage the land as coal mining does, and it often uses marginal or unproductive land in deserts or old industrial areas; and unlike wind turbines that offend many people by their visual intrusion, ground-mounted PV is hardly ever visually aggressive or unattractive. Finally there may be some risk when PV comes to the end of its useful life, but most of the materials involved are benign and the industry is very aware of its environmental credentials and the need for recycling and careful disposal; we discuss this in Chapter 7. All in all the negative impacts of PV seem relatively modest and containable.

Much of PV's magic is due to its elegance and simplicity. A solar cell turns sunlight directly into electricity without fuel, moving parts, or waste products. Made from a thin slice or layer of semiconductor material, it is literally a case of "photons in, electrons out." By contrast a fossil fuel or nuclear power station working on a classic thermodynamic cycle turns heat from fuel combustion or nuclear reaction into high pressure steam and then uses the steam to drive a turbine coupled to an electrical generator. This complex chain of events produces undesirable by-products, including spent fuel and a great deal of waste heat, and in the case of fossil fuels also a lot of carbon dioxide and other greenhouse gases. The high pressures and temperatures at which modern plant is operated put great stresses on materials and components. Small-scale electricity production using diesel generators has similar disadvantages. Meanwhile the solar cell works silently and effortlessly, a model of operational simplicity. Place it under sunlight and you can tap electricity directly from its terminals.

Not that PV is simple science. As we shall see, solar cells are high-tech products based on more than half a century of impressive research in universities, companies, and government institutes around the world. Their manufacture demands very high standards of precision and cleanliness. And they are strongly related to another modern technology that has a certain magic for many people—semiconductor electronics and computers.

Figure 1.11 A certain magic: "sunflowers" in Korea (*Source:* Reproduced with permission of IEA-PVPS).

1.5 A Piece of History

Light has fascinated some of the world's greatest scientists. One of the most famous of them all, Isaac Newton (1642–1727), thought of it as a stream of particles rather like miniature billiard balls. But in the early 19th century, experiments by the English polymath Thomas Young and French physicist Augustin Fresnel demonstrated interference effects in light beams, which include the bands of colors often seen on the surface of soap bubbles. This suggested that light acts as a wave rather like the ripples on a pond—a theory reinforced by James Clerk Maxwell's work in the 1860s, showing visible light to be part of a very wide spectrum of electromagnetic radiation.

Yet Newton's "billiard ball" theory refused to go away. The German physicist Max Planck used it to explain the characteristics of blackbody radiation, and it subsequently proved central to Albert Einstein's famous work on the photoelectric effect in 1905 for which he got the Nobel Prize, in which he proposed that light is composed of discrete miniature particles or packets of energy known as *quanta*. The subsequent development of quantum theory was one of the great intellectual triumphs of the 20th century. So our modern view is that light has an essential *duality*: for some purposes we may think of it as a stream of particles and, for others, as a type of wave. The two aspects are complementary rather than contradictory.

The earliest beginnings of PV go back to 1839 when the young physicist Edmond Becquerel, working in his father's laboratory in France, discovered the PV effect as he shone light onto an electrode in an electrolyte solution. By 1877 the first solid-state PV cells had been made from selenium, and these were later developed as light meters for photography. Although a proper understanding of the phenomena was provided by quantum theory, practical application to useful PV devices had to await the arrival of semiconductor electronics in the 1950s. Thus, there was a gap of over a hundred years between Becquerel's initial discovery and the development of PV as we know it today.

The story of modern PV has been expertly reviewed in an article by solar cell pioneer Joseph Loferski,[4] formerly a professor at Brown University in the United States. Although we must leave aside the technical details of his account, the following broader points will serve well to bring the PV story up toward the 21st century.

Figure 1.12 Isaac Newton, Edmond Becquerel, and Albert Einstein (Wikipedia).

The modern PV age may be said to have begun in 1954 with the work of researchers at the Bell Telephone and RCA laboratories, who reported new types of semiconductor devices, based on silicon and germanium, that were an order of magnitude more efficient than previous cells at converting radiation directly into electricity. The fledgling PV community hoped that this would lead to new applications for solar cells including electrical power generation. However their hopes were not realized, in part because that decade was a time of great expectations for nuclear energy. Sceptics believed that solar energy was too diffuse and intermittent, and the new devices far too expensive. At that moment in history, PV looked rather like a solution in search of a problem.

What changed the situation almost overnight was the launch of the first Earth satellite, the USSR's *Sputnik*, in 1957. Satellites and solar cells—even expensive ones—were made for each other. The early satellites needed only a very modest amount of electricity, and the weight and area of solar panels needed to produce this were acceptable to satellite designers. Also, the types of cell made in 1954 were proving reliable and seemed likely to operate in the space environment for many years without significant deterioration. The first US research satellite using a PV power supply was launched in 1958. It was the size of a large grapefruit. Its solar cells covered an area of about $100\,cm^2$ and produced just a few tens of milliwatts. In 1962 the first-ever commercial telecommunication satellite, *Telstar*, was launched with sufficient solar cells to produce 14 W from the sun (Figure 1.13). By the early 1970s, space satellites powered by solar cells had become quite commonplace. PV in space had already made its mark.

The possibilities for "bringing PV down to Earth" depended crucially on lowering the price of solar cells. In 1970 the US price was around $300/W_p$ (we normally quote solar cell power in peak watts (W_p), being the rated power at a standard insolation of $1000\,W/m^2$). This was acceptable for extremely expensive space satellites, but hopeless for terrestrial electricity production on a significant scale. What encouraged PV researchers to hold on to their dream was the realization that prices would almost certainly fall dramatically as production levels rose, in accordance with the well-known "learning curve" concept. Experience had shown that, for every doubling of cumulative production of a wide range of manufactured products, price tended to drop between 10 and 30%. For mature technologies such as steel or electric motors, such doubling, given the high current production levels, would require many decades. But a fledgling technology like PV had many doublings of cumulative production to look forward to, so major reductions in cost could be expected over a comparatively short timescale. Indeed, it was predicted that by the time cumulative PV cell production reached gigawatt levels (1 GW being equal to 1000 MW), as required for serious terrestrial application, the price would have dropped nearer to $1/W_p$. As with other technologies, new inventions and manufacturing systems that could not be visualized would arise, unpredictable political and economic factors would occur, and the price would be driven down.

This apparently bold prognosis, which helped lift the gloom and spurred the PV community on to ever-greater efforts, proved farsighted. The actual "learning curve" for world PV production over the period 1987–2016 is shown in Figure 1.14, plotted on logarithmic scales in terms of US dollars per peak watt against cumulative peak megawatts (MW_p). We see that as cumulative production advanced from $100\,MW_p$ in 1987 to $1000\,MW_p$ in 1999, the cost per peak watt fell from about $15 to $7. Another tenfold

Figure 1.13 Telstar 1, the first commercial telecommunication satellite; diameter approximately 0.8 m; weight 77 kg (*Source:* Courtesy of NASA).

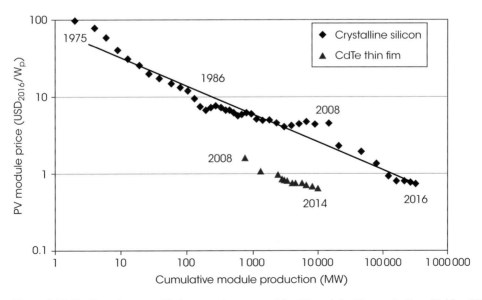

Figure 1.14 The "learning curve" (price-experience curve) for PV modules (*Source:* A. Jäger-Waldau; PV Status Report 2016; EUR 28159 EN).

advance to $10\,000\,MW_p$ was achieved in 2007 and the cost reduced to around $3.0. At that time a new technology, thin-film cadmium telluride (CdTe), was introduced in the market and quickly became the lowest production cost technology combining quick processing and economy on semiconductor materials. The market growth continued, accounting to annual growth rates of 44% in the period of 2000–2015.

As of 2017, the cumulative production of crystalline Si PV has grown to $200\,000\,MW$ (200 GW), and the lowest module cost, for modules made in China, has decreased to $0.40/W of rated power. CdTe PV modules made in the United States and Malaysia have been also produced at such low costs. This outstanding trend corresponds to an average cost reduction of about 20% for every doubling of cumulative production. PV has become truly competitive with conventional electricity generation in several regions. The costs of the balance of system components (i.e., power electronics, structures, installation) were not reduced as rapidly as the module costs with the total installed system costs being in the range of $1–1.5/W for large utility systems and $2–4/W for residential rooftop systems. We expect that PV costs will continue to decrease as technologies and systems integration improves and their cost is reaching parity with conventional electricity in regions of tropical and subtropical climates. These are exciting times!

Returning now to our historical review, the renewed optimism of the PV community, based on "learning curve" predictions, was bolstered by the first "oil shock" in 1973 when oil-producing countries decided greatly to increase the price of crude oil and exert more control over its supply. Funding for PV research and development (R&D) in the United States under President Jimmy Carter then increased dramatically. Unfortunately, government support was subsequently cut back hard by President Ronald Reagan's administration from 1980 onward, but great advances had already been made, and major PV research programs in Germany and Japan were adding their own important contributions. The efficiencies of solar cells were constantly being raised; new PV materials and cell structures were being investigated, and, on the applications front, a range of PV power plants emerged with multi-megawatt capacity. In 1985 Professor Martin Green's group at the University of New South Wales in Australia pushed the efficiency of new design of silicon cells above the 20% barrier— some four times higher than the cells that had heralded the arrival of the modern PV age in 1954. By 1990 the same group achieved efficiencies of 24% in the lab, and it has continued as a major pacesetter for crystalline silicon cells, the long-term workhorse of the PV industry. Also during the 1980s, Professor Richard Swanson's group at Stanford developed the point-contact solar cell, with laboratory cell records of 28% in concentrator cells and 23% in one-sun cells. Dr. Swanson founded SunPower in 1985, which today is the world's leader in high-efficiency solar cells offering modules of 21.5% efficiency, composed of 24% efficient cells.

The early 1990s proved also to be crucial for thin-film PV with CdTe leading the way. Photon energy in Colorado was first to raise the CdTe cell efficiency above 12%. Ting and Shirley Chu—University of South Florida professors—improved it to 14.6%; and in the summer of 1992, their former students Christos Ferekides and Jeff Britt raised it to near 16%. This record was held for nearly 10 years with the National Renewable Energy Laboratory increasing the record research cell efficiency to 16.5% in 2001. The technology was commercialized by First Solar, which in 2007 was able to produce 10% efficient modules at a cost of only a dollar per watt. First Solar led the global PV industry in module cost reductions and as of 2016 they produced 16.6% efficient modules at only $0.40/W_p$, while on schedule for further improvements toward their record efficiency of 21.5%.

The pace of PV research, development, and application for all technologies continues unabated today. Growing awareness of global warming and the vital role of the renewable energies in combating it have ensured that governments around the world appreciate the need to encourage and stimulate PV, and we are now in the multigigawatt era. This must have been almost unimaginable half a century ago.

At the end of his 1993 article, Joseph Loferski noted that the blossoming PV edifice was destined to grow manyfold again. The small band of researchers who had ushered in the modern PV age in the 1950s had multiplied into "a band of brothers and sisters, we happy few," who shared the dream that solar PV electricity was destined for an evergreater future in the service of humanity. From today's perspective, his vision and optimism seem entirely justified.

1.6   Coming Up to Date

How can we summarize the current status of a technology such as PV that has been, and still is, experiencing dramatic growth? Today's research and development, novel PV installations, and global statistics will very soon seem history. But fortunately certain trends that have developed over the past 15 or 20 years seem likely pointers to the future. We can discuss these trends more easily by dividing PV systems into two broad categories: *grid-connected systems* (also called *grid-tied systems*) that feed any surplus PV electricity into a grid and accept electricity from the grid when there is a solar deficit and *stand-alone systems* that are self-contained and not tied to a conventional electricity grid. These categories may usefully be subdivided as follows:

Grid-connected systems

1) Building-integrated photovoltaics (BIPV) and rooftop systems on houses, offices, factories, and other commercial premises (or on adjacent land)—covering a wide power range from $1\,kW_p$ to several MW_p
2) Land-based PV power plants, often remote from individual electricity consumers— typically $10–150\,MW_p$, with a few up to $550\,MW_p$

Stand-alone systems

3) Low-power solar home systems (SHSs), supplying small amounts of electricity to individual homes in developing countries—typically $30–100\,W_p$

4) Higher-power systems for isolated homes and buildings in the developed world— typically $1–20\,kW_p$

5) PV systems for a wide range of applications, including water pumping and irrigation, isolated telecommunications equipment, marine buoys, traffic control signs, and solar-powered cars and boats

In the 1980s PV started to make a major contribution, supplying small amounts of electricity to the millions of families in "sunshine countries" of the developing world with no access to, or promise of, an electricity grid. This was rightly seen as a noble social objective as well as a commercial opportunity that would increase PV's international market. However it became increasingly evident that SHSs (item 3 earlier) could not, by themselves, boost global production toward the levels dreamed of by the PV community. A typical SHS requires only one small PV panel, but even so the "up-front" costs cannot easily be afforded by individual families in developing countries without effective government financing schemes that are not always forthcoming. And maintenance problems (generally with batteries or other system components rather than the PV itself) can easily reduce reliability. So although the SHS market is socially important and continues to grow, it no longer represents a major plank of the global PV industry.

Various other types of stand-alone system were steadily developed in the 1980s, often providing valuable PV electricity in remote locations that would otherwise need diesel generators. In addition a number of grid-connected PV power plants were commissioned, mainly in the United States, by electric utilities keen to assess the commercial possibilities and reliability of the new technology. However, the limited number and scale of all these systems offered little prospect for the exciting expansion of PV needed to make it a major source of electricity.

What really changed the outlook for global PV production was the emphatic shift toward grid-connected systems in the developed world that got under way in the 1990s. It was the citizens of richer countries that would provide the up-front costs and market stimulus to propel PV faster along its "learning curve," leading to price reductions as

Figure 1.15 This PV module powers a solar home system in Bolivia (*Source:* Reproduced with permission of EPIA/BP Solar).

cumulative production really took off. This policy shift was supported by increasing awareness among governments of the importance of renewable energy for combating climate change, and by the growing enthusiasm of individuals and companies to "do their bit" by installing roof-top systems, even though the price of solar electricity was not yet competitive. Electricity utilities began to accept that the flow of electricity was not all "one-way," allowing customers to be providers as well as consumers and introducing tariffs for feeding electricity back into the grid.

As far as governments are concerned, the price support mechanisms devised for grid-connected systems have proved crucial. PV is similar to other renewables such as hydroelectric and wind in having high up-front capital costs and very low running costs. But this can make it hard for families and organizations to find the initial capital, and even harder if they are not guaranteed an attractive price for surplus PV electricity fed back into the grid. In recent years many governments have provided capital grants to encourage people to install domestic PV systems, and the more farsighted ones have introduced *feed-in tariffs* that offer long-term guaranteed payments for renewable electricity. Countries that have given PV a big boost with feed-in tariffs—especially Germany—have stimulated their home markets and, by doing so, have pushed PV decisively along its "learning curve" into the multi-gigawatt era. As cumulative world production surges and the price comes down, poorer families in sunshine countries are more likely to get their SHSs.

Figure 1.16 A 9 kW$_p$ grid-connected PV system in Northern Italy (*Source:* Reproduced with permission of IEA-PVPS).

Not that feed-in tariffs and other forms of government support are universally popular. Some politicians tend to regard them with suspicion, arguing that market forces alone should determine the price of PV and other renewables, forgetting that power production from fossil fuels had and still have government subsidies. Others are more likely to vote for taxpayers' or consumers' money being used to support a new and promising technology that will, in due course, benefit the whole of society as well as the planet. Such differences tend to produce stop–go support for PV when governments change, causing confusion and clouding investment decisions. Yet in spite of these drawbacks, the clear trend is toward support by governments regardless of their political hue, mainly because of near-universal agreement that global warming must be checked and renewable energy championed.

The emphasis on grid-connected systems in the developed world continues today, making them far more important than stand-alone systems in terms of the total volumes of PV required. Huge numbers of rooftop installations are being installed on homes, offices and commercial buildings increasingly use PV on their roofs and facades, and large factory rooftops are being fitted with PV, sometimes retrospectively. The market for power plants is also developing rapidly, with Germany, China and the United States prominent in pressing ahead with ever larger installations. Several PV plants of 200–550 MW capacities were constructed in the southwest of the United States and more are planned for China, India, southern Europe, Chile, and the Middle East.

Figure 1.17 115 kW$_p$ rooftop installation of the Ford Motor Company in London (*Source:* Reproduced with permission of IEA-PVPS).

So where has all this activity got us? We previously noted that world cumulative PV production passed the $10\,GW_p$ landmark in 2007 and it exceeded $350\,GW_p$ in the middle of 2017. PV electricity generation has grown to $47\,GW$ in the United States, with most deployment in the country happening during the last 7 years. Reflecting back in time, in 2008, Fthenakis with collaborators James Mason and Ken Zweibel published in Scientific American[5] and in Energy Policy[6] a "Solar Grand Plan" demonstrating the feasibility of renewable energy in providing 69% of the US electricity demand by 2050 while reducing CO_2 emissions by 60% from 2005 levels; the PV contribution to this plan was assessed to be $250\,GW$ by 2030 and $2900\,GW$ by 2050. For 2020 we projected a module manufacturing cost of \$0.50/W and installed PV power plant cost of \$1.20–1.30/W; as of 2016 the industry has already reached these 2020 targets. The US Department of Energy (DOE) more detailed SunShot vision study, released in 2012, showed the possibility of having $300\,GW$ of PV installed in the United States by 2030 and $630\,GW$ by 2050.[7] As of 2017, the PV industry has already accomplished the efficiency improvements and price reduction projections targeted for 2020.

The industry is now in an exciting new phase, with multi-gigawatt annual production set to challenge fossil fuel and nuclear plants and achieve "grid parity." Crystal ball gazing is always risky, but if current and projected increases in cumulative production are maintained, it seems possible that we will be approaching $1000\,GW_p$ of PV installed around the world by 2020 or soon after. Recalling that a large conventional power station generates about $1\,GW$, it is clear that renewable electricity on this scale would make a serious contribution to global supplies.

This raises an interesting question: what total area will be required to accommodate all this PV? After all, sunlight is not a highly concentrated energy source and $1000\,GW_p$ of installed capacity would take up a large area. Will Planet Earth be smothered with solar cells? An approximate answer may be found by noting that $1\,kW_p$ of 16% efficient solar modules has an area of about $6\,m^2$. However, modules cannot generally be crammed together, especially in large ground-mount installations where space may be needed to allow servicing and prevent shading, so we might allow $15\,m^2/kW_p$. This means that $1\,000\,GW_p$, equal to $1\,000 \times 10^6\,kW_p$, would need around $15\,000 \times 10^6\,m^2$, which could be provided by $1\,220 \times 1\,220\,km^2$ of land, roughly two times the area taken up by London and its suburbs, or by Paris. In other words, our projected global PV scenario for 2020 might require a total area comparable with just two large modern cities—but spread right around the globe. When we consider that huge arid regions and deserts of the world, marginal and ex-industrial land, and hundreds of millions of rooftops on houses and commercial buildings are all candidates for PV, there appears to be plenty of space, and as we discuss in Chapter 7 this space is less than coal mining uses!

Our brief summary of recent developments and likely trends has so far ignored one of the most important aspects—R&D of solar cells, modules, and the additional items that go to make up a complete PV system. In fact, the past 20 years has seen extraordinary R&D activity by teams in universities, government institutes, and PV companies. Solar cells are constantly being improved, new types of cell invented, and system components improved in reliability as well as reduced in price. However, we will be in a better position to consider such topics after covering some of the basic science of solar cells in the next chapter.

Figure 1.18 PV power plant in Colorado, USA (*Source:* Reproduced with permission of IEA-PVPS).

Appendix 1.A   Energy Units and Conversions

Thermal (primary) energy	Electricity
Joule(J); kilojoule(kJ), $1\,kJ = 10^3\,J$; Megajoule(MJ), $1\,MJ = 10^6\,J$; Exajoule(EJ), $1\,EJ = 10^{16}\,J$	Wh; kWh; MWh; GWh, $1\,GWh = 10^9\,Wh$; TWh, $1\,TWh = 10^{12}\,Wh$
Btu; QBtu(Quad), $1\,QBtu = 10^{15}\,Btu$	$1\,kWh = 1000\,W \times 3600\,s = 3.6\,MW\ s = 3.6\,MJ_{electricity}\ (MJ_e)$

$1\,MJ_e = {\sim}3\,MJ$ (conversion of electricity to primary energy assuming an average grid conversion factor of 0.33)

CO_2 Emissions per Fuel Type

Fuel type	kg of CO_2 per unit of consumption
Grid electricity	40–60/kWh (depending on grid mix)
Natural gas	3142/Mt
Diesel fuel[a]	2.68/l (l)
Petrol[b]	2.31/l
Coal	2419/Mt

[a] One liter of diesel weighs 0.83 kg. Diesel consists 86% of carbon, or 0.72 kg of carbon per liter. According to the stoichiometry of the combustion reaction, $C + O_2 = CO_2$ 12 kg of C produces 3.6 times more (44 kg) of CO_2. Thus, combustion of one liter of diesel would produce approximately 2.6 kg of CO_2.
[b] One liter of petrol weighs 0.75 kg. Petrol consists 87% of carbon, or 0.65 kg of carbon per liter of petrol. In order to combust this carbon to CO_2, 1.74 kg of oxygen is needed. The sum is then 2.39 kg of CO_2 per liter of petrol.

CO_2 Emissions in Transportation

Vehicle type	Miles/gal (mpg)	km/l	kg CO_2/km
Medium hybrid car[a]	50	18	0.11
Medium gasoline car[a]	28	12.5	0.20
Diesel car[b]	47	20	0.12
Trailer truck, diesel engine		3	2.68
Rail			0.06/person
Air, short haul (500 km)			0.18/person
Air, long haul			0.11/person
Shipping			0.01/ton

[a] A car rated at 20 km/l (47 mpg) uses 5 l/100 km; this corresponds to $5 l \times 2.6$ kg/l/(per 100 km), thus 0.13 kg CO_2/km.
[b] An average consumption of 5 l/100 km then corresponds to $5 l \times 2.39$ g/l/(per 100 km) = 0.12 kg CO_2/km.

Self-Assessment Questions

Q1.1   Explain the difference between *capital* and *income* as applied to the Earth's energy resources.

Q1.2   Using Figure 1.7, estimate the approximate annual mean insolation in W/m^2 just outside the Earth's atmosphere and at the Earth's surface for (a) Oslo, Norway; (b) Lisbon, Portugal; and (c) the Caribbean Islands.

Q1.3   Explain why solar spectrum AM0 is relevant to PV on Earth satellites, whereas spectrum AM1.5 is generally used for terrestrial PV.

Q1.4   What is meant by the *direct, diffuse, and albedo* components of sunlight received by solar cells? Would you expect direct or diffuse radiation to be more significant in the following locations: (a) Phoenix, Arizona; (b) Glasgow, Scotland; and (c) Atacama desert, Chile.

Q1.5   Solar panels on the *Telstar* 1 telecommunication satellite (see Figure 1.13) were rated at 14 W, sufficient for its entire power requirements. To put this into an everyday perspective, how long would a 14 W heating element take to raise the temperature of 0.3 l of water from 20 to 90°C, sufficient for a generous mug of coffee? Neglect heat losses.

Q1.6   During the last decade global cumulative PV capacity has been increasing at an average annual rate of about 40%. What is the equivalent percentage for your nation (or country of residence), and what factors (climatic, economic, political) seem responsible for any difference?

Q1.7   What is the difference between primary energy and electricity and what losses are accounted for in converting the first to the latter?

Q1.8 What are (a) the total annual electricity demand and (b) the total annual energy demand for your nation (or country of residence) expressed in kWh, GWh, and TWh for electricity and in MJ, EJ, and quads for primary energy.

Q1.9 Consider a ground-mounted $10\,MW_p$ solar farm on flat land. If the modules are 17% efficient, and the total area allocated to the solar farm is 2.5 times larger than the area covered by the modules to allow access and prevent shading, what is approximately the total area of the solar farm, expressed in (a) square meters, (b) hectares, and (c) acres?

Problems

1.1 The current US annual electricity demand is approximately 4000 TWh; what is the primary energy corresponding to this demand if the energy mixture is 37% natural gas, 30% coal, 20% nuclear, 10% hydro and 3% wind and solar? How would the primary energy demand change if the renewable energy contribution to the mixture increases to 70% by 2050 (according to the Solar Grand Plan) from the current 13% and the balance of 30% is satisfied by 10% each of coal, natural gas and nuclear? Assume a 10,000 TWh/year scenario. Assume the following primary to electrical energy conversion factors: natural gas 0.4; coal 0.28; nuclear 0.33; hydro 1.0, wind and solar 0.1.

1.2 What is the height required for a flow of one ton of water per second to produce 1 kWh of electricity, in a hydrodynamic plant with efficiency of 100%?

1.3 Derive the conversion of Joule to units of mass, length, and time in the metric systems using the kinetic energy equation.

1.4 How many tons of CO_2 are produced annually by a typical internal engine car and by a hybrid car?

1.5 Burning a metric ton of coal with 67% carbon content generates about 2.4 tons of CO_2 and burning a liter of petroleum generates about 2.7 kg of CO_2.
a) If a total of 20 million tons of coal are burned every day, worldwide, calculate the weight of carbon dioxide emitted into the atmosphere from burning coal and associated ppm increase, ignoring the absorption of CO_2 on land and the sea.
b) Calculate the same for the total combustion of petroleum assuming a global consumption of 70 million barrels per day (a barrel of petroleum contains 42 gallons and a gallon is 3.78 l).
c) Calculate the same for 20,000 tons consumption of natural gas.

1.6 The United States with a population of 300 million accounts for about 18% of the 550 quad (quadrillion Btu) global energy demand. (a) What would be the increase in the global demand if all the 7.4 billion people on the globe used the same per capita energy as the US population and (b) what would be the corresponding CO_2 emissions assuming that the composition of the energy source mix doesn't change?

Use the following US energy mix composition: 30% natural gas; 37% petroleum; 14% coal; 9% nuclear; 10% wind + solar. (see Appendix 1.A for emission factors per fuel).

1.7   A square silicon solar cell with $5\,cm \times 5\,cm$ and thickness of $250\,\mu m$ has an electric power input of $22\,W$. The bottom surface of the cell is exposed to a coolant whose temperature is 20°C. If the convection heat transfer coefficient between the cell and the coolant is $180\,W/m^2/K$, determine the surface temperature of the cell.

Answers to Questions

Q1.1   Fossil fuel resources are finite so the income provided by using them can not last for ever. In contrast solar and wind are renewable energy resources.

Q1.2   (a) 220, 100 (b) 320, 180 (c) 400, 240

Q1.3   AMO corresponds to irradiation at the top of the atmosphere whereas AM1.5 corresponds to average daily irradiation at the surface of the earth.

Q1.4   (a) direct (b) diffuse (c) direct

Q1.5   105 minutes

Q1.6   The answer is country-specific.

Q1.7   Primary is the energy embedded in a fuel and there are heat losses during combustion and other processing that converts primary energy to electricity.

Q1.8   The answer is country specific; conversion factors are listed in Appendix 1.A.

Q1.9   (a) 147,058 (b) 14.7 (c) 36.3

References

1 E.F. Schumacher. *Small Is Beautiful*, 1st edition, 1973, republished by Hartley & Marks: London (1999).
2 A. Gore. *An Inconvenient Truth*, Bloomsbury Publishing: London (2006).
3 H. Scheer. *A Solar Manifesto*, James & James: London (2005).
4 J.J. Loferski. The first forty years: a brief history of the modern photovoltaic age. *Progress in Photovoltaics: Research and Applications*, 1, 67–78 (1993).
5 K. Zweibel, J. Mason and V. Fthenakis, A solar grand plan, *Scientific American*, 298(1), 64–73 (2008).
6 V. Fthenakis , J. Mason and K. Zweibel, The Technical, Geographical and Economic Feasibility for Solar Energy to Supply the Energy Needs of the United States, *Energy Policy*, 37, 387–399 (2009).
7 U.S. Department of Energy, *SunShot Vision study* (2012). https://energy.gov/sites/prod/files/2014/01/f7/47927.pdf (Accessed on September 18, 2017).

2

Solar Cells

2.1 Setting the Scene

We are now ready to discuss the underlying principles and operation of the invention central to our story—the modern solar cell. To help set the scene, we shall also say a few words about photovoltaic (PV) modules, reserving a detailed discussion for the next chapter. It will be helpful to start this chapter with a brief account of the main types of solar cell and module in widespread use today.

Silicon solar cells have been the workhorse of the PV industry for many years and currently account for well over 80% of world production. Modules based on these cells have a long history of rugged reliability, with guarantees lasting 20 or 25 years that are exceptional among manufactured products. Although cells made from other materials are constantly being developed and some are in commercial production, it will be hard to dislodge silicon from its pedestal. The underlying technology is that of semiconductor electronics: a silicon solar cell is a special form of semiconductor diode. Fortunately, silicon in the form of silicon dioxide (quartz sand) is an extremely common component of the Earth's crust and is essentially nontoxic. There is a further good reason for focusing strongly on silicon cells in this chapter: in its *crystalline* form silicon has a simple lattice structure, making it comparatively easy to describe and appreciate the underlying science.

There are two major types of crystalline silicon solar cell in current high-volume production:

- *Monocrystalline.* The most efficient type, made from a very thin slice, or wafer, of a large single crystal obtained from pure molten silicon. The circular wafers, often 5 or 6 inches (15 cm) in diameter, have a smooth silvery appearance and are normally trimmed to a pseudo-square or hexagonal shape so that more can be fitted into a module—see Figure 2.1. Fine contact fingers and bus bars are used to conduct the electric current away from the cells, which have a highly ordered crystal structure with uniform, predictable properties. However, they require careful and expensive manufacturing processes, including "doping" with small amounts of other elements to produce the required electrical characteristics. Typical commercial module efficiencies fall in the range 16–20%. The module surface area required is about 5–$6\,m^2/kW_p$.

Electricity from Sunlight: Photovoltaic-Systems Integration and Sustainability, Second Edition.
Vasilis Fthenakis and Paul A Lynn.
© 2018 John Wiley & Sons Ltd. Published 2018 by John Wiley & Sons Ltd.
Companion website: www.wiley.com/go/fthenakis/electricityfromsunlight

Figure 2.1 Each of these PV modules contains 72 monocrystalline silicon solar cells (*Source:* Reproduced with permission of EPIA/Phoenix Sonnenstrom).

- *Multicrystalline*, also called *polycrystalline.* This type of cell is also produced from pure molten silicon, but using a casting process. As the silicon cools it sets as a large irregular multicrystal that is then cut into thin square or rectangular slices to make individual cells. Their crystal structure, being random, is less ideal than with monocrystalline material and gives slightly lower cell efficiencies, but this disadvantage is offset by lower wafer costs. Cells and modules of this type often look distinctly blue, with a scaly, shimmering appearance, as in the building façade shown in Figure 2.2. Multicrystalline modules exhibit typical efficiencies in the range 13.5–16% and have overtaken their monocrystalline cousins in volume production over recent years. The module surface area is about 6–7 m^2/kW_p.

You have probably already gathered that the *efficiency* of any solar cell or module, the percentage of solar radiation it converts into electricity, is considered one of its most important properties. The higher the efficiency, the smaller the surface area for a given power rating. This is important when space is limited and also because some of the additional costs of PV systems—especially mounting and fixing modules—are area related. Monocrystalline silicon cells, when operated in strong sunlight, have the highest efficiencies of all cells commonly used in terrestrial PV systems, plus the promise of modest increases as the years go by due to improvements in design and manufacture. But it is important to realize that other types of cell often perform better in weak or diffuse light, a matter we shall return to in later sections.

Research laboratory cells achieve considerably higher efficiencies than mass-produced cells. This reflects the ongoing R&D effort that is continually improving cell design and leading to better commercial products. In some applications where space is

Figure 2.2 The façade of this cable-car station in the Swiss Alps is covered with multicrystalline silicon PV modules (*Source:* Reproduced with permission of IEA-PVPS).

limited and efficiency is paramount—for example, the famous solar car races held in Australia—high-quality cells made in small batches are often individually tested for efficiency before assembly.

Module efficiencies are slightly lower than cell efficiencies because a module's surface area cannot be completely filled with cells and the frame also takes up space. It is always important to distinguish carefully between cell and module efficiencies.

There is one further type of silicon solar cell in common use:

- *Amorphous.* Most people have met small amorphous silicon (a-Si) cells in solar-powered consumer products such as watches and calculators that were first introduced in the 1980s. Amorphous cells are cheaper than crystalline silicon cells, but have much lower efficiencies, typically 7–9%. Nowadays, large modules are available and suitable for applications where space is not at a premium, for example, on building façades. The surface area required is about $16\,\mathrm{m^2/kW_p}$. We shall discuss amorphous silicon in Section 2.3.

We focus initially on crystalline silicon solar cells for two main reasons: their comparatively simple crystal structure and theoretical background and their present dominant position in the terrestrial PV market. Their wafer technology has been around for a long time and is often referred to as "first generation"; they are the cells you are most likely to see on houses, factories, and commercial buildings.

However, it is important to realize that many other semiconductor materials can be used to make solar cells. Most come under the heading of *thin film*—somewhat confusing because a-Si is also commonly given this title—and involve depositing very thin layers of semiconductor on a variety of substrates. Thin-film products are generally regarded as the ultimate goal for terrestrial photovoltaics (PV) since they use very small amounts of semiconductor material and large-scale continuous production processes without any need to cut and mount individual crystalline wafers. Thin-film modules based on the compound semiconductors *cadmium telluride (CdTe)* and *copper indium diselenide (CIS)* are in commercial production, and the former is the technology with the lowest production cost. Often referred to as "second generation," they started with efficiencies lower than those of crystalline silicon, but they are currently catching up with multicrystalline silicon efficiencies. We will discuss them, and several types of specialized cells and modules, later in this chapter.

2.2 Crystalline Silicon

2.2.1 The Ideal Crystal

A large single crystal of pure silicon (Figure 2.3) forms the starting point for the monocrystalline silicon solar cell—the most efficient type in common use. As we shall see, the simple and elegant structure of such crystals makes it comparatively easy to explain the basic semiconductor physics and operation of PV cells. We are talking here of silicon refined to very high purity, similar to that used by the electronics industry to

Figure 2.3 Chunks of silicon (*Source:* Reproduced with permission of EPIA/Photowatt).

make semiconductor devices (diodes, transistors, and integrated circuits including computer chips). Its purity is typically 99.99999%. This contrasts with the far less pure metallurgical-grade silicon, produced by reducing quartzite in electric arc furnaces, that is used to make special steels and alloys.

The *Czochralski (CZ)* method of growing silicon crystals is quite easy to visualize. Chunks of pure silicon with no particular crystallographic structure are melted at 1414°C in a graphite crucible. A small seed of silicon is then brought into contact with the surface of the melt to start crystallization. Molten silicon solidifies at the interface between seed and melt as the seed is slowly withdrawn. A large ingot begins to grow both vertically and laterally with the atoms tending to arrange themselves in a perfect crystal lattice.

Unfortunately, this classic method of producing crystals has a number of disadvantages. Crystal growth is slow and energy intensive, leading to high production costs. Impurities may be introduced due to interaction between the melt and the crucible. And in the case of PV, the aim is of course to produce thin solar cell wafers rather than large ingots, so wire saws are used to cut the ingot into thin slices, a time-consuming process that involves discarding valuable material. For these reasons the PV industry has spent a lot of R&D effort investigating alternatives, including pulling crystals in thin sheet or ribbon form, and some of these are now used in volume production. Whatever method is employed, the desired result is pure crystalline silicon with a simple and consistent atomic structure.

Semiconductors, such as Si, are made up of individual atoms bonded together in a structure where each atom is surrounded by eight electrons. The electrons surrounding each atom of Si are part of a covalent bond, consisting of two atoms "sharing" a single electron. Each Si atom forms four covalent bonds with the four surrounding atoms. This is illustrated in Figure 2.4(a). Each line connecting the atoms represents an electron being shared between the two. Since each atom has four valence electrons that are not tightly bound to its nucleus, a perfect lattice structure is formed when each atom forms bonds with its four nearest neighbors (which are actually at the vertices of a three-dimensional tetrahedron, but shown here in two dimensions for simplicity). The structure has profound implications for the fundamental physics of silicon solar cells.

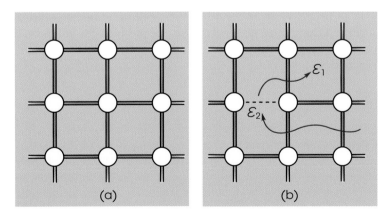

Figure 2.4 (a) Silicon crystal lattice; (b) electrons and holes.

Silicon in its pure state is referred to as an *intrinsic* semiconductor. It is neither an insulator like glass nor a conductor like copper, but something in between. At low temperatures its valence electrons are tightly constrained by bonds, as in part (a) of the figure, and it acts as an insulator. But bonds can be broken if sufficiently jolted by an external source of energy such as heat or light, creating electrons that are free to migrate through the lattice. If we shine light on the crystal, the tiny packets, or *quanta*, of light energy can produce broken bonds if sufficiently energetic. The silicon becomes a conductor, and the more bonds are broken, the greater its conductivity.

Figure 2.4(b) shows an electron ε_1 that has broken free to wander through the lattice. It leaves behind a broken bond, indicated by a dotted line. The free electron carries a negative charge and, since the crystal remains electrically neutral, the broken bond must be left with a positive charge. In effect it is a positively charged particle, known as a *hole.* We see that breaking a bond has given rise to a pair of equal and opposite charged "particles," an electron and a hole. At first sight the hole might appear to be an "immovable object" fixed in the crystal lattice. But now consider the electron ε_2 shown in the figure, which has broken free from somewhere else in the lattice. It is quite likely to jump into the vacant spot left by the first electron, restoring the original broken bond, but leaving a new broken bond behind. In this way a broken bond, or hole, can also move through the crystal, but as a positive charge. It is analogous to people sitting in a theater row and one by one moving one seat to the side. The moving people resemble the electrons and the emptied seats resemble the holes that appear as moving in the opposite direction.

We see that the electrical properties of intrinsic silicon depend on the number of mobile electron–hole pairs in the crystal lattice. At low temperatures, in the dark, it is effectively an insulator. At higher temperatures, or under sunlight, it becomes a conductor. If we attach two contacts and apply an external voltage using a battery, current will flow—due to free electrons moving one way and holes on the other. We have now reached an important stage in understanding how a silicon wafer can be turned into a practical solar cell.

Yet there is a vital missing link: remove the external voltage and the electrons and holes wander randomly in the crystal lattice with no preferred directions. There is no tendency for them to produce current flow in an external circuit. A pure silicon wafer, even under strong sunlight, cannot *generate* electricity and become a solar cell. What is needed is a mechanism to propel electrons and holes in opposite directions in the crystal lattice, forcing current through an external circuit and producing useful power. This mechanism is provided by one of the great inventions of the 20th century, the semiconductor p–n junction.

2.2.2   The p–n Junction

A conventional monocrystalline solar cell has a silvery top surface surmounted by a fine grid of metallic fingers forming one of its electrical contacts. What is less obvious is that the cell actually consists of two different layers of silicon that have been deliberately *doped* with very small quantities of impurity atoms, often phosphorus and boron, to form a *p–n junction.* The addition of such *dopants* is absolutely crucial to the cell's operation and provides the mechanism that forces electrons and holes generated by sunlight to do useful work in an external circuit.

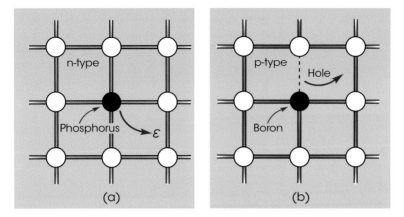

Figure 2.5 (a) A phosphorus atom in n-type silicon provides an extra free electron; (b) a boron atom in p-type silicon provides an extra hole.

The p–n junction may be regarded as the basic building block of the semiconductor revolution that began back in the 1950s. It is perhaps a little surprising that an invention normally associated with mainstream electronics should also form the basis of PV technology; but a silicon solar cell is essentially a form of p–n junction specially tailored to the task of converting sunlight into electricity.

We have already noted that heating or shining light on pure silicon can alter its electrical properties, progressively converting it from an insulator into a conductor. Another extremely important way of modifying its properties is to add small amounts of dopants. For example, if phosphorus is added to molten silicon, the solidified crystal contains some phosphorus atoms in place of silicon. While the latter has four valence electrons able to form bonds with neighboring atoms, phosphorus has five. The extra one is only weakly bound to its parent atom and can easily be enticed away, as shown in Figure 2.5(a). In other words, silicon doped with phosphorus provides plenty of free electrons, known as the *majority carriers*. Generally, there are also a few holes present due to thermal generation of electron–hole pairs, as in intrinsic silicon, and these are called *minority carriers*. The material is a fairly good conductor and is referred to as negative-type or *n-type*.

A complementary situation arises if silicon is doped with boron, which has only three valence electrons loosely bound to its nucleus, illustrated in part (b) of the figure. Each boron atom can only form full bonds with three neighboring silicon atoms, so boron introduces broken bonds into the crystal. In this case holes are the majority carriers and electrons the minority carriers. Once again, the material becomes a conductor; it is referred to as positive-type or *p-type*.

We see that n-type material has many surplus electrons and p-type material has many surplus holes. The next step is to consider what happens when the two materials are joined together to form a p–n junction, illustrated in Figure 2.6(a).

Near the interface, free electrons in the n-type material start diffusing into the p-side, leaving behind a layer that is positively charged due to the presence of fixed phosphorus atoms. Holes in the p-type material diffuse into the n-side, leaving behind a layer that is negatively charged by the fixed boron atoms. This diffusion of the two types of majority

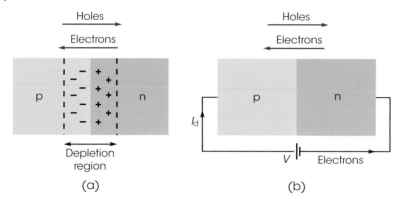

Figure 2.6 (a) A p–n junction; (b) applying forward bias.

carriers, in opposite directions across the interface, has the extremely important effect of setting up a strong electric field, creating a potential barrier to further flow. Equilibrium is established when the tendency of electrons and holes to continue diffusing down their respective concentration gradients is offset by their difficulty in surmounting the potential barrier. In this condition there are hardly any mobile charge carriers left close to the junction and a so-called *depletion region* is formed.

The depletion region makes the p–n junction into a diode, a device that conducts current easily in one direction only. Figure 2.6(b) shows an external voltage *V* applied to the diode, making the p-type material positive with respect to the n-type, referred to as *forward bias*. In effect the external voltage counteracts the "built-in" potential barrier, reducing its height and encouraging large numbers of majority carriers to cross the junction—electrons from the n-side and holes from the p-side. This results in substantial forward current flow (note that conventional positive current is actually composed of negatively charged electrons flowing the other way; we may think of them as going right around the circuit through the battery and back into the n-type layer). Conversely if the external voltage is inverted to produce a *reverse bias*, the potential barrier increases and the only current flow is a very small *dark saturation current* (I_0). This is because a bias that increases the potential barrier for majority carriers decreases it for minority carriers—and at normal temperatures there are some of these present on both sides of the junction due to thermal generation of electron–hole pairs.

The practical result of these movements of electrons and holes is summarized by the diode characteristic in Figure 2.7. Diode current *I* increases with positive bias, growing rapidly above about 0.6 V; but with negative bias the reverse current "saturates" at a very small value I_0. Clearly this device only allows current flow easily in one direction. Mathematically the curve is expressed as

$$I = I_0 \left[\exp\left(\frac{qV}{kT}\right) - 1 \right] \tag{2.1}$$

where *q* is the charge on an electron, *k* is Boltzmann's constant (1.3807×10^{-23} J/K), and *T* is the absolute temperature (K).

Equation (2.1) is called the **diode** or **Shockley** equation.

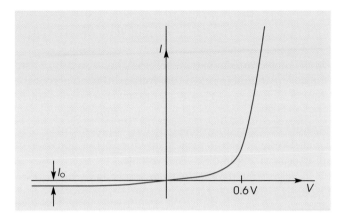

Figure 2.7 The voltage–current characteristic of a silicon diode.

You are perhaps beginning to wonder what all this has to do with solar cells, because we have not so far discussed the effects of shining light on the diode and it is not obvious what these will be. However, rest assured that understanding the aforementioned discussion of electrons and holes, majority and minority carriers, and potential barriers is essential for unraveling the mysteries of PV!

2.2.3 Monocrystalline Silicon

2.2.3.1 Photons in Action

We are now close to understanding how a monocrystalline silicon wafer, doped to create a semiconductor diode, can work as a power-generating solar cell. The basic scheme in Figure 2.8 shows a small portion of such a cell. At the top, several metallic contact fingers form part of the cell's negative terminal. Next comes a thin layer of n-type material interfacing with a thicker layer of p-type material to produce the crucial p–n junction. And finally there is a back contact that acts as the positive terminal. For clarity the cell's thickness is exaggerated in the figure; it is actually a very slim wafer, normally less than 0.3 mm from top to bottom.

A stream of photons containing minute packets or *quanta* of energy shines on the cell. Their numbers are staggering: under strong sunlight a 6-inch (15-cm) cell receives more than 10^{19} photons every second. Various possible fates await them, some productive and others fruitless, and we show a few important examples in the figure.

Unfortunately, there is some loss of photons by optical reflection back from the conducting fingers, top surface, and rear surface (nos. 1, 2, and 3 in the figure). The rest enter the cell body, but only those with a certain minimum energy, known as the *bandgap*, have any chance of creating an electron–hole pair and contributing to the cell's electrical output. The most productive ones, for reasons explained in the following text, create electron–hole pairs in the p-type layer or in the n-type layer very close to the junction (4 and 5). Less productive, on average, are the ones that travel further into the p-type material (6). Successful cell design involves producing as many electron–hole pairs as possible, preferably close to the junction. But even high-quality cells are subject to theoretical limits dictated by the spectral distribution of sunlight, nature of light absorption in silicon, and quantum theory. We shall discuss these topics a little later.

First comes the big question: what happens to the electron–hole pairs generated within the cell by sunlight, and how do they produce current flow in an external circuit?

As we have seen, majority carriers (electrons in n-type material, holes in p-type) are the main players in a conventional semiconductor diode. By initial diffusion across the p–n junction, they set up a depletion layer and create a potential barrier. Forward-biasing the diode reduces the height of the barrier, making it easier for them to cross the junction and produce substantial current. In reverse bias the barrier increases and current flow is severely inhibited. Diode action is principally due to the behavior of majority carriers under the influence of an applied external voltage.

With solar cells, however, it is light-generated *minority carriers* that take center stage in creating electric current. The basic reason may be simply stated: a potential barrier that inhibits transfer of majority carriers across a p–n junction positively encourages the transfer of minority carriers. Whereas majority carriers experience "a hill to climb," minority carriers see "a hill to roll down." With luck they are swept down this hill, *collected* at the cell terminals, and produce an output current proportional to the intensity of the incident light.

Let us consider the three photons in Figure 2.8 that successfully create electron–hole pairs in the crystal lattice. Number 4 produces a pair in the p-type region, close to the junction. Its free electron, a minority carrier in p-type material, is easily swept across the junction and collected. So is the hole produced in the n-type region by number 5, which is swept across the junction in the opposite direction. Both these minority carriers should contribute to the light-generated current.

Photon 6 also creates an electron–hole pair, but well away from the junction and its associated electric field. The free electron does not immediately experience "a hill to roll

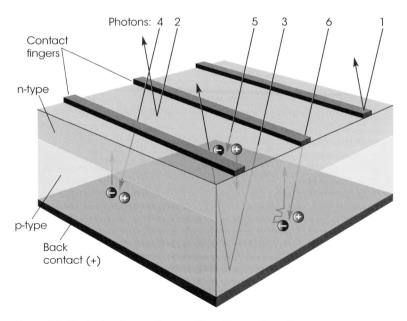

Figure 2.8 The basic scheme of a crystalline silicon solar cell.

down," but instead starts wandering randomly through the silicon lattice. In the figure it is shown eventually reaching the junction and being swept away to success. But the journey is a dangerous one: it may instead encounter a hole and be annihilated. Although such *recombination* is not illustrated in the figure, unfortunately it occurs not only in the main body of the cell (*bulk recombination*) but also even more importantly at the edges and metal contacts due to defects and impurities in the crystal.

The longer a minority carrier wanders around, the greater the distance traveled through the crystal and the more likely it is to be lost by recombination. Two measures are used to describe the risk. The *carrier lifetime* is the average amount of time between electron–hole generation and recombination (the bigger, the better), which for silicon is typically 1 μs. The *diffusion length* is the average distance a carrier moves from the point of generation until it recombines, for silicon typically 0.2 mm that is comparable with the thickness of the monocrystalline wafer. This again emphasizes the value of electron–hole pairs generated close to the junction.

We have now covered some fundamental aspects of solar cell operation, including the key role played by light-generated minority carriers. The next task is to consider the voltage–current characteristics of the cell as measured at its output terminals.

2.2.3.2 Generating Power

We have seen solar photons at work, creating minority carriers that speed toward the solar cell's output terminals under the magical influence of the p–n junction. But how is all this internal activity reflected in the cell's power generation, and what voltages and currents are produced at its terminals? Figure 2.9(a) helps answer the question with an equivalent circuit summarizing the cell's behavior as a circuit component. It consists of a diode representing the action of the p–n junction together with a current generator representing the light-generated current I_L.

In dark conditions I_L is zero and the cell is quiescent. If an external voltage source is connected, the cell behaves just like a semiconductor diode with the characteristic shown in part (b) of the figure (this has the same form as Figure 2.7). We choose to define the current I as flowing into the circuit and, in the dark, it must be the same as the diode current I_D. Note also that since a diode is a *passive* device that dissipates power, the cell's dark characteristic lies entirely in the first and third quadrants (I and V either both positive or both negative). But if sufficient sunlight falls on the cell to turn it into an *active* device delivering power to the outside world, the current I must reverse

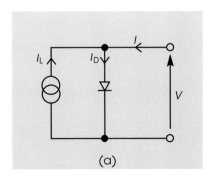

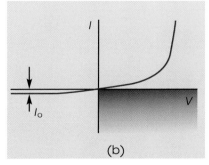

Figure 2.9 (a) The equivalent circuit of a solar cell; (b) its *I–V* characteristic in the dark.

and the characteristic will shift into the fourth quadrant (*I* negative, *V* positive) shown shaded in the figure.

In sunlight the generator produces a current I_L proportional to the level of insolation. It is effectively superimposed on the normal diode characteristic, and we may write

$$I = I_D - I_L \tag{2.2}$$

Substituting for the diode current using Equation (2.1) gives

$$I = I_0\left[\exp\left(\frac{qV}{kT}\right) - 1\right] - I_L \tag{2.3}$$

This equation confirms that the diode *I–V* characteristic is shifted down into the fourth quadrant by an amount equal to the light-generated current I_L. This is shown in Figure 2.10(a).

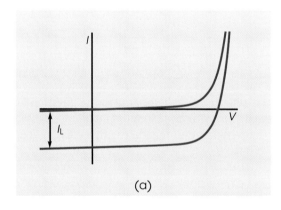

(a)

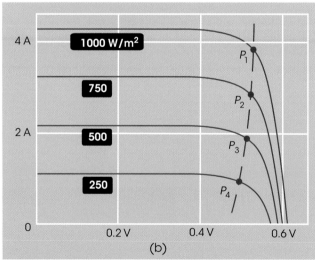

(b)

Figure 2.10 (a) The light-generated current shifts the cell's characteristic into the fourth quadrant; (b) a family of *I–V* curves for a 2 W_p solar cell.

Most people are unfamiliar with curves in the fourth quadrant, so for convenience the *I–V* characteristics of a solar cell are normally "flipped over" to the first quadrant. This is equivalent to plotting *V* against –*I*. Part (b) of the figure illustrates a family of such curves for a typical crystalline silicon cell rated by the manufacturer at $2\,W_p$. Each curve represents a different strength of sunlight and hence a different value of I_L. You will recall that PV cells and modules are normally rated in peak watts (W_p), indicating the maximum power they can deliver under standard conditions (insolation $1000\,W/m^2$, cell temperature 25°C, AM1.5 solar spectrum). Therefore, we should first consider how the rated power of $2\,W_p$ relates to the $1000\,W/m^2$ *I–V* curve.

In general, the cell's power output equals the product of its voltage and current. No power is produced on open circuit (maximum voltage, zero current) or short circuit (maximum current, zero voltage). The full rated power is obtained by operating the cell slightly below maximum voltage and current at its *maximum power point* (*MPP*), shown as P_1 against the $1000\,W/m^2$ curve, and corresponding to about 4 A at 0.5 V or 2 W. We can only obtain the promised output power by operating the cell at its MPP. Three other curves are shown for lower insolation values of 750, 500, and $250\,W/m^2$; each has its own MPP (P_2, P_3, P_4), indicating the maximum power available from the cell at that particular strength of sunlight.

Note that the maximum voltage produced by a silicon solar cell is about 0.6 V, considerably less than the 1.5 V of a dry battery cell. This means that it is essentially a low-voltage, high-current device, and many cells must be connected in series to provide the higher voltages required for most applications. For example, the PV module previously illustrated in Figure 2.1 has 72 individual cells connected in series, giving a DC voltage of about 35 V at the MPP. Higher voltages may be obtained by connecting a number of modules in series.

The *I–V* characteristics suggest another important aspect of the solar cell—it is helpful to think of it as a *current source* rather than a *voltage source* like a battery. A battery has a more or less fixed voltage and provides variable amounts of current; but at a given insolation level, the solar cell provides a more or less fixed current over a wide range of voltage.

The maximum voltage of the cell, its *open-circuit voltage* V_{oc}, is given by the intercept on the voltage axis and lies in the range 0.5–0.6 V. It does not depend greatly on the insolation. The close relationship between the diode characteristic of the p–n junction and the *I–V* characteristics in sunlight, illustrated in Figure 2.10(a), means that the open-circuit voltage is similar to the forward voltage of about 0.6 V at which a silicon diode starts to conduct heavily.

The maximum current from the cell, its *short-circuit current* I_{sc}, is given by the intercept on the current axis and is proportional to the strength of the sunlight. Other things being equal, it is also proportional to the cell's surface area. It represents the full flow of minority carriers generated by the sunlight and successfully "collected" after crossing the p–n junction.

The aforementioned parameters are further illustrated in Figure 2.11. The blue curve shows a typical *I–V* characteristic at $1000\,W/m^2$ insolation, labeled with the short-circuit current, open-circuit voltage, and MPP. The red curve shows how power output varies with voltage; the maximum value is $P_{mp} = I_{mp} \times V_{mp}$. Since the current holds up well over most of the voltage range, it follows that the cell's output power is roughly proportional to voltage up to the MPP. This emphasizes once again the importance of operating the cell close to the MPP if its power output potential is to be realized.

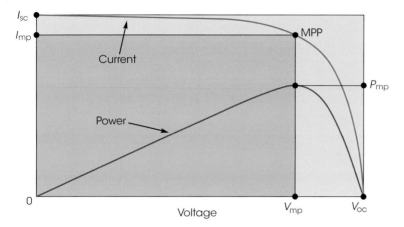

Figure 2.11 Current and power at standard insolation.

A widely used measure of performance that reflects the overall quality of the cell is its *fill factor* (*FF*) given by

$$FF = \frac{I_{mp}V_{mp}}{I_{sc}V_{oc}}$$
(2.4)

An "ideal" cell in which the current held right up to the short-circuit value, then reduced suddenly to zero at the MPP, would have an FF of unity. Needless to say, practical cells do not achieve this; the $I-V$ characteristics in the figure have an FF of about 70%. Equation (2.4) shows that graphically it is equal to the ratio between the areas of the small and large shaded rectangles in the figure.

So far we have not considered the effects of temperature on cell performance, but actually they are quite important, especially in the case of crystalline silicon. Many people imagine that solar cells are more efficient if operated at elevated temperatures, perhaps thinking of the type of solar–thermal panel used for water heating. But solar PV cells like to be kept cool—they do very well in strong winter sunshine in the Swiss Alps! In hot climates cell temperatures can reach 70°C or higher and system designers often go to considerable lengths to ensure adequate ventilation of PV modules to assist cooling.

The main effect of temperature on a cell's $I-V$ characteristics is a reduction in open-circuit voltage, illustrated in Figure 2.12. We have repeated the $1000\,W/m^2$ curve for the $2\,W_p$ cell already shown in Figure 2.10(b) for the standard temperature of 25°C and added two further curves for 0 and 50°C. The open-circuit voltage changes by about 0.1 V between these extremes, corresponding to 0.33% per °C. Note that the *temperature coefficient* is negative; in other words the voltage decreases as the temperature rises. There is a much smaller effect on the short-circuit current. Generally, the cell loses power at elevated temperatures, a more serious effect with crystalline silicon than most other types of solar cell.

You have probably noticed one major omission from this discussion—an explanation of efficiency. At the start of this chapter, we noted that commercial multicrystalline silicon modules have typical efficiencies in the range 13–16%, but we have not so far

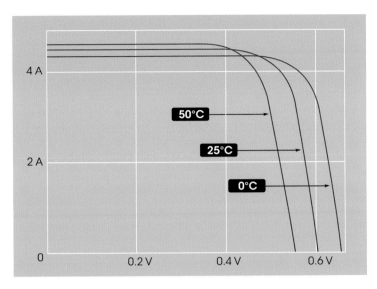

Figure 2.12 Effects of temperature on the *I–V* characteristic.

explained the reasons for this apparently rather disappointing performance. Returning for a moment to Figure 2.10(b), it is not clear from our discussion why this cell, which probably receives up to about $14\,W_p$ of incident solar energy, only manages to convert $2\,W_p$ into electrical output. Where does the rest go, and why cannot the efficiency be dramatically improved by better design? This raises some fundamental issues that we tackle in the next section.

2.2.3.3   Sunlight, Silicon, and Quantum Mechanics

It may seem a little surprising to find "quantum mechanics" mentioned in an introductory book on PV—and possibly unnerving in view of a quotation by Richard Feynman (1918–1988), latterly a professor at the California Institute of Technology, who received a Nobel Prize in Physics in 1965 for his work on quantum mechanics and famously declared: "I think I can safely say that nobody understands quantum mechanics."

So it is clear we must tread lightly, leaving the great body of 20th-century quantum theory undisturbed. Yet not entirely, for it contains precious nuggets relating to the nature of sunlight and imposes fundamental limits on the efficiency of solar cells.[1,2]

Back in Section 1.4 we noted that certain eminent physicists, from Isaac Newton in the 17th century to Albert Einstein in the 20th, viewed light as a stream of minute particles carrying discrete packets of energy. And in Section 2.2.3.1 we stated—without explanation—that a light quantum or photon needs to have a certain minimum energy, known as the *bandgap*, if it is to have any chance of creating an electron–hole pair in a silicon crystal lattice. It is now time to bring these ideas together with the help of a little quantum theory.

The human eye is sensitive to visible light—all the colors of the rainbow from violet to red. The corresponding range of wavelengths is about 0.4–0.8 μm. The complete solar spectrum, previously shown in Figure 1.6, also contains significant energy at ultraviolet (UV) and especially infrared (IR) wavelengths. A key concept of quantum theory is that the energy content of a photon is related to wavelength by a surprisingly simple equation:

$$E = \frac{hc}{\lambda} \tag{2.5}$$

where E is the photon energy, h is Planck's constant ($6.62607004 \times 10^{-34}\,\mathrm{m^2\,kg/s}$), c is the velocity of light ($\sim 3 \times 10^8\,\mathrm{m/s}$), and λ is the wavelength. This means that the packet of energy or quantum is about twice as large for a violet photon as for a red photon. And as Einstein proposed in 1905, quanta can only be generated or absorbed as complete units.

When E is expressed in electron volts (eV) and λ in micrometers (μm),

$$E(\mathrm{eV}) = \frac{1.24}{\lambda}(\mu\mathrm{m}) \tag{2.6}$$

A second key point is that solar cells based on semiconductors are essentially quantum devices. An individual solar photon can only generate an electron–hole pair if its quantum of energy exceeds the bandgap of the semiconductor material, also known as its *forbidden energy gap*. This is illustrated in Figure 2.13 and the energy bandgaps of various semiconductors are shown in Figure 2.14.

You may recall that the creation of an electron–hole pair involves jolting a valence electron to produce a broken bond in the crystal lattice. The electron moves from the *valence band* to the *conduction band*, leaving behind an equal, but oppositely charged hole. However, the energy levels of an electron in the two bands are separated by a discrete energy gap. Moving from one band to another requires a "quantum leap"—it is all or nothing, and intermediate levels are forbidden. Long-wavelength IR and red photons do not generally have the necessary amount of energy. Conversely most photons toward the violet end of the spectrum have more than enough and the excess must be dissipated as heat. These fundamental considerations, taken in conjunction with the sun's spectral distribution, reduce the theoretical maximum efficiency of a silicon solar cell at an insolation of $1000\,\mathrm{W/m^2}$ to about 30%. The figure does not take account of various

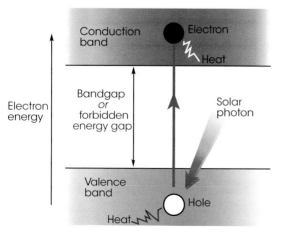

Figure 2.13 Quantum effects in solar cells.

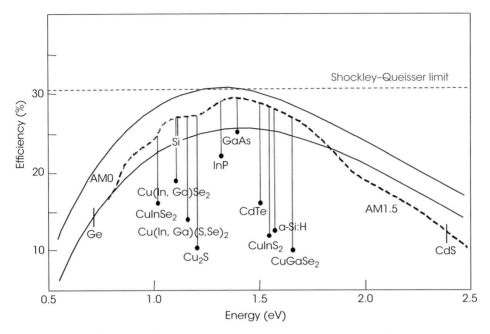

Figure 2.14 The theoretical efficiencies of various solar cell types as a function of energy bandgap.

other loss mechanisms and practical design considerations, some of which were illustrated in Figure 2.8. So it is not hard to appreciate why cells made in research laboratories do well to reach 30% and why most commercial, mass-produced cells achieve less than 20%.

The Shockley–Queisser limit of 30% shown in Figure 2.14 corresponds to (i) single junction, (ii) unconcentrated light, (iii) $T = 25°C$, and (iv) thermal relaxation so that energy above the bandgap is dissipated as thermal energy instead of contributing to energizing more electrons. To go above the Shockley–Queisser, we need to use concentrated light, more than one junction so a stack of cells one at the top of the other, and/or materials and configurations that can give better quantum efficiency.

We can now appreciate why the size of the bandgap is a very important influence on solar cell efficiency. It represents the minimum energy needed to free an electron from its valence band to the conduction band so that it can move within the crystal and conduct electricity. The energy bandgap of semiconductors tends to slightly decrease as the temperature is increased because interatomic spacing increases with the amplitude of atomic vibrations increasing due to the increased thermal energy. An increased interatomic spacing decreases the potential seen by the electrons in the material, which in turn reduces the energy bandgap. The levels shown in Figure 2.14 correspond to 0°C (273.15°K).

It becomes obvious now why there are several types of solar cells based on different semiconductor materials. Using a semiconductor with a wide bandgap yields a device with a high voltage, but photons are "thrown away." Using a semiconductor with a narrow bandgap yields a device with a high current (as more photons are absorbed) but low voltage (Figure 2.15). Since $P = V \times I$, there is not an optimal bandgap that corresponds

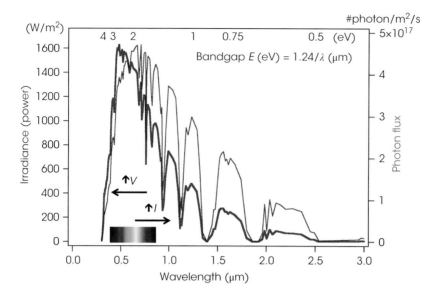

Figure 2.15 Solar spectrum wavelength and its relationship with semiconductor energy bandgap.

to the maximum product of current and voltage, although most efficient harvesting of the sun's energy requires bandgaps in the range 1.0–1.6 electron volts (eV). Silicon's bandgap of 1.1 eV is fairly good in this respect. Certain other semiconductor materials (e.g., CdTe, GaAs) have bandgaps closer to the middle of the range, and we will discuss them later.

Unfortunately, not all photons with the necessary energy are readily absorbed. Most solar cell materials, the *direct-bandgap* semiconductors, act as good light absorbers within layers just a few micrometers thick. But crystalline silicon, an *indirect-bandgap* material, is not so effective. It absorbs high-energy blue photons quite easily, close to the cell's top surface, but low-energy red photons generally travel much further before absorption and may exit the cell altogether. The basic problem is that successful generation of conduction electrons in silicon requires additional quantum lattice vibrations that complicate the process, so that layers less than about 0.1 mm thick are not good light absorbers. Special *light-trapping* techniques may be used to increase the path length of light inside the cell and give a better chance of electron–hole generation. These are described in the next section.

To summarize, it would be helpful if every photon entering a solar cell produced an electron–hole pair and contributed to power generation, in other words if the *quantum efficiency*, thus the number of electrons divided by the number of photons, was 100%. But quantum theory tells us this is impossible. Photons are all-or-nothing packets of energy that can only be used in their entirety. Some are too feeble in their energy content, while others are unnecessarily strong, placing fundamental limits on solar cell efficiency. Disappointing though this may seem, we should always remember that sunlight is "free" energy to be used or not as we wish. Photons are not wasted if untapped— at least not in the sense of an old-fashioned power station burning fossil fuel that effectively discards around 60% of its precious fuel as waste heat.

2.2.3.4 Refining the Design

Solar cell designers are constantly striving to improve conversion efficiencies and have used their ingenuity over many years to refine crystalline silicon cells beyond the basic scheme already illustrated in Figure 2.8. Some of the constraints on efficiency are caused by fundamentals of light and quantum theory, others by the properties of semiconductor materials or the problems of practical design.

One important point should be made at the outset. Researchers use various sophisticated techniques to achieve "record" efficiencies and can select their best cells for independent testing and accreditation. But PV companies engaged in large-scale production have an additional set of priorities: simple, reliable, and rapid manufacturing processes and high yield coupled with minimal use of expensive materials, all aimed at lower costs. Manufacturers are certainly interested in the commercial advantages of high cell efficiency and over the years have incorporated many design advances coming out of research laboratories, but cost must always be a big consideration and there are often significant time lags.

Figure 2.16 summarizes the main factors determining the efficiency of a typical, commercial, crystalline silicon solar cell operated at or near its MPP. On the left the incident solar power is denoted by 100%. Successive losses, shaded in blue, reduce the available power to around 15–20% at the cell's output terminals—its rated efficiency value. We will now discuss each loss category in turn.

Quantum Theory

We emphasized the fundamental limitations imposed by quantum theory[3] in the previous section. They represent the biggest loss of efficiency in a solar cell based on a single p–n junction. One way of reducing the problem is to stack together two or more junctions with different bandgaps, creating a *tandem cell.* A well-known example, which has been exploited commercially for many years, is based upon amorphous rather than crystalline silicon, and we shall mention this again in Section 2.3.

Optical Losses

Optical losses affect the incoming sunlight, preventing absorption by the semiconductor material and production of electron–hole pairs. The small section of solar cell shown in Figure 2.17 illustrates three main categories of optical loss: blocking of the light by the top contact (i); reflection from the top surface (ii); and reflection from the back contact without subsequent absorption (iii).

Shadowing by the top contact can obviously be minimized by making the total contact area as small as possible. This area comprises not only the metallic contact fingers

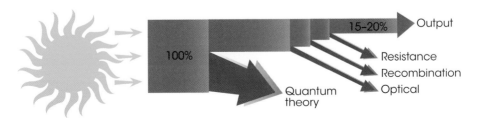

Figure 2.16 Solar cell losses.

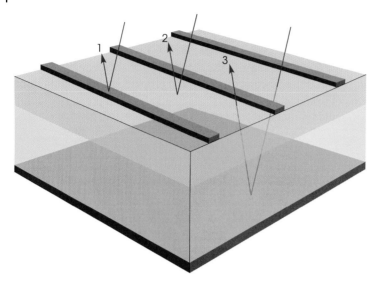

Figure 2.17 Optical losses.

shown in the figure (and previously in Figure 2.8) but also wider strips known as *bus bars* that join many fingers together and conduct current away from the cell. Clearly a well-spaced grid of very fine fingers and narrow bus bars helps reduce optical loss, but the disadvantage is increased electrical resistance. As always, practical design involves compromise.

The photo in Figure 2.18 shows the top surface of a monocrystalline silicon cell, surrounded by its neighbors in a PV module. This example has very simple grid geometry, consisting of 49 fine vertical fingers and two horizontal bus bars, giving a shadowing loss of about 11%. The fingers have constant width; a more efficient design would taper them to account for the increasing current each carries as it nears a bus bar. The bus bars are slightly tapered toward the low-current end; it would be better to taper them along their length as they pick up current from more and more fingers. Ideally the cross

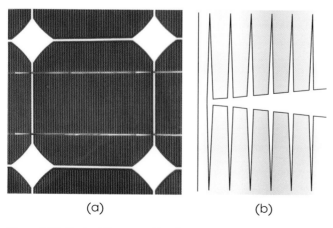

(a) (b)

Figure 2.18 Contact fingers and bus bars.

sections of fingers and bus bars should be roughly proportional, at each point, to the current carried. To illustrate this, a small section of a more efficient finger–bus bar design is shown in part (b) of the figure.

The *metallization pattern* of fingers and bus bars, as well as having its own inherent resistance to current flow, introduces contact resistance at the semiconductor interface. This may be reduced by heavy doping of the top layer of semiconductor material, at the risk of forming a significant dead region at the surface that reduces the collection efficiency of blue photons.

Conventional top contacts are made from very thin metallic strips formed using a screen-printing process. A metallic paste is squeezed through a mask, or screen, depositing the desired contact pattern that is then fired. The shading loss, typically between 8 and 12%, represents a significant drain on cell efficiency. A major design improvement, pioneered in the 1990s at the University of New South Wales,[4] uses laser-formed grooves to define a metallization pattern with narrower but deeper fingers just below the cell's surface. Such *buried contact solar cells* offer valuable gains in efficiency compared with normal screen-printed designs.

The second category of optical loss illustrated in Figure 2.17 is reflection from the cell's top surface. Two main design refinements are commonly employed. The first is to apply a transparent dielectric *antireflection coating* (ARC) to the top surface, illustrated in Figure 2.19. If the coating is made a quarter-wavelength thick, the light wave reflected from the ARC/silicon interface is 180° out of phase with that reflected from the top surface, and when the two combine, the resulting interference effects produce cancelation. This condition is met when

$$d = \frac{\lambda}{4n} \tag{2.7}$$

where d is the thickness, n is the refractive index of the coating material, and λ is the wavelength (interestingly, we are temporarily considering light as a wave rather than a stream of particles, a good example of the dual nature of light first mentioned in Section 1.4). Clearly, exact cancelation can only occur at one value of λ, normally chosen

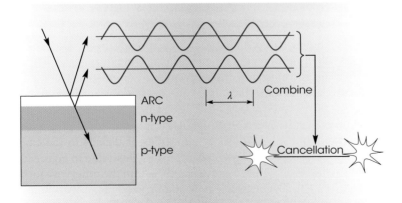

Figure 2.19 An antireflection coating reduces reflection from the top surface by cancelation.

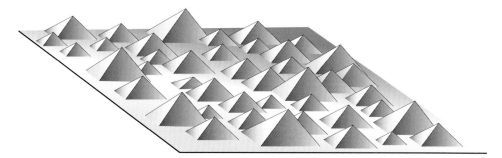

Figure 2.20 Texturization by raised pyramids.

to coincide with the peak photon flux about 0.65 µm. The antireflection performance falls off to either side of this value. For optimum performance the refractive index of the ARC material should be intermediate between that of the materials on either side, usually silicon or either air or glass.

The second design refinement involves *texturizing* the top surface so that light is reflected in a fairly random fashion and has a better chance of entering the cell. Almost any roughening is helpful, but the crystalline structure of silicon offers a special opportunity because careful surface etching can be used to create a pattern of minute raised pyramids, illustrated in Figure 2.20. Light reflected from the inclined pyramidal faces is quite likely to strike adjacent pyramids and enter the cell.

The third type of optical loss is reflection of light from the back of the cell, without subsequent absorption. This may be reduced by an uneven back surface that reflects the light in random directions, trapping some of it in the cell by total internal reflection. The technique is referred to as *light trapping*[4] and is very important in crystalline silicon cells because silicon is a relatively poor light absorber, especially of longer-wavelength (red) light. It is illustrated in Figure 2.21.

It is difficult to put precise figures on the efficiency losses caused by these various optical effects. However, a cell that includes carefully designed metallization, ARC, texturization, and light trapping can give major improvements compared with the basic structure first illustrated in Figure 2.8.

Recombination Losses

The undesirable process known as recombination has already been discussed in Section 2.2.3.1. It occurs when light-generated electrons and holes, instead of being swept across the p–n junction and collected, meet up and are annihilated. The wastage of charge carriers adversely affects both the voltage and current output from the cell, reducing its efficiency.

Some recombination takes place as electrons and holes wander around in the body of the cell (*bulk recombination*), but most occurs at impurities or defects in the crystal structure near the cell's surfaces, edges, and metal contacts, as illustrated in Figure 2.22. The basic reason is that such sites allow extra energy levels within the otherwise forbidden energy gap (see Figure 2.13). Electrons are now able to recombine with holes by giving up energy in stages, relaxing to intermediate energy levels before finally falling back to the valence band. In effect they are provided with stepping stones to facilitate the quantum leaps necessary for recombination.

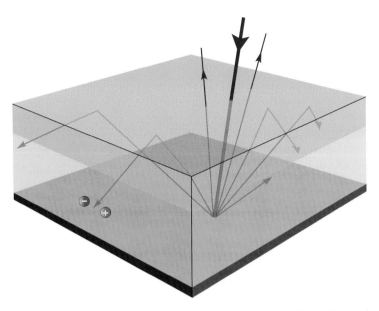

Figure 2.21 Light trapping helps keep incoming light within the cell by total internal reflection.

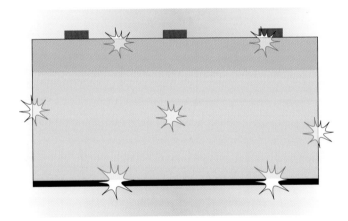

Figure 2.22 Typical recombination sites. The central one represents bulk recombination, the others occur close to surfaces, edges, and contacts.

What can be done to reduce recombination? Three important techniques may be briefly mentioned here. The first involves processing the cell to create a *back surface field* (*BSF*). Although the details are subtle,[3] the tendency of long wavelength photons to recombine at the back of the cell may be reduced by including a heavily doped aluminum region, which also acts as the back contact. Next, it is possible to reduce recombination at the external surfaces by chemical treatment with a thin layer of *passivating oxide*. And finally, regions adjacent to the top contacts may be heavily doped to create "minority carrier mirrors" that dissuade holes in the n-type top layer from approaching the contacts and recombining with precious free electrons.

Resistance Losses

The final efficiency loss shown in Figure 2.16 is due to electrical resistance. We previously noted that a solar cell is best thought of as a current generator. As with other current generators, it is desirable to minimize resistance in series with the output terminals and maximize any shunt resistance that appears in parallel with the current source. A low shunt resistance, which could happen due to manufacturing defects, would cause power losses in solar cells by providing an alternate current path for the light-generated current.[5] Figure 2.23 shows two equivalent circuits similar to that previously used for a solar cell (Figure 2.9) but modified to include a series resistance R_1 and a shunt resistance R_2. Ideally, R_1 would be zero and R_2 infinite, but, needless to say, we cannot expect these values in practice.

The physical interpretation of R_1 is straightforward. It represents the resistance to current flow offered by the bus bars, fingers, contacts, and the cell's bulk semiconductor material. A well-designed cell keeps R_1 as small as possible. R_2 is more obscure, relating to the nonideal nature of the p–n junction and impurities near the cell's edges that tend

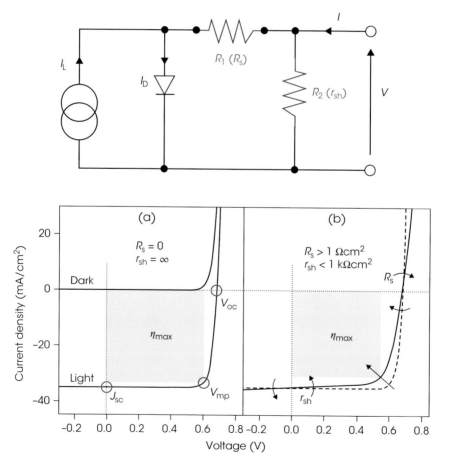

Figure 2.23 Equivalent circuits and *I–V* characteristics of a solar cell: (a) ideal solar cell; (b) "less than ideal" actual solar cell with series (R_s) and shunt (r_{sh}) resistances (courtesy Markus Gloeckler).

to provide a short-circuit path around the junction. In practical designs both resistors cause losses, but it is simpler to appreciate their effects if we treat them separately.

The I–V, also termed J–V, characteristic shown in (a) is for $R_1 = 0$ and $R_2 =$ infinity, which is the ideal case. Part (b) in Figure 2.23 shows characteristics for a cell with finite values of R_1 and R_2. Let us first consider the open-circuit condition, $I = 0$. In this case there is no current through R_1 and no voltage drop across it, so the open-circuit voltage V_{oc} must be the same as for the reference cell. We assumed that series resistance due to a cell's bus bars, fingers, contacts, and semiconductor material has no effect on the open-circuit voltage. However, full circuit analysis shows that it causes a small reduction in short-circuit current and a loss of FF, as indicated. R_1 (R_s) causes a slight rotation of the I–V curve around the (V_{oc}, $I = 0$) point.

To consider the effects of shunt resistance, it is helpful to consider the short-circuit condition, $V = 0$. In this case there is no voltage across R_2 and no current through it, so the short-circuit current I_{sc} must be the same as for the reference cell. Thus, a finite shunt resistance due to imperfections in and around the cell's p–n junction has no effect on the short-circuit current. However, it has a minor effect on the open-circuit voltage and a considerable one on the FF. The shunt resistance R_2 (r_{sh}) causes a slight rotation of the I–V curve around the (I_{sc}, $V = 0$) point. To conclude, a practical cell with both series and shunt resistance losses is expected to suffer small reductions in both V_{oc} and I_{sc}; but the most serious effect is generally degradation of the FF (defined in Equation (2.4)).

This interdependency can be quantified by replacing the simplified model shown in Figure 2.9 and Equation (2.3) by a more standard model that includes both the series resistance R_1 and the shunt resistance R_2/(shown in Figure 2.23). Then for deriving the characteristic I–V curves, the current becomes $I = I_1 - I_D - I_2$, where I_2 is estimated from Equation (2.7):

$$I_2 = \frac{V_D}{R_2} = \frac{V + I \cdot R_1}{R_2} \tag{2.8}$$

Then the characteristic I–V equation is given as

$$I = I_1 - I_0 \cdot \left(e^{\frac{(V + I \cdot R_1)q}{k_T}} - 1 \right) - \frac{V + I \cdot R_1}{R_2} \tag{2.9}$$

We have now covered the main categories of efficiency loss in crystalline silicon solar cells. The techniques for counteracting them have been conceived and enhanced over many years in R&D laboratories around the world, leading to continuous improvements in cell and module efficiencies. Of course, the degree to which they are employed in a commercial product depends upon the manufacturer's expertise and judgment; the number and complexity of processing steps have a big impact on cost and there is inevitably a trade-off between cost and performance.

2.2.4 Multicrystalline Silicon

In most respects multicrystalline silicon, also referred to as *polycrystalline silicon* or more simply as *poly-Si*, produces solar cells that are very similar to their monocrystalline cousins. The theoretical background is shared, even though the initial stage of

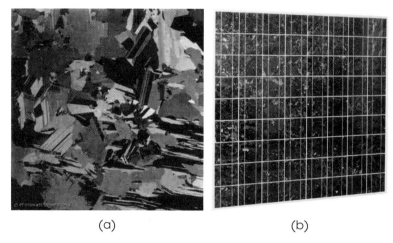

(a)   (b)

Figure 2.24 (a) Multicrystalline silicon wafer; (b) module (*Source:* Reproduced with permission of EPIA/Photowatt).

manufacture is different. As first mentioned in Section 2.1, multicrystalline cells also start life as pure molten silicon, but the material is cast in substantial blocks, cut into smaller bricks, and finally made into thin wafers. The casting process produces a multi-grain crystal structure that is less ideal than monocrystalline material and gives cell and module efficiencies typically 1% (absolute) lower, but this disadvantage is offset by lower wafer costs. And since the cells are cut square or rectangular, rather than "pseudo-square" as with monocrystalline cells, they can be packed closely in modules. They have a scaly, shimmering appearance. The façade exhibited in Figure 2.2 shows that the modules tend to have a distinctly blue appearance due to their ARC, a property often appreciated by architects.

As the molten silicon cools, crystallization occurs simultaneously at many points, producing crystal grains with random sizes, shapes, and orientations. After cutting into thin wafers, the material has the surface appearance in Figure 2.24(a). Within each grain the crystal structure is highly regular, but the many grain boundaries represent imperfections and provide unwelcome sites for electron–hole recombination. The problem is reduced if grains are at least a few millimeters across and extend from front to back of the wafer. As part (b) of the figure shows, a multicrystalline module tends to present a uniform, shimmering appearance without the gaps between cells associated with the "pseudo-square" shape of monocrystalline cells.

On the whole there is little to choose between the performance of monocrystalline and multicrystalline PV modules. From a user's point of view, efficiency and cost differences may not be decisive and the choice often comes down to appearance, availability, and the manufacturer's reputation and guarantee.

2.3 Second-Generation Photovoltaics

The crystalline silicon PV technologies that comprise interconnected small cells forming PV modules are the first generation of PV in the market. The second-generation includes technologies based on the deposition of thin films on large substrates and then dividing those to form the cells and their interconnects. There are three types of commercial "thin-film" technologies, amorphous and thin-film silicon, CdTe, and copper indium gallium diselenide (CIGS). These technologies entail a lower manufacturing cost than crystalline silicon at the expense, at least until recently, of lower efficiencies.

2.3.1 Amorphous and Thin-Film Silicon

Amorphous silicon (a-Si) was the first *thin-film* technology used in PV. Small a-Si cells in consumer products such as watches and calculators have introduced solar cells to millions of people since the 1980s. The tiny amounts of power required by such products make the comparatively low efficiency of their cells unimportant, and in any case they are rarely used outdoors under strong sunlight! Ease of manufacture and low cost are their strong points. What is not so generally realized is that a-Si technology has been developed in recent years and scaled up for higher-power applications. Although it only accounts for a few percent of world production, it is no longer confined to consumer products. A good example is building façades; a-Si modules can serve as attractive cladding and may well be competitive with other types of PV module. PV cladding is not necessarily more expensive than traditional high-quality materials and may be chosen for its aesthetic appeal or as an environmental statement. If the façade also generates electricity, so much the better. Efficiency is not the only criterion.

In any case the question of efficiency needs further discussion. We noted at the start of this chapter that a-Si module efficiencies typically fall in the range 6–9%, about half that of crystalline silicon. But efficiencies quoted by PV manufacturers invariably relate to standard insolation ($1000\,W/m^2$, 25°C) and tell only part of the story. While crystalline silicon modules are impressive under strong sunlight, their performance in weak or diffuse light is often inferior to thin-film products and is more adversely affected by high temperatures. In recent years there have been many reports of thin-film modules, both a-Si and CdTe, outperforming crystalline silicon in terms of annual energy yield, especially in climates with significant cloud cover and plenty of diffused light.

Amorphous silicon is also a far better light absorber than crystalline silicon, so extremely thin layers of semiconductor may be used—of the order $1\,\mu m$. Like other thin-film technologies, it offers more advantages:

- Relatively simple fabrication at low temperatures using inexpensive substrates and continuous "production line" methods
- Integrated, monolithic design obviating the need to cut and mount individual wafers
- Potential for manufacturing flexible, lightweight products

The word *amorphous*, derived from ancient Greek, means "without form or shape." a-Si, which may be deposited as a thin film on a variety of substrates, does not exhibit a regular lattice structure. The distances and angles between the silicon atoms are randomly distributed, giving rise to incomplete bonds and a high concentration of defects. The result is a high density of allowed energy states within the nominal energy gap, in stark contrast to crystalline silicon. In effect, the extra energy states act as stepping stones, allowing conduction electrons to relax back into the valence band and recombine. There is also a problem of low charge-carrier mobility within the semiconductor material (referred to as poor *carrier transport*). Fortunately, early research into a-Si solar cells suggested two effective ways of countering these difficulties.

Figure 2.25 Amorphous silicon PV modules on a building façade (*Source:* Reproduced with permission of EPIA/Schott Solar).

First, it was discovered that introducing hydrogen into amorphous silicon could passivate incomplete bonds, also known as *dangling bonds* (*DBs*), greatly reducing the number of excess energy states within the bandgap. The modified material is referred to as a-Si(H) to denote its hydrogen content and is illustrated in Figure 2.26. This shows the irregular arrangement of silicon atoms, a DB, and a DB that has been passivated by a hydrogen atom (H). Using this approach it is possible to make effective n-type and p-type materials by doping with phosphorus or boron, resulting in a direct bandgap semiconductor with an energy gap of about 1.75 eV.

The second problem, poor carrier transport, is reduced by introducing an intrinsic layer (which, in practice, is usually slightly n-type) into the p–n junction giving the p–*i*–n structure shown in Figure 2.27. This *i*-layer greatly increases the width of the depletion region and the associated electric field that sweeps minority carriers

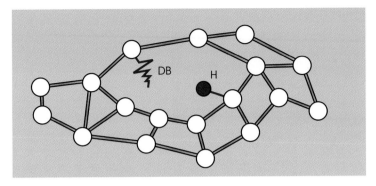

Figure 2.26 Irregular structure and bonding in a-Si(H).

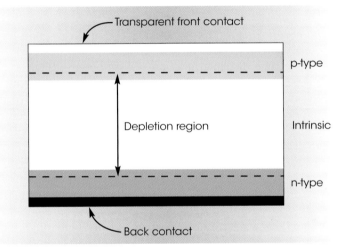

Figure 2.27 The basic structure of a single-junction a-Si(H) solar cell.

across the junction. Assuming the *i*-layer is in fact lightly doped n-type, the highest electric field occurs at the p–*i* interface and it is therefore best to design the cell so that light enters through a transparent front contact into a very thin, heavily doped p-type layer. This ensures that most charge carriers are created near the top of the cell and successfully collected.

Unfortunately, the introduction of an *i*-layer has its drawbacks. During initial exposure to strong sunlight, absorption by the *i*-layer creates additional defects that aid recombination and reduce cell efficiency. The phenomenon, known as the *Staebler–Wronski effect*, depends on the total number of photons absorbed and therefore on the intensity and duration of the light and the thickness of the *i*-layer. Building up over a timescale of months, it results in final or "stabilized" efficiencies significantly lower than the initial values. In the past this has given single-junction a-Si(H) cells a rather doubtful reputation. But most PV clouds have a silver lining. In the case of Staebler–Wronski, the initial loss of efficiency can be largely overcome using multi-junction or stacked cell structures in which light absorption is shared between two or more much thinner *i*-layers. Furthermore, by stacking cells with different bandgaps, it is possible

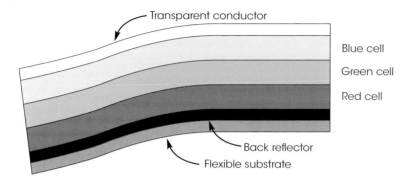

Figure 2.28 A triple-junction amorphous silicon solar cell.

to capture a bigger percentage of solar photons and achieve relatively good levels of efficiency and stability, especially in weak or diffuse sunlight.

The basic scheme for one type of triple-junction cell is shown in Figure 2.28. It depends on the ability of a-Si to form good alloys with germanium, producing semi-conductor material with smaller bandgaps. The top a-Si "blue cell" is effective at cap-turing high-energy blue photons with its bandgap of about 1.75 eV. Next comes the "green cell," based on amorphous silicon–germanium alloy containing about 15% ger-manium with a bandgap of around 1.6 eV. And finally the bottom "red cell," designed to capture low-energy red and IR photons, uses an alloy with about 50% germanium giving a bandgap of around 1.4 eV. Photons that are not absorbed on the first pass through the cells are returned by the back reflector which may be texturized to encourage light trapping.

The supporting substrate does not have to be flexible, but flexibility offers exciting possibilities during production and also for the user. The production process can be continuous "roll to roll," the various layers being deposited on an extremely long thin sheet of stainless steel or plastic as it travels between rollers in the manner of a magic carpet (Figure 2.29). This was the dream of solar cell pioneers back in the 1950s! Sheet thickness is typically a small fraction of a millimeter, with sheet lengths up to an amazing kilometer. Individual solar cells are automatically scribed and interconnected as a monolithic circuit. From the user's point of view, flexibility tends to go hand in hand with lightness and allows easy mounting on curved or awkward surfaces.

The lack of a crystal structure in amorphous silicon ultimately prevents it from matching the efficiency of crystalline silicon, and its commercial production has been phased out. However recent years have seen much R&D effort directed toward a new microcrystalline form of silicon that, like other thin-film materials, can be deposited in extremely thin layers of about 1 μm onto various substrates including glass. Crystalline silicon's comparatively poor light absorption means that success depends upon highly effective light trapping to keep incident light within the film. The hope is that microcrystalline silicon will rival wafer technology for ruggedness and electrical stability while at the same time using minimal amounts of cheap and plentiful raw materials, improving efficiency above amorphous products, and greatly reducing costs.

Figure 2.29 Roll-to-roll manufacture of a-Si solar cells (*Source:* Reproduced with permission of IEA-PVPS).

Silicon and germanium may be the best-known semiconductors, but they are certainly not the only ones. Many compounds incorporating rather unfamiliar chemical elements also display electrical properties midway between insulators and conductors. Some readily absorb solar photons to produce electron–hole pairs and may be doped to make n-type or p-type material and deposited as thin layers on a variety of substrates. In other words they are candidates for "second-generation" thin-film cells that surpass amorphous silicon's efficiency and challenge crystalline silicon's usage of materials and production costs. Of the various possibilities, two materials—*CIGS* and related compounds and *CdTe*—have a highly significant presence in the terrestrial PV market and are set to lead PV decisively into a new era.

Not that crystalline silicon cells will be easily displaced. Global production continues apace. Gigawatts of wafer-based modules are already installed and will generate electricity for many years to come, catching the public eye as ambassadors for PV around the world. However, CdTe thin-film technology has been over the last 10 years the leader in manufacturing cost reductions and may offer the best chance of achieving grid parity with conventional electricity generation on a large scale.

In the following text, we discuss CIGS and CdTe, and in Section 2.5 we will discuss even more exotic third-generation technologies.

2.3.2 Copper Indium Gallium Diselenide (CIGS)

To be successful, inorganic crystalline solar cell materials need two essential properties. They must be good light absorbers, turning solar photons into electron–hole pairs; and they must include an efficient junction to sweep light-generated minority carriers across the junction and force current through an external circuit.

Many years ago it was discovered that the compound semiconductor copper indium selenide (CIS) offers excellent light absorption in small-grained layers a micrometer or two thick. Although the electronic and chemical properties of CIS and related compounds are subtle and complex,[6,7] a few key points can be made here. First, and unlike silicon, CIS cannot be doped to form an efficient p–n junction on its own (it cannot form a *homojunction*); but it can be interfaced with another semiconductor, *cadmium sulfide (CdS)*, to produce an effective *heterojunction.* CIS and CdS are well matched and do not suffer excessive recombination at the interface. Since CdS can only be successfully doped as n-type material, the CIS must be doped p-type. It is rather difficult to make good metallic contact with CIS; gold is effective, but expensive, so molybdenum is normally used as a back contact.

There is a further important twist to the story. In the 1970s it was discovered that the rather low bandgap of CIS (about 1.1 eV) may be increased by substituting some gallium in place of indium. By varying the gallium content, a range of bandgaps relevant to PV cells can be obtained, from about 1.1 eV (no gallium) up to 1.7 eV. In addition, the low open-circuit voltage of CIS is raised toward 0.5 V, comparable with crystalline silicon, meaning that fewer cells need be interconnected to achieve useful module voltages. The modified material, *copper indium/gallium diselenide (CIGS)*, has achieved many cell efficiency records (it is worth noting that the initials CIS and CIGS tend to be used interchangeably, which can lead to a certain amount of confusion). CIGS passed the 20% efficiency milestone for laboratory cells in 2008. At that stage commercial module efficiencies were already attaining 10–12%, comfortably beating amorphous silicon and within aiming distance of crystalline silicon.

The basic scheme of a typical CIGS cell is shown in Figure 2.30. Light enters the cell via a transparent conducting layer acting as the top contact. Next comes an extremely thin layer of CdS that forms a p–n heterojunction with the thicker (but still very thin!) CIGS absorber. A metallic layer, normally molybdenum, provides the back contact and completes the electronic design. The doping of the p-type absorber is often graded, being lightest near the junction. This extends the depletion region and its associated electric field well into the absorber where most charge carriers are generated and helps

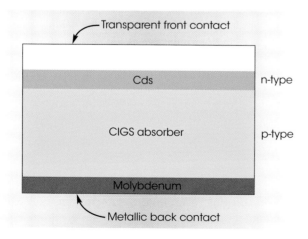

Figure 2.30 The basic scheme of a CIGS solar cell.

sweep them across the junction. Not shown in the figure is the necessary supporting substrate, which may be rigid or flexible and made of glass, metal, or plastic.

As our attention moves from silicon cells with their superabundance of cheap raw material to thin-film cells based on unfamiliar elements, it is time to question cost and availability of supplies. Cost is not generally seen as a problem, given the tiny amounts of material used in thin-film cells compared with silicon wafers; indeed one of thin-film technology's main promises is to make PV ever more affordable. But the situation could change if production levels continue to increase dramatically. The indium used in CIS and CIGS cells is a case in point: indium is a comparatively rare element of the Earth's crust, in demand for electronic products other than solar cells. Availability may become a problem in large penetration scenarios and one advantage of partially substituting gallium into CIGS is a decreased demand for indium. This topic is further discussed in Chapter 7.

As thin-film solar cells contribute more and more to "second-generation" PV technology and challenge the pole position occupied for so long by crystalline silicon, we will become used to seeing CIS and CIGS modules with the smooth, dark grey/black appearance (shown in Figure 2.31) often favored by architects. There is also intensive development of semitransparent modules that act as windows, allowing a portion of light to enter a building while at the same time generating electricity. The possibilities for exciting and innovative PV products are enormously increased by thin-film techniques.

Figure 2.31 An array of CIS solar modules in Austria (*Source:* Reproduced with permission of EPIA/ Shell Solar).

2.3.3   Cadmium Telluride (CdTe)

Cadmium telluride (CdTe) is another important semiconductor material for thin-film solar cells, its direct bandgap of 1.45 eV being close to optimum for capturing the sun's spectrum using a single-junction device. Also its high optical absorption coefficient allows light to be fully captured using only a 1.5-µm-thick layer. Like many II–VI compounds, CdTe sublimes congruently; it vaporizes homogeneously and the compound's thermodynamic stability simplifies the deposition of layers of stoichiometric CdTe.[7] However, initially there was considerable concern among environmental groups about the commercialization of CdTe cells and modules because of cadmium's toxicity. However, these fears have been largely allayed as the life cycle of CdTe PV has been critically reviewed by expert panels in more than 12 countries and all concluded that the technology is safe and friendly to the environment.[8–10] CdTe has not got the toxicity of its individual constituents Cd and Te. Cadmium is commonly obtained as a byproduct of zinc mining and smelting, so removing it from the environment for use in solar cells offers an environmental benefit, especially when modules are recycled at the end of their useful life. This issue is discussed in Chapter 7. Cadmium and tellurium are more abundant elements than the indium used in CIS/CIGS products, so availability is not so big an issue. However, the market is growing strongly. CdTe modules accounted for about 10% of world production in 2016, a lot more than any other thin-film technology, and they are finding large-scale application in PV power plants. Comparatively simple production processes mean that CdTe modules are currently about the cheapest on the market in terms of price per peak watt. Furthermore their conversion efficiencies of around 16.5% (2016 average) look set to advance toward 20% in the next few years.

The rationale behind a thin-film CdTe solar cell results in a scheme very similar to that for CIS and CIGS. The essential layers in the thin-film "sandwich" are a transparent top contact, a CdS/CdTe p–n heterojunction and absorber, and a metallic back contact, as shown in Figure 2.32. Also required is a suitable supporting substrate of glass, metal, or plastic, which determines whether cells are rigid or flexible. Bear in mind that

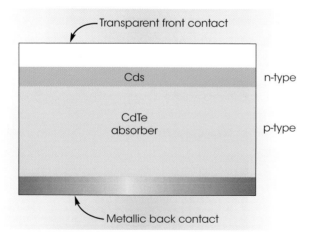

Figure 2.32   A CdTe solar cell (simplified cross section).

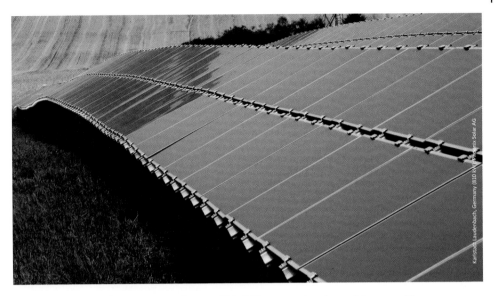

Figure 2.33 Farming the Sun; part of an 810 kW$_p$ CdTe power plant in rural Germany (*Source: Reproduced with permission of First Solar/Phoenix Solar*).

although the figure represents the cell as rather thick and narrow, it is actually manufactured as part of an extremely thin sheet.

As worldwide thin-film PV production grows, and it seems that cadmium telluride will continue its important contribution. An installation that nicely illustrates the possibilities for "farming sunshine" alongside conventional crops is shown in Figure 2.33. Further up the power scale, two 550 MW$_{ac}$ power plants with CdTe modules made by First Solar have been commissioned in south California (Figure 2.34); such sizes were almost unimaginable 10 years ago.

2.4   Cell Efficiency and Module Cost

As discussed in the first chapter, government subsidies for PV have enabled the scales and the developments that brought the cost of PV down to levels equal to those of fossil-fuel-based electricity in the sunniest regions of the world. As shown in Figure 2.14, all the single-junction solar cells have about the same (~28–30%) theoretical (Shockley–Queisser) limit of photon to electron conversion efficiency. Crystalline Si solar cells have approached this limit with record cell efficiencies of 25%, whereas thin-film solar cells are slightly behind, reflecting the earlier and higher investment in c-Si semiconductors for both integrated circuits and PV. The record efficiencies for these and other emerging cell types are shown in Figure 2.35.

The module efficiencies are well behind those of the record cells for a number of reasons. First, some record cells use expensive materials and processing that are not cost effective in commercial production; second, there are interconnection and area losses between the cells and the modules (mostly in c-Si) and film quality challenges as the area increases (in thin-film modules).

Figure 2.34 Large-scale farming of the Sun; part of the Desert Sunlight 550 MW$_{ac}$ CdTe power plant in California (First Solar).

This efficiency gap is shown in Figure 2.36. In the United States where the electricity prices are among the lowest among the developed nations, the expectation is that within a decade, the cost of PV electricity will be in parity with electricity from the grid, making additional subsidies unnecessary.

Up to a few years ago, first-generation monocrystalline Si PV held the commercial PV module efficiency record, and second-generation cadmium telluride held the record of the lowest module production cost. This was exemplified with SunPower modules exceeding 20% module efficiencies at a production cost of about $1.75/W$_p$, and First Solar thin-film PV modules produced at costs of 75 ¢/W$_p$ and efficiencies of 12.5%. By 2017, this distinction was less apparent as the efficiency of First Solar modules reached 16.5% and their cost fell to 40 ¢/W$_p$, whereas the cost of the SunPower modules fell to approximately $1/W. Both companies continue to improve their products and have produced record, not yet commercial, module efficiencies of 18.6% and 22.8% correspondingly. However, a quest of accomplishing both high efficiency and low production cost has brought up what are called "third-generation" technologies that promise bridging the goals of high efficiency and low cost. Figure 2.37 shows the current costs of producing first- and second-generation technology PV modules and the projected costs of producing third-generation technologies, assuming the same low per-area manufacturing costs as those of thin films. The figure shows power conversion efficiency and manufacturing costs per unit of area and the resultant module cost (diagonal lines).

Of course the module costs are only one parameter in the system cost equation and higher efficiency modules would have lower mounting structure and installation costs, which are proportional to the area required. This is the reason that every PV manufacturer tries to increase the efficiencies of the modules they produce by progressively decreasing the efficiency gap between the cell and the module.

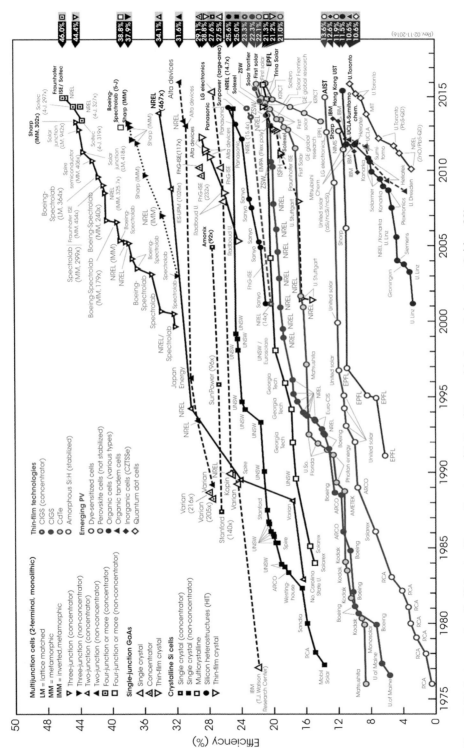

Figure 2.35 Best research-cell efficiencies (*Source*: Courtesy of National Renewable Energy Laboratory, Golden, CO).

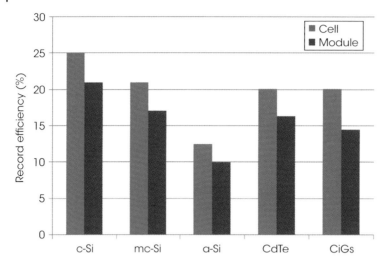

Figure 2.36 Efficiencies of record cells and commercial PV modules (Wolden *et al.*[11]).

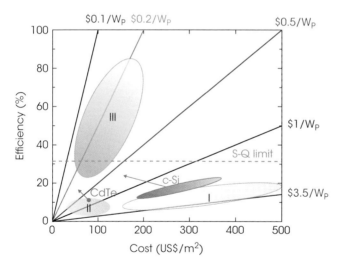

Figure 2.37 Classification of PV technologies superimposed with the current status (solid) and projected evolution (arrows) of c-Si and CdTe PV technologies (*Source:* Wolden *et al.*[11]. Reproduced with permission of AIP Publishing LLC).

2.5   Third-Generation Solar Cells

In this section we will discuss notable advances of third-generation technologies, namely, organic cells and nanostructures, dye-sensitized cells (DSCs), and multi-junction III/V cells. The underlining promise of most third-generation technologies is that the thermodynamic Shockley–Queisser efficiency limit (discussed in Section 2.2.3.3) of single-junction solar cells will be surpassed. In first-generation crystalline silicon and second-generation a-Si, CdTe, and CIGS solar cells, only photons within a narrow wavelength, corresponding to the semiconductor bandgap, are effectively absorbed. Photons with energy lower than the semiconductor bandgap are not absorbed and their energy is not used for carrier generation, whereas photons with energy larger than the bandgap are absorbed but excess energy is lost to heat, negatively affecting the voltage generation of the devices. Primarily for this reason, the thermodynamic Shockley–Queisser limit of single-junction cells is only about 31%, with spectral losses being as large as 50%. Several approaches have been proposed to reduce or eliminate spectral losses, for example, multi-junction cells, intermediate bandgaps, multiple exciton generation, quantum dot concentrators, down- and up-converters, and downshifters.[12,13] The first approach is already commercialized, with multi-junction solar cells based mainly on GaAs with efficiencies as high as 46% being produced in the lab and 38% efficient cells being deployed in terrestrial concentrator systems. The others involve the generation of more than one electron from a highly energetic photon, and the upshifting of low energy photons so that their energy is also utilized in electron production. Nanotechnology is essential in realizing these concepts. Nanotechnology-enabled organic photovoltaics (OPV), in addition to enabling high photon conversion efficiencies, have the potential to further lower the production cost of PV by using inexpensive materials. Dye-sensitized solar cells seem especially suited for low irradiation regions, and their efficiencies can also be augmented with nanostructures. The newest addition to this next generation PV is perovskites, which have shown in the laboratory phenomenal efficiency increases over just a few years (see Figure 2.35).

2.5.1 Gallium Arsenide (GaAs) Multi-Junctions

Gallium is one of the elements in Group III of the periodic table; arsenic is in Group V. So gallium arsenide (GaAs) is often referred to as a *Group III–V* semiconductor. GaAs and associated compounds have two claims on our attention as specialized, but important, PV materials: for making solar cells used in spacecraft and for their use in terrestrial concentrator systems that focus sunlight using mirrors or lenses.

In the early years of space exploration, silicon solar cells were the main source of electricity for spacecraft, reaching efficiencies of about 15% by 1970. Since then GaAs has made a big impact, for two main reasons. First, it is less susceptible than silicon to damage by radiation in space, a key consideration on long missions where the performance and reliability of electricity supply is paramount. Second, its direct bandgap of 1.42 eV (compared with 1.1 eV for silicon) allows a greater percentage of the solar spectrum to be harvested, giving better conversion efficiencies. Since the 1980s solar cell designers have learned how to deposit thin films on crystalline germanium wafers, producing lightweight multi-junction devices of even higher efficiency. Triple-junction modules have gained a high reputation for their reliability and light weight. And although the material and processing costs of GaAs cells are high, this is hardly a major consideration for vastly expensive space projects.

A typical scheme for triple-junction GaAs cells is shown in Figure 2.38. Like the triple-junction amorphous silicon cell described earlier, it is a "sandwich" of three stacked cells with different bandgaps designed to capture different portions of the sun's spectral energy. For space applications the relevant spectrum corresponds to *air mass zero* (*AM0*), received by solar cells outside the Earth's atmosphere (refer back to Figure 1.8). Each cell includes n-type and p-type crystalline layers. The top cell, with a bandgap of

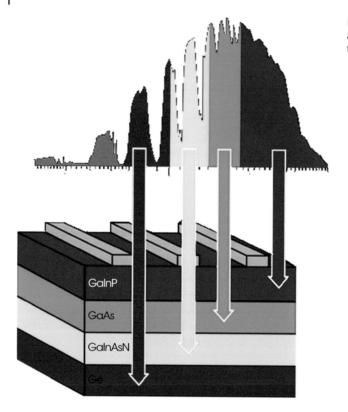

Figure 2.38 Photon absorption in a triple-junction cell.

about 1.9 eV obtained using the alloy *gallium indium phosphide (GaInP)*, is very effective at absorbing high-energy UV/blue photons. The GaAs cell in the middle has a bandgap of 1.42 eV; and the bottom cell, based on germanium that also provides the supporting substrate, has a bandgap of 0.7 eV to absorb IR photons.

Although such triple-junction devices come in the general category of "gallium arsenide," we see that they actually use carefully controlled proportions of several III–V elements plus Group IV germanium to achieve bandgap control. These highly specialized solar cells are built up monolithically, with many layers being grown on top of one another with optimal thickness and doping. All this requires expensive materials and very advanced processing. But the technical rewards are high: the best laboratory cells have efficiencies of 40% and commercial cells over 35%. Such impressive efficiencies pose an interesting question. Can gallium arsenide be "brought down to Earth" and make a significant contribution to terrestrial PV generation? Success depends upon effective *concentration* of sunlight using mirrors or lenses, focusing the light onto cells of far smaller area with correspondingly reduced material and processing costs. For example, increasing the light intensity 1000 times ("1000 Suns") should allow the cell area to be reduced 1000 times for the same power output. Indeed, it is rather better than this because the efficiency of many solar cells improves under concentrated sunlight. Triple-junction GaAs concentrator cells have already passed the 44% landmark in the laboratory, with commercial cells not far behind.

Successful PV concentration systems must aim to reduce cell costs sufficiently to offset the expense of focusing the light and tracking the sun across the sky on its daily journey. Not surprisingly, there are sceptics; yet PV concentration is being intensively researched and developed, with many systems in commercial production.

2.5.2 Dye-Sensitized Cells

Some of the new PV concepts and materials introduced in recent years would have astounded early PV pioneers whose attention was entirely focused on inorganic semi-conductors, principally silicon and germanium. We are now moving into an era where artificial organic materials seem certain to play an important role in converting sunlight directly to electricity. They are seen as part of PV's "third generation." Of many possible approaches, Dye-Sensitized Cells (DSCs) and *organic cells* are presently in the vanguard of development and commercial application.

DSCs are a hybrid organic–inorganic technology that uses small-molecule absorber dyes absorbed onto an electron-accepting material, such as *titanium dioxide (TiO_2)*, along with an electrolyte to regenerate the dye. They are also called the "Graetzel cells" after Michael Graetzel who with Brian O'Regan at the Federal Polytechnic in Lausanne, Switzerland, found that a 10 μm thin film of TiO_2 could work as an effective solar cell if coated with an organic dye, immersed in an electrolyte, and provided with electrical contacts. Most importantly, the TiO_2 was made in the form of a nanoporous "sponge" of minute particles just tens of nanometers (nm) across, propelling PV into the modern field of nanotechnology. And since titanium dioxide (also known as *titania*) is an inor-ganic semiconductor, whereas the dye and electrolyte are organic, the *Graetzel cell* is sometimes referred to as an organic–inorganic thin-film device.

But why *dye sensitized*? Unlike conventional cells in which the absorption of light and transport of light-generated charges takes place within the same semiconductor, in a DSC these roles are split. The dye acts as light absorber, generating electrons that it *injects* into the conduction band of the semiconductor. In other words the dye acts as a "sensitizer" of the TiO_2, which would not be effective on its own because of its large bandgap. Another key aspect of the Graetzel cells is their use of new organic dyes able to absorb a wide solar spectrum. And the use of TiO_2 nanoparticles, rather than larger crystals, hugely increases the surface area of the adsorbed dye coating and hence the efficiency of light absorption.

Most people, meeting DSCs for the first time, find their detailed operation very complex—certainly more so than crystalline silicon cells. Although it involves many of the same basic concepts[12]—photon absorption, charge generation and transport, recombination, optical and resistance losses—the electrochemical terminology is unfa-miliar and the names of the organic materials can seem unreasonably long! So we restrict ourselves here to a brief summary.

The basic scheme of a DSC is illustrated in Figure 2.39. Light enters the cell via a transparent front contact and is absorbed by the organic dye covering the TiO_2. Excitation electrons are injected into the conduction band of the TiO_2, causing oxida-tion of the dye. They are efficiently transported through the semiconductor by diffusion and reach the electrical contact. Assuming the cell is connected to an electrical load, the electrons now pass through the external circuit and reenter the cell via the back contact

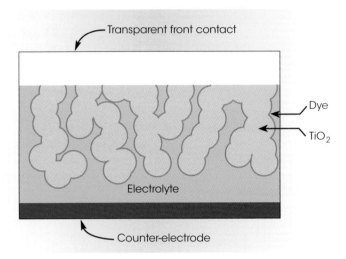

Figure 2.39 A dye-sensitized solar cell.[14]

or *counter electrode.* Here they provide the negative charges required to restore the dye to its original (unoxidized) state with the help of the intervening electrolyte. The circuit is completed.

Unfortunately some recombination does occur, but not in the same manner as in conventional silicon cells. Although electrons are injected by the dye into the conduction band of the semiconductor, holes are not formed in its valence band; so there is no generation of electron–hole pairs, or subsequent annihilation. But electrons can recombine with the oxidized dye. Fortunately, electron injection and transport in the semiconductor is extremely fast compared with the recombination process, so effective charge separation does in fact take place. Overall, the photon-to-electron generation process in a DSC is analogous to photosynthesis in leaves and plants where chlorophyll acts as the sensitizer.

Early Graetzel cells achieved very respectable efficiencies of up to about 10% in standard insolation conditions (1000 W/m^2, 25°C). A great deal of ongoing research has since improved performance, raising efficiency above amorphous silicon and within sight of other thin-film technologies. However, efficiency in bright sunshine is probably not the main criterion for DSCs. They work well in low diffused light and in high ambient temperatures, indoors and out. Flexible modules can easily be made using plastic substrates. They use nontoxic and plentiful materials (TiO$_2$ is a widely used chemical, e.g., in paints and toothpastes) and their relatively simple manufacturing techniques include fast roll-to-roll production. Unusual and exciting possibilities are opening up for building-integrated photovoltaics (BIPV), including roofing products, transparent and semitransparent tinted windows, partitions, and decorative features. Instead of restricting architects to standard rectangular PV modules, DSC products can be tailor made to particular sizes, shapes, and aesthetic design criteria; see, for example, Figure 2.40. The wide range of applications promises an exciting future.

Figure 2.40 Innovative and flexible: dye-sensitized solar cells in Australia (*Source:* Reproduced with permission of Greatcell Solar Limited).

2.5.3 Organic Solar Cells

In addition to the inorganic semiconductors we examined before, there are organic compounds that can absorb photons or carry electric charges, functioning as semiconductors. These are oligomers or polymers made up by carbon and hydrogen atoms with, sometimes, added nitrogen, sulfur, and oxygen. Like silicon, in general they are insulators but become semiconducting when doped, or by photoexcitation. OPV cells use long-chained molecular systems for the electron-donating material (e.g., P3HT), along with fullerenes as the electron-accepting system (e.g., PC60BM); see the box in the next page for a basic description of fullerene structure.

They attracted interest for PV applications because they use inexpensive materials and can be fabricated with low-cost, solution-based processing. Soluble organic molecules enable roll-to-roll processing techniques and allow for low-cost manufacturing. Also these materials can be applied to flexible substrates enabling a wide variety of uses.

Progress in the development of OPV has been fast; as shown in Figure 2.35, OPV record cell efficiencies have gone from 3 to 11% within 14 years of development, and tandem cells that have the potential for exceeding the Shockley–Queisser limit reached the same record efficiency within only 6 years (2008–2015). They have been commercialized, amid small volumes, by Konarka and Heliatek since the potential for new applications counterbalances their low efficiency. The area where OPV is expected to have the largest impact is in BIPV where they offer shape and design flexibility since they can be tuned to the desirable colors by slightly changing their chemical properties, thereby allowing solar cells to be an integral part of the design. Currently only Heliatek produces OPV; these are used for retrofits on window and building exteriors. Their vision is that eventually OPV could be used to create solar-coated cars and homes.

Fullerene Structure

A fullerene is a molecule of carbon in the form of a hollow sphere, tube, or other shape. Spherical fullerenes are also called buckminsterfullerene (buckyballs), and they resemble the balls used in soccer. The first fullerene molecule to be discovered, buckminsterfullerene (C60), was prepared in 1985 by Richard Smalley and coworkers at Rice University. The name was a homage to the architect Buckminster Fuller, whose geodesic domes it resembles. The suffix "-ene" indicates that each C atom is covalently bonded to three others.

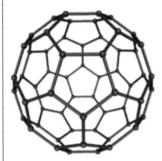 

The fullerene (C60) structure and its crystalline form (Wikipedia, Fullerene, April 10, 2016).

Let's now see how OPV functions. Polymer-based OPV cells use long-chained molecular systems for the electron-donating material (e.g., poly 3-hexylthiophene (P3HT)), along with fullerenes as the electron-accepting system (e.g., C60PCBM). The absorber is used in conjunction with an electron acceptor, such as a fullerene, which has molecular orbital energy states that facilitate electron transfer. Photon absorption in OPV does not lead directly to an electron and a hole, but it first generates an exciton, a state where the two charges are bound together. The exciton then migrates to the interface (heterojunction) between the absorber material and the electron acceptor material. At the interface, the energetic mismatch of the molecular orbitals provides sufficient driving force to split the exciton and create free-charge carriers (an electron and a hole).

The most common device structure for OPV uses a mixture of donor and acceptor materials referred to as a bulk heterojunction (BHJ) that resides between two electrodes. Figure 2.41 (a) illustrates a BHJ (green-blue colors) packed between an electron blocking layer (EBL) and a hole blocking layer (HBL), which are in contact with an indium tin oxide (ITO) and a silver electrode.[14] As depicted, the PV effect follows the following steps: the illumination of an organic semiconductor donor (1) generates excitons (2) with a binding energy of about 0.4 eV. To separate into free charges, the exciton must diffuse until it reaches a donor/acceptor interface (3) with a difference in electron affinities and an ionization potential large enough to overcome the binding energy. The energy cascade required for charge extraction is illustrated in Figure 2.41(b). The free charges then can travel (4) through either the donor or acceptor material (5), and then

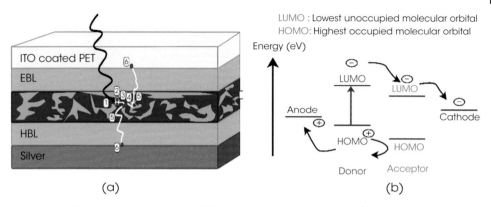

Figure 2.41 (a) Photovoltaic effect in a bulk heterojunction organic solar cell[14] (b) conditions for charge transfer in a donor/acceptor photovoltaic device [14] (*Source:* Anctil, Fthenakis http://cdn.intechopen.com/pdfs/32591/InTech-Life_cycle_assessment_of_organic_photovoltaics.pdf. CC BY 3.0).

are collected at the electrodes (6). The overall efficiency of the device therefore is determined by the optical absorption and the efficiency of each of those steps.

Various challenges have limited the use of OPV; in particular, the large bandgap of most organic polymers is responsible for low power-conversion efficiency because a large portion of the solar spectrum is unabsorbed. In theory, in an optimal solar cell, the acceptor bandgap should be around 1.4 eV, wherein the maximum efficiency would be 31% under 1 sun AM1.5. Most early-generation semiconducting polymers have bandgaps higher than 2 eV (corresponding to a wavelength of 620 nm), so limiting their maximum efficiency. There are two alternatives to increase the devices' efficiency: lowering the bandgap to absorb a maximum of photons in one layer or using a multi-junction approach where two different materials absorb in a different region of the solar spectrum.

The low bandgap approach has received considerable interest in the last few years and produced the current record efficiency for polymer devices. By lowering the bandgap from 2 to 1.5 eV, the maximum theoretical efficiency increases from 8 to 13%. To increase efficiencies further, the multi-junction approach illustrated in Figure 2.42 is necessary. For OPV this approach not only is advantageous to capture a broader range of the solar spectrum but also helps overcome the poor charge carrier mobility and lifetime of carriers, which prevents the fabrication of a thick absorbing layer. In comparison with inorganic material, organic semiconductors absorb only a narrow portion of the spectrum and therefore a combination of multiple materials is necessary to enhance photon absorption.

Also note that because various absorbers can be used to create colored or transparent OPV devices, this technology is particularly appealing to the BIPV market. OPV has achieved efficiencies near 11%, but efficiency limitations as well as long-term reliability remain significant barriers; these challenges present opportunities for further research, which we discuss in the following text.

Research Directions
The low efficiencies of OPV cells are related to their small exciton diffusion lengths and low carrier mobilities. These two characteristics ultimately result in the use of thin

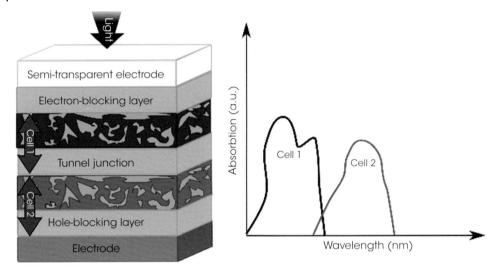

Figure 2.42 Organic photovoltaic with two sub-cells having different complementary absorption spectra[14] (*Source:* Anctil, http://cdn.intechopen.com/pdfs/32591/InTech-Life_cycle_assessment_of_organic_photovoltaics.pdf. CC BY 3.0).

active layers that affect overall device performance. Furthermore, the operational lifetime of OPV modules remains significantly lower than for inorganic devices.

BHJs are most commonly created by forming a solution containing the two components, casting (e.g., drop casting and spin coating) and then allowing the two phases to separate, usually with the assistance of an annealing step. The two components will self-assemble into an interpenetrating network connecting the two electrodes. They are normally composed of a conjugated molecule-based donor and fullerene-based acceptor. The nanostructural morphology of BHJ tends to be difficult to control, but is critical to PV performance.

Current research focuses on increasing device efficiency and lifetime. Substantial efficiency gains have been achieved already by improving the absorber material, and research is being done to further optimize the absorbers and develop organic multijunction architecture. Improved encapsulation and alternative contact materials are being investigated to reduce cell degradation and push cell lifetimes to industry-relevant values.

2.5.4 Perovskites

Perovskites are the latest addition to the promise of third-generation PV. Figure 2.35 shows the extremely rapid efficiency improvements in perovskite cells. In just 5 years, lead-based organometal halide perovskite solar cells have shown in the lab efficiencies up to 22%. This matches the best efficiencies of CdTe and CIGS thin-film device technologies that have been developed over several decades. However, high efficiencies were observed only at very small (i.e., $1\,cm^2$ or less) sizes that become unstable within minutes, whereas PV is supposed to last 30 years. Perovskites have the crystallographic structure ABX3, where A is a large cation such as methyl-ammonium, B is typically lead

(Pb), and X is halogen or a mixture of halogens (I, Br, and Cl). They are appealing because they have large absorption coefficients and offer a range of bandgaps for both single-junction devices and Si-based tandem cells. Moreover, perovskites are less sensitive than other technologies to structure defects and show long carrier lifetimes and diffusion lengths. The rapid improvement in device performance has been largely based upon improved understanding and control of composition, microstructure, and complementary charge transport layers in the device structure. Early reports of rapid improvements in efficiency led to an explosion of research efforts, with well over 1000 publications on the topic in 2015.

Despite great excitement, commercialization of perovskite PV technology requires overcoming three critical challenges: demonstrating long-term operational stability of PV modules, scaling up to large areas, and avoiding or mitigating real and perceived risks associated with the use of toxic lead. Regarding stability and size, early lab-scale perovskite solar cells have exhibited significant degradation of performance on timescales of minutes although their size is as small as a fingernail. The causes of this degradation are only partially understood. Proper encapsulation will address some of the stability issues related to moisture exposure, but work to modify the device layers themselves has also shown promise.

The other challenge to commercialization is the use of toxic lead. The intensity and risks associated with lead in perovskite solar cells are subjects of controversy. Some reports have indicated that any amount of lead is unacceptable, whereas other reported that the amount of lead potentially used in PV would be negligibly small compared with lead usage in from other industries such as in electronic solder, aviation fuel, and coal-fired electricity production.

Research Directions

Avoiding lead altogether seems the most desirable solution. However, Pb-free alternatives such as Sn-based perovskites have shown much lower efficiencies and worse stability than Pb-based ones and Sn has toxic properties as well. Hazard and risk characterization studies are necessary to address the issue in a comparable context accounting for the whole life cycle of the product from cradle to grave, or recycling. These issues are discussed in Chapter 7.

Self-Assessment Questions

Q2.1   Describe the main differences between monocrystalline, multicrystalline, and amorphous silicon solar cells.

Q2.2   Explain why pure silicon, even under strong sunlight, cannot generate electricity. How may dopants such as phosphorus and boron be used to convert it into a solar cell?

Q2.3   Why are minority carriers so central to the operation of a solar cell?

Q2.4   How large are the energy bandgaps and theoretical efficiencies of the following types of solar cell: (a) crystalline silicon and (b) cadmium telluride?

Q2.5 Note approximate ranges of energies for different regions of the electromagnetic spectrum (Figure 2.15). Specifically, give ranges for ultraviolet, visible, and infrared in terms of commonly used units of energy (eV), wavelength (nm and μm), and frequency (GHz and THz).

Q2.6 Crystalline Si has an energy bandgap of 1.12 eV and Ge has a bandgap of 0.66 eV; what are the corresponding wavelengths of the solar spectrum that can excite these semiconductors?

Q2.7 Gallium arsenide has an energy bandgap of 1.39 eV and Ge has a bandgap of 0.66 eV. How do you stack those two (thus which would face the sun) in order to maximize the photon absorption?

Q2.8 Why do antireflection coatings and surface texturization improve the efficiency of silicon solar cells?

Q2.9 What process advantages have cadmium telluride over crystalline silicon in the manufacture of photovoltaics?

Q2.10 What is the difference between shunt resistance and series resistance?

Q2.11 What are the potential advantages, disadvantages, and challenges of nanostructured solar cells?

Q2.12 What are perovskites and why are they interesting materials for solar cells?

Q2.13 What is a record efficiency of monocrystalline–Si solar cells in the lab, and what is the efficiency of high-performance mono-c–Si modules in the market?

Q2.14 Note the gap between record cell efficiencies and commercial module efficiencies for all currently available commercial PV technologies.

Q2.15 What is the effect of connecting solar cells in series and of connecting them in parallel?

Q2.16 What measures can be taken to minimize reflection from the top surface of a solar cell?

Q2.17 Does the level of solar irradiation on a solar cell affect mainly its (mark all that apply):
a) Power
b) Voltage
c) Current

Q2.18 What are typical n-dopants and p-dopants in silicon solar cells?

Q2.19   The energy bandgap of c-Si is 1.12 eV and that of CdTe is 1.45 eV. Which of the two is expected to have a lower open-circuit voltage, and why?

Problems

2.1   The current–voltage characteristic of a silicon solar cell may be expressed as

$$I = I_0 \left[\exp\left(\frac{qV}{kT}\right) - 1 \right] - I_L$$

What is the value of the cell current I when the applied voltage V is (a) large and negative; (b) zero?

2.2   What is meant by the *maximum power point* (*MPP*) of a solar cell? Figure 2.10(b) shows a family of I–V curves for a 2 W_p silicon solar cell. What value of resistance would you have to connect to the cell to extract maximum power from it when the insolation is (a) 1000 W/m^2 and (b) 250 W/m^2? What are the implications for operating the cell at its MPP in variable sunlight?

2.3   We want to connect two slightly different silicon solar cells in series or in parallel. The first has under standard conditions a $V_{oc} = 0.56$ V and $I_{sc} = 25$ mA/cm^2 and the second has $V_{oc} = 0.6$ V and $I_{sc} = 22$ mA/cm^2. Using the diode equation (eqn 2.1) find the values of V_{oc} and I_{sc} in each of the two configurations.

2.4   Estimate the voltage associated with the maximum power point of a solar cell with $I_L = 2.2$ A, $I_0 = 0.0001$ mA under standard temperature of 298° K. (Hint: assume ideal conditions, so no series resistance and infinite shunt resistance, and use equation 2.3 for I_L; multiply with voltage and differentiate the resulted power equation in respect to V).

2.5   Using a computer, plot I–V characteristics using a reasonable set of physical parameters. Show how the shape of I–V curves changes for different values of

a) Light intensity
b) Shunt resistance
c) Series resistance
d) Reverse saturation current, I_0

Describe why the curves change the way they do in relationship to the physics of the solar cell. State the values you used for the fixed and varied parameters. Choose an appropriate range such that you see significant changes. As a baseline, you can use $I_{sc} = 32.6$ mA/cm^2, $I_0 = 5 \times 10^{-8}$ mA/cm^2, zero series resistance, and infinite shunt resistance. Use the standard model, which describes the electrical losses in the solar cell.

2.6   Why is a solar cell best thought of as a current source, whereas a battery is normally considered a voltage source?

2.7   Estimate the fill factor of the solar cell shown in Figure 2.10(b) when the insolation is $750 \, \text{W/m}^2$.

2.8   a-Si solar cell is under a lamp producing monochromatic light with wavelength of $0.810 \, \mu\text{m}$ and intensity of $20 \, \text{mW/cm}^2$. Given that the energy bandgap is $1.12 \, \text{eV}$, and assuming an 80% quantum efficiency, estimate the short current of a 36-inch solar cell.

2.9   For the same quantum efficiency and solar cell area, calculate the short current if the cell was made of CdTe (energy bandgap of $1.45 \, \text{eV}$).

2.10   A silicon cell with active area of $100 \, \text{cm}^2$ gives under rated irradiation conditions ($1 \, \text{kW/m}^2$, $25°\text{C}$) an open circuit voltage of $680 \, \text{mV}$ and a short current of $3.5 \, \text{A}$. Calculate the cell's photon to electron conversion efficiency assuming that a) the cell has a zero series resistance and an infinite shunt resistance and b) the cell has a zero series resistance and a shunt resistance of $5 \, \text{Ohm}$. Assume a Filling Factor of 0.80.

2.11   A silicon solar cell has $V_{oc} = 0.6 \, \text{V}$ and $I_{sc} = 25 \, \text{mA/cm}^2$ under standard rated ($1 \, \text{kW/m}^2$, $T = 298 \, \text{K}$) conditions corresponding to one sun irradiation. Estimate the V_{oc} expected under illumination by 100 suns and by 500 suns correspondingly.

Answers to Questions

Q2.1   Crystallinity and efficiency

Q2.2   It needs a junction to propel the electrons and the holes to different directions so that they do not recombine. One p- and one -n dopant are needed.

Q2.3   Because they go across the junction to the electrodes and create current.

Q2.4   (a) $1.1 \, \text{eV}$, 27% (b) $1.45 \, \text{eV}$, 28%.

Q2.5   Use $E(\text{ev}) = 1.24/ \, \text{wavelength(micron)}$ for the conversion

Q2.6   For Si: $\lambda \leq 1.1 \, \mu\text{m}$, for Ge: $\lambda \leq 1.9 \, \mu\text{m}$.

Q2.7   The material with energy bandgap corresponding to longer wavelengths goes to the bottom as such photons can penetrate deeper; so in our case GaAs goes on the surface facing the sun is Ge goes on the bottom.

Q2.8   They reduce optical losses.

Q2.9   Fewer processing steps.

Q2.10   A high series resistance hinders the movement of electrons to the collecting electrodes; a low shunt resistance creates alternative electron pathways thereby causing losses.

Q2.11   They provide control of morphology, grain and chemistry and potentially collection of electrons in bulk formations. The later is the challenge.

Q2.12   Described in 2.5.4; they have high absorption coefficients and can harvest a wide part of the solar spectrum.

Q2.13   See Figure 2.35 and Google for updates.

Q2.14   See Figure 2.36 and Google for updates.

Q2.15   A connection in series increases voltage; a connection in parallel increases current.

Q2.16   Texturing, antireflective coating, reducing width of contacts, taking contacts to the back of the cell

Q2.17   Current and correspondingly power

Q2.18   P, As and B, Al

Q2.19   The lower bandgap semiconductor can effectively capture more energetic photons and those can create a higher voltage; see figure 2.15.

References

1 T. Markvart (ed.). *Solar Electricity*, 2nd edition, John Wiley & Sons, Ltd: Chichester (2000).
2 P. Wuerfel. *Physics of Solar Cells*, Wiley-VCH: Weinheim (2005).
3 A Luque and S. Hegedus (eds). *Handbook of Photovoltaic Science and Engineering*, 2nd edition, John Wiley & Sons, Ltd: Chichester (2011).
4 S.R. Wenham *et al. Applied Photovoltaics*, 2nd edition, UNSW Centre for Photovoltaic Engineering: Sydney (2009).
5 C. Honsberg and S. Bowden, University of Delaware (2013). www.PVEducation.org (Accessed on August 24, 2017).
6 J. Poortmans and V. Arkhipov. *Thin Film Solar Cells*, Wiley-VCH: Weinheim (2006).
7 M. Pagliaro *et al. Flexible Solar Cells*, Wiley-VCH: Weinheim (2008).
8 Fthenakis V. Life Cycle Impact Analysis of Cadmium in CdTe Photovoltaic Production, *Renewable and Sustainable Energy Reviews*, 8, 303–334 (2004).
9 European Commission, DG JRC, Peer Review of Major Published Studies on the Environmental Profile of Cadmium Telluride (CdTe) Photovoltaic (PV) Systems. http://iet.jrc.ec.europa.eu/remea/sites/remea/files/report_summary-peer_review.pdf (Accessed on August 24, 2017).

10 Scientific Comment Fraunhofer to Life Cycle Assessment of CdTe Photovoltaics. http://www.csp.fraunhofer.de/presse-und-veranstaltungen/details/id/829/ (Accessed on August 24, 2017).

11 C. Wolden, *et al.* Photovoltaic Manufacturing: Present Status and Future Prospects, *Journal of Vacuum Science and Technology A*, 29(3), 030801-1–030801-16 (2011).

12 G. Crabtree and N. Lewis, Solar energy conversion, *Physics Today*, 60(3), 37–42 (2007).

13 V. Fthenakis (ed.) *Third Generation Photovoltaics*, InTech (open access): Rijeka (2012).

14 A. Anctil and V. Fthenakis, Life Cycle Assessment of Organic Photovoltaics, in V. Fthenakis (ed.). *Third Generation Photovoltaics*, InTech (Open Access): Rijeka (2012).

3

PV Modules and Arrays

3.1   Introduction

Modules and arrays present photovoltaic (PV)'s face to the world, as well as the sun, and the technology's reputation depends crucially on their technical performance, reliability, and appearance. They must be designed and manufactured for a long and trouble-free life. The solar cells they contain need careful encapsulation to provide mechanical strength and weatherproofing, and the electrical connections must remain robust and corrosion-free.

Most PV modules are provided with aluminum frames to give extra protection and simplify mounting on a roof or support structure. Modules without frames, known as *laminates*, are sometimes preferred for aesthetic reasons, for example, on the façade of a building where reflections from metal frames would be unwelcome. A group of inter-connected modules working together in a PV installation is referred to as an *array*. We mentioned PV modules briefly in Section 2.1, noting the module areas required for a given power output using different cell technologies, and discussed cell and module efficiencies. In this chapter we will focus mainly on electrical characteristics and effective mounting to capture the available sunlight. But first, a few words about module sizes and designs.

For a given level of solar cell efficiency, the rated power output of a module is proportional to its surface area. As we noted in Section 2.1, about $7-8\,m^2$ of surface area is required to generate $1\,kW_p$ using crystalline silicon cells, about $16\,m^2$ using amorphous silicon, and intermediate areas for thin-film technologies such as CIGS and CdTe. As more and more PV installations including power plants move into the megawatt range, huge arrays and module numbers are involved. For example, the 550 MW power plant, part of which is shown in Figure 2.3.4, uses more than 5 million CdTe modules each of $0.72\,m^2$ area. Recently, First Solar, the manufacturer of these modules, announced a new product, with three times larger area, to speed up installation. The larger the size of a module, the lower the number of electrical interconnections and the associated mounting costs. In response PV manufacturers have steadily increased module sizes and power ratings that now range up to several hundred peak watts. However, such advances must be weighed against the difficulty of handling the larger, and therefore heavier, modules. In addition, for thin-film PV, the module size

Electricity from Sunlight: Photovoltaic-Systems Integration and Sustainability, Second Edition.
Vasilis Fthenakis and Paul A Lynn.
© 2018 John Wiley & Sons Ltd. Published 2018 by John Wiley & Sons Ltd.
Companion website: www.wiley.com/go/fthenakis/electricityfromsunlight

Figure 3.1 A large array of PV modules on a rooftop in Switzerland (*Source:* Reproduced with permission of EPIA/BP Solar).

is also constrained by the challenge of depositing high-quality semiconductor films over large areas.

Figure 3.2 shows a cutaway view of the edge of a typical module containing crystalline silicon solar cells. The cells, which are brittle, are cushioned by encapsulation between two layers of *ethylene vinyl acetate* (*EVA*) and adhesive that also holds the module parts together. On top is a cover of tempered glass that is sometimes treated with an antire-flection coating (ARC) to maximize light transmission. Underneath is a sheet of *Tedlar*, a light synthetic polymer, acting as a barrier to moisture and chemical attack. The whole "sandwich" is located in a slot in the aluminum frame and fixed with sealant. Other PV module types (e.g., CdTe) are sandwiched between two sheets of glass and do not have frames. This construction must withstand up to 25–30 years of outside exposure in a variety of climates that include desert sands, alpine snows, wind, rain, pollutants, and extremes of temperature and humidity—a highly demanding specification. When things go wrong, it is often due to ingress of moisture or corrosion of electrical contacts rather than faults in the solar cells.

PV module design is by no means static, especially with the new thin-film technolo-gies. In most cases the films, deposited on glass or other substrates, are scribed to produce the complete pattern of solar cells and interconnections, avoiding the need to handle and mount individual semiconductor wafers. Not only does this reduce man-ufacturing costs, but it also promises improved electrical reliability within the module. Another area where thin films are having a big impact is in flexible products. The his-toric market dominance of rigid, relatively heavy glass-covered modules is increasingly challenged by flexible lightweight designs more easily tailored to the shapes of awkward roofs and unusual structures or the aesthetic demands of architects.

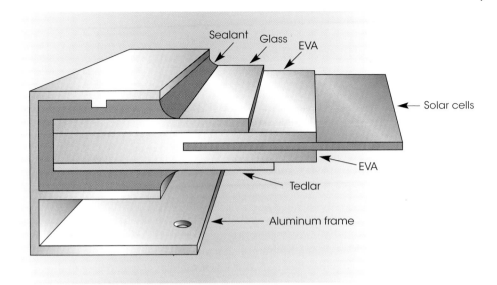

Figure 3.2 Typical construction of a conventional crystalline Si PV module.

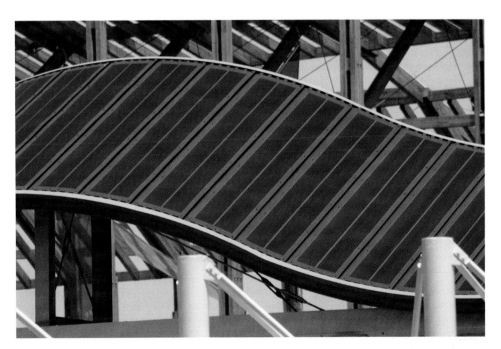

Figure 3.3 Innovative design: an example from Japan (*Source:* Reproduced with permission of IEA-PVPS).

3.2 Electrical Performance

3.2.1 Connecting Cells and Modules

Individual solar cells are hardly ever used on their own. A cell is essentially a low-voltage, high-current device with a typical open-circuit voltage of around 0.5 V, far lower than the operating voltage of most electrical loads and systems. So it is normal for a PV module to contain many *series-connected cells, raising the voltage* to a more useful level. For example, many manufacturers offer modules with 36 crystalline silicon cells connected in series, suitable for charging 12 V batteries. These modules have an open-circuit voltage V_{oc} of around 20 V and a voltage at the maximum power point (MPP) V_{mp} of about 17 V, giving a good margin for battery charging, even in weak sunlight. As PV moves increasingly toward high-power grid-connected systems, the trend is for more cells per module giving higher output voltages—for example, the modules previously shown in Figure 2.1 each contain 72 cells producing about 35 V at the MPP.

Of course the peak power of a module is also one of its key characteristics. The surface area of a monocrystalline silicon solar cell is limited by the diameter of the original ingot, which in turn restricts its power output. Many cells must therefore be interconnected to produce substantial module power: for example, 72 cells, fitted into a module of about $1.5\,m^2$ area, can yield about $200\,W_p$. Multicrystalline silicon cells, being cut from large cast blocks of silicon, are less restricted in area; thin-film cells are even less so. But the modules must still incorporate many cells to achieve useful voltage levels and have a substantial surface area to give a reasonable power output.

What happens when many cells are connected in series? The answer would be very straightforward assuming that all the cells are identical and exposed to the same strength of sunlight: with n cells in series, the module voltage would be n times the cell voltage, and the module current would be the same as the cell current. But in practice cells are not functioning exactly the same. There could be small manufacturing tolerances and small temperature differences, depending on where cells are located in the module. If a module becomes partially shaded by buildings or trees, some cells receive more sunlight than others. In all cases the module's output is limited by the cell with the lowest output—the "weakest link in the chain." The resulting loss of power is referred to as *mismatch loss*.

Small mismatch losses are to be expected in commercial modules and are covered by manufacturing tolerances. They need not normally concern us. But significant losses can easily be caused by shading, which should obviously be avoided where possible. The situation can worsen dramatically if one cell in a string becomes truly "bad" and fails to generate current. It then acts as a load for the other cells and starts to dissipate substantial power, which can lead to breakdown in localized areas of its $p–n$ junction. Severe local overheating occurs, possibly causing cracking, melting of solder, or damage to the encapsulating material. This is known as *hot-spot formation*.

The hot-spot condition is illustrated in Figure 3.4. At the top in part (a) is a string of n cells of which $n-1$ are "good" and one is "bad." The string is shown short-circuited, which is the worst-case scenario. Since the cells are in series, the current must be the same for all. But whereas the good cells happily generate a solar current I_L, the bad cell cannot do so and is forced into reverse bias. With the string short-circuited, the bad cell is subjected to the full voltage and power output of the good cells, leading to breakdown and hot-spot formation.

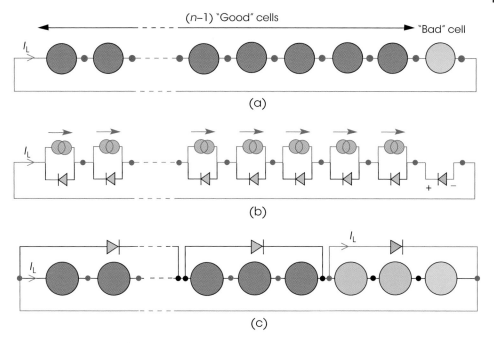

Figure 3.4 (a) A string of cells including one "bad" cell; (b) equivalent circuit; (c) addition of bypass diodes.

Part (b) of the figure clarifies the situation from a circuit point of view. The complete current path is shown in red. Each good cell is represented by the simple equivalent circuit previously shown in Figure 2.9(a), consisting of a semiconductor diode in parallel with a current generator. Since the bad cell's current generator is inactive, the circuit current I_L must pass through its diode in a reverse direction (of course, a diode is not supposed to pass reverse current; but in this case the voltage produced by all the good cells in series is sufficient to cause breakdown and make it conduct). The power produced by the good cells must now be absorbed by the bad cell since none is dissipated in an external load. In conditions other than short circuit, the situation is less severe, but hot spots may still occur.

Hot-spot failure is avoided by incorporating additional diodes known as *bypass diodes*, shown in part (c). Here red indicates circuit elements carrying the main current, green indicates inactive (but serviceable) elements, and blue indicates the bad cell. The bypass diodes offer an easy current path around any bad cells. Ideally there would be one bypass diode for each solar cell, but this is rather expensive. Many PV modules therefore incorporate one diode per small group of cells. In our example there are three cells per group. The disadvantage is that each bad cell takes two good cells out of action, preventing them from contributing to the power output of the string. In practice the maximum number of cells in a group that prevents damage is normally reckoned about 10. Bypass diodes are often built into commercial modules, but if they are not, care should be taken to avoid short circuits, especially when there is partial shading by trees, buildings, or other structures.

Application of Ohm's Law to Cell or Module Connections in Series and in Parallel

$$I(\text{Ampere}) = \frac{V(\text{volt})}{R(\text{Ohm})}$$

Series: (resistors R_1, R_2, R_3 in series)

$$R_{\text{equivalent}} = R_1 + R_2 + R_3 + \cdots$$

$$R_{\text{equivalent}} = \frac{V}{I} = \frac{V_1 + V_2 + V_3 + \cdots}{I} = \frac{V_1}{I_1} + \frac{V_2}{I_2} + \frac{V_3}{I_3} + \cdots = R_1 + R_2 + R_3 + \cdots$$

Series key idea: The current is the same in each resistor by the current law.

Parallel: R_1, R_2, R_3

$$\frac{1}{R_{\text{equivalent}}} = \frac{1}{R_1} + \frac{1}{R_2} + \frac{1}{R_3} + \cdots$$

$$\frac{V}{R_{\text{equivalent}}} = I = I_1 + I_2 + I_3 + \cdots = \frac{V_1}{R_1} + \frac{V_2}{R_2} + \frac{V_3}{R_3} + \cdots$$

$$\frac{V}{R_{\text{equivalent}}} = \frac{1}{R_1} + \frac{1}{R_2} + \frac{1}{R_3} + \cdots$$

Parallel key idea: The voltage is the same across each resistor by the voltage law.

Many of the ideas we have developed for solar cells also apply to modules. For example, a module, like an individual cell, may be characterized by its open-circuit voltage, short-circuit current, and MPP. Indeed we may think of a complete module as a type of "supercell" with higher voltage and power ratings.Modules are connected together—sometimes in large numbers—to form arrays. Whereas the cells in a single module are usually series-connected to raise the voltage as much as possible, modules in an array may be connected in series, parallel, or a mixture of the two. This is illustrated by the array of six modules shown in Figure 3.5, consisting of two strings in *parallel to amplify the current*, each string containing three modules in *series to amplify the voltage*. According to Ohm's law ($I = V/R$), if the modules are perfectly matched, this arrangement produces an array voltage three times the module voltage, an array current twice the module current, and an array power six times the module power.

In practice the array performance will be slightly reduced by mismatch losses between the various modules. It is also worth noting that modules from different manufacturers should not be mixed together in an array, even if they are nominally similar, because differences in *I–V* characteristics and spectral response are likely to cause extra mismatch losses.

The figure also shows a number of diodes. Those colored green are bypass diodes, one in parallel with each module to provide a current path around the module if it fails

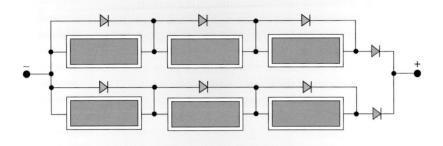

Figure 3.5 An array of six modules.

or becomes "bad." The two diodes colored red are referred to as *blocking diodes*, one in series with each string to ensure that current only flows out of the modules. They are generally used in battery charging systems to prevent the batteries from discharging back through the modules at night. Some manufacturers include blocking diodes within their modules.

An array installed on a domestic roof might typically contain 10–20 modules. A single module is normally sufficient for the type of solar home system (SHS) that provides modest amounts of electricity for a family in a developing country. But the huge PV systems and power plants now being installed in industrialized countries incorporate hundreds of thousands of modules and involve major decisions about how they should be interconnected. A complete *PV generator* is, of course, a DC generator, and since most large systems feed their solar electricity into an AC grid, it is important to design arrays that can be interfaced safely and efficiently. We will return to this topic in the next chapter.

3.2.2 Module Parameters

Not surprisingly, most of the electrical parameters of a PV module closely reflect those of its solar cells. However the efficiency of a module, measured in standard conditions of bright sunlight (1000 W/m^2 at 25°C, AM1.5 spectrum), is slightly less than that of the constituent cells because the cells do not completely fill the module's area and there are small power losses as sunlight passes through the top cover and encapsulant. If blocking diodes are built in, these too will produce small power losses. Average (measured under standard conditions) efficiencies for the most widely used terrestrial modules are 18% for monocrystalline silicon, although modules with efficiencies as high as 21.5% are also in the market, 17% for multicrystalline silicon, approximately 15% for CIS, and 16.5% for CdTe; efficiencies of *a*-Si PV are below 10%. Note that these module efficiencies reflect second quarter of 2017 and module efficiencies may be higher by the time you read this text as the PV industry continues to invest in R&D. But as we have emphasized previously, efficiencies in bright sunlight do not tell the whole story. Crystalline silicon modules tend to lose their advantage in weak or diffuse light, or high temperatures, and there is accumulating evidence that the newer thin-film modules may produce higher annual yields in regions with substantial cloud cover.

We have already described monocrystalline silicon solar cells in some detail and illustrated typical I–V characteristics for a 2 W_p cell at various levels of insolation in Figure 2.10(b). As we pointed out in the previous section, the cells in a module are normally series-connected, raising the power and voltage (but not the current). As an example we now consider a module containing 72 monocrystalline cells with a peak power rating of 180 W_p—a popular size and power rating. The following module parameters are at the low end of the range of commercial modules in the market today:

Nominal power	180 W_p
Open-circuit voltage	43.8 V
Short-circuit current	5.50 A
Voltage at maximum power	35.8 V
Current at maximum power	5.03 A
Power reduction per °C	0.45%
Voltage reduction per °C	0.33%
Length	1600 mm
Width	804 mm
Weight	18 kg
Efficiency	14.0%

As expected, the nominal power of 180 W_p equals the product of voltage and current at the MPP. The efficiency is given by the module power in kW_p divided by the module area in square meters. And the reduction of power at elevated cell operating temperatures, 0.45% per °C, is typical of crystalline silicon—and more serious than for thin-film technologies. It means, for example, that if the cell temperature is allowed to rise to 65°C, the power output will fall by about 18%, emphasizing the need to keep this type of module as cool as possible with adequate ventilation. For comparison, the loss of power due to heat buildup in CdTe PV is 0.34% per °C and the voltage reduction is 0.29% per °C.

A family of I–V curves for the module is shown in Figure 3.6. Their form is very similar to those for the 2 W_p cell of Figure 2.10(b). The top curve, labeled 1000 W/m^2, refers to standard insolation and corresponds to the parameters in the previous table. The other three curves confirm that as the level of insolation reduces, the current falls in proportion. Each has its own MPP, labeled P_1–P_4, the operating point at which maximum power output may be obtained.

Other aspects of module performance closely mirror those of the constituent solar cells. For example, since the module is effectively a current source, the actual power output is closely proportional to the voltage at which it is operated, up to the MPP, and the main effect of temperature rise on the I–V curves is a reduction in open-circuit voltage. You may like to refer back to Figures 2.11 and 2.12 for a discussion of these points.

We have so far concentrated on monocrystalline silicon and you may be wondering how module parameters and I–V characteristics differ for other cell technologies. In fact the differences are very slight for multicrystalline silicon, the main effect being a small decrease in module efficiency (and increase in module area) for a given power rating. The sizes and power ratings of both types of crystalline silicon modules have tended to rise steadily in recent years; most current production is in

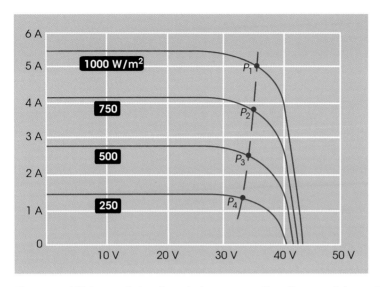

Figure 3.6 *I–V* characteristics of a typical monocrystalline silicon module rated at 180W$_p$.

the range 150–300 W$_p$. In part this is due to advances in manufacturing processes to satisfy a market increasingly slanted toward large grid-connected systems. The first commercial module to achieve 500 W$_p$ appeared a few years ago.

When we come to the newer thin-film modules, the situation changes in two main respects. First, manufacturers generally start testing the market for new cell technologies with relatively small modules. It takes a lot of skill and experience to produce large modules with consistent performance and reliability. CIS and CdTe modules, for example, have worked up from small to medium sizes over a number of years, but maximum power ratings are still behind those of monocrystalline silicon. Large a-Si modules, a more mature technology, offer power levels into the hundreds of watts, but at amorphous silicon's relatively low efficiency. Second, the manufacture of thin-film cells offers considerable flexibility over module voltages. Cells can be scribed in a wide range of sizes within a module; a few large cells give high currents at low voltages, while many small cells give high voltages at low currents. Higher voltages are often preferred for interfacing to an electricity grid and for reducing cable losses when modules are interconnected, a significant factor in large PV systems.

Figure 3.7 illustrates this flexibility with *I–V* characteristics for four commercial modules based on different technologies; multicrystalline silicon (mc-Si), CIS, CdTe, and a-Si. The modules all have the same nominal peak power of 75 W$_p$ but very different voltage and current levels. The curves are all drawn for standard insolation (1000 W/m^2), and the MPP for each module is shown by a dot. The figure is meant to be indicative, not definitive, and the situation is of course developing rapidly. For example, the latest Series 4 CdTe PV rated at 117.5 W (16.3% efficiency) gives at PMax both higher voltage and current, 71.2 V and 1.65 A correspondingly. The main point is the increasing range of power levels offered by thin-film deposition and scribing techniques. Chapter 6 on PV manufacturing discusses processes used by the industry to increase current and voltage of solar cells.

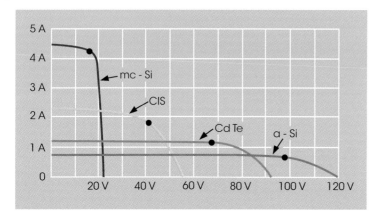

Figure 3.7 *I–V* characteristics in strong sunlight (1000 W/m²) of four 75 W$_p$ modules.

3.3 Capturing Sunlight

The sun is a tremendous fusion reactor, with a volume 1.3 million times larger than of our Earth, at a safe distance from the Earth (Figure 3.8). Forty minutes of sunshine striking the Earth on a clear day is equal to 1 year of global consumption of energy.

Solar radiation at normal incidence received at the surface of the Earth is subject to variations due to change in the extraterrestrial radiation and to two more significant phenomena: (i) atmospheric scattering and (ii) atmospheric absorption by O_3, H_2O, and CO_2.

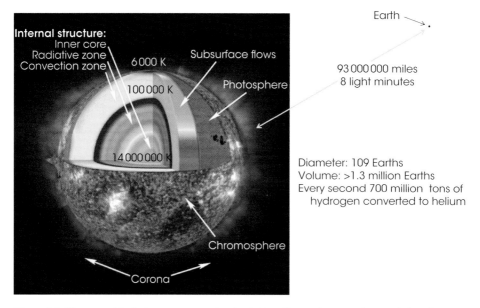

Figure 3.8 The sun, a big fusion reactor on the sky, is 93 million miles away, but sunlight takes only 8 light minutes to reach the Earth (University of Oregon website).

The effects of sunlight scattering and absorption on the spectral distribution of solar beam irradiance were shown in Figures 1.8 and 1.10. Absorption and scattering levels change as the constituents of the atmosphere change. Clouds are the most familiar example of change; they can block most of the direct radiation. Also seasonal variations and trends in ozone layer thickness have an important effect on terrestrial ultraviolet level. Fortunately not only the direct light from the sunlight but also the light that diffuses into the atmosphere and the light that is reflected from surfaces can be useful for creating electrons in photovoltaics.

Also, as discussed in Chapter 2, solar cells respond differently to the different wavelengths, or colors, of light, depending on the energy bandgap of the semiconductor materials they employ. For example, crystalline silicon can use most of the visible spectrum, plus some part of the infrared spectrum. But energy in part of the infrared spectrum, as well as longer-wavelength radiation, is too low to produce electrons. Higher-energy radiation can produce electrons, but much of this energy is likewise not usable. In summary, light that is too high or low in energy is not usable by a cell to produce electricity and is converted into heat.

Now to measure all this sunlight, we need instruments that are sensitive across the whole solar spectrum, and they can measure all the components of solar irradiation; as we discussed in Chapter 1, these are the direct (beam), diffuse, and albedo (reflected on the ground) solar components. Their total is called global horizontal irradiation (GHI) as is typically measured on horizontal surfaces by instruments called **pyranometers**. The name is derived from the ancient Greek words pyr "fire" + ouranos "sky," implying an instrument that measures the strength of the sun's "fire" from the sky. A pyranometer uses a large number of thermocouples that are connected in series in a pile facing upwards, with each absorbing a part of the solar spectrum. It thus measures voltage induced by thermal energy absorbed at the tip of each thermocouple. A double-glass dome prevents condensation and reduces heat losses, while its hemispheric shape allows the global irradiation to enter with minimum reflections. Many pyranometers have a cartridge of silica gel to absorb moisture, and this need to be replaced periodically. Also the sensitivity of a pyranometer can decline with time, and it may need to be recalibrated every 2 years or so.

Since pyranometers are rather expensive (costing $1500–2000), irradiation sensors made of solar cells, called **reference cells**, are an economic alternative. These are precisely calibrated silicon solar cells that are short-circuited, so the short-circuit current is converted into a voltage that is being measured in conjunction with temperature measurements with an enclosed thermocouple. Their cost is low, ranging from $150 to 500, but their accuracy is not better than ±5% or ±10%.

When using solar cells with concentrators on sun-tracking devices, we need to measure the direct (normal on the surface of the PV module) irradiation (DNI). DNI is measured by instruments called **pyrheliometers**, which track the sun. The name is also derived from the ancient Greek words pyr "fire" + helios "sun" + meter "measure." Similarly to the pyranometers, sunlight enters the instrument through a window and is directed onto a stack of thermocouples that convert heat to voltage. The signal voltage is amplified and is reported as power per area (W/m^2). In contrast to a pyranometer, a pyrheliometer has a very small window allowing only direct sunlight to enter, and it is used with a solar tracking system to keep the instrument aimed at the sun.

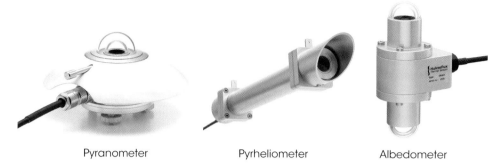

Pyranometer   Pyrheliometer   Albedometer

Figure 3.9 Instruments measuring solar irradiance (*Source:* Reproduced with permission of Hukseflux thermal Sensors B.V.).

Albedo, also called solar reflectance, is defined as the ratio of the reflected to the global irradiation. The solar albedo depends on the directional distribution of incoming radiation and on surface properties at ground level. Albedos of typical surfaces range from about 4% for asphalt and 15% for grass to 90% for snow. This can be measured by an **albedometer** composed of two identical pyranometers, one facing the sky and one facing the ground (Figure 3.9). Albedo is especially important if the modules are composed of bifacial solar cells; thus solar cells can absorb sunlight from both sides. When conventional PV modules are placed horizontally or with small angles to the horizontal, then the albedo contribution is negligible.

The performance, thus the cost-effectiveness, of a PV system depends crucially on positioning its solar array to capture as much sunlight as possible. We must therefore appreciate how the sun's apparent path across the sky varies according to the time of year and the latitude of the site. Of course the variation is the result of the rotation of the Earth around its axis and its trajectory around the sun. Figure 3.10 shows this trajectory, whereas Figure 3.11 shows how a stationary observer on the northern hemisphere sees this trajectory in daily cycles. The sun appears to be at its lowest height from the ground at the *winter solstice*, around December 22 in the northern hemisphere, and at its highest at the *summer solstice*, around June 21. In between are the two mid-season *equinoxes* around March 22 and September 23, when the sun rises due east and sets due west, giving equal hours of day and night.

The *declination* shown in Figure 3.10 is the angle between the equator plane and the line connecting the centers of the Earth and the sun. It varies seasonally due to the tilt of the Earth on its axis of rotation and the movement of the Earth around the sun. If the Earth was not tilted on its axis of rotation, the declination would always be 0°. However, the Earth is tilted by 23.45° and the declination angle varies between −23.45° and +23.45°. Only at the spring and fall equinoxes is the declination angle equal to 0°.

The high point is always reached at *solar noon* when the sun is in the south. In winter, sunrise occurs south of east and sunset occurs south of west; in summer both veer toward the north. In other words the angular span as well as the height of the trajectory varies with the season. This all applies equally well to the southern hemisphere if we swap north for south and interchange the dates of the winter and summer solstices.

Time measured by the apparent motion of the sun is called *solar time* and fluctuates slightly around the time given by a conventional clock. This is because the Earth's journey

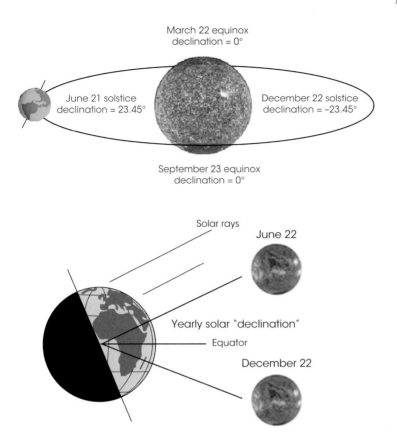

Figure 3.10 The Earth trajectory around the sun and its rotation around its axis. Declination is the angle that direct (beam) irradiation from the sun hits the equator.

around the sun is slightly elliptical and our distance to the sun varies with the time of year. In this chapter we always refer to solar time (e.g., *solar noon*) because we are, of course, interested in the apparent movement of the sun in relation to PV installations on Earth. However, the time system used in our everyday lives and shown by our clocks and watches averages out the fluctuations to make every day of equal length and is referred to as *mean time.* The best-known example is *Greenwich Mean Time*, being the time measured at the Greenwich Observatory in London. This is on the *prime meridian*, reckoned as 0° longitude, thus the longitude where we start measuring east and west. Meridian originates from the Latin word medius "middle" + dies "day." The use in astronomy is due to the fact that the sun path crosses a local meridian at noon. Fortunately the difference between local solar time and mean time, described by the so-called equation of time, never exceeds about 17 minutes at any time of year. This is only really significant when designing PV systems that use highly concentrated sunlight and must track the sun very accurately across the sky. We shall meet them later in the chapter.

Latitude also has a big effect—the further we are from the equator, the lower the sun's path through the sky. On the equinox dates, its elevation angle, also called *solar altitude*, above the horizon at solar noon, *labeled a* in Figure 3.11, is equal to 90° minus the latitude. For example, in Madrid, latitude 40°N, the solar altitude at noon on March 22 and

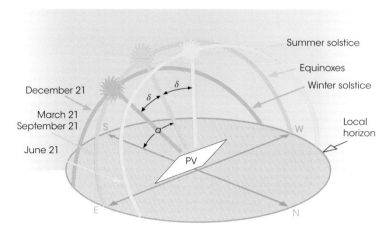

Figure 3.11 Solar trajectories over the northern hemisphere, from a stationary observer on Earth.

September 23 is 50°; in more northerly Berlin, latitude 52°N, it is 38° (and at the North Pole, latitude 90°N, the sun is on the horizon). At the summer solstice, June 21, the noon elevation increases by an angle δ equal to 23.45° and is at its annual peak. At the winter solstice it is reduced by the same amount. In general, $a = 90° + \delta - \varphi$

Thus, at a location of latitude φ,

$$a = 113.45° - \varphi \text{ is the maximum } \textit{solar altitude} \text{ in the summer, and}$$
$$a = 66.55° - \varphi \text{ is the minimum } \textit{solar altitude} \text{ in the winter.}$$

(3.1)

So in Madrid the summer and winter solstice elevations are 73.45° and 26.55°, respectively (intrepid explorers at the North Pole for the winter solstice, in total darkness, are perhaps unaware that the sun is 23.45° *below* the horizon). These seasonal variations are caused by the offset angle between planet Earth's axis of rotation and its plane of revolution around the sun.

3.3.1 Aligning the Array

In the previous section we saw how the sun's trajectory varies according to the time of day, the season, and the latitude. This information suggests how a fixed PV array should be aligned to capture as much sunshine as possible. First, for installations in the Northern hemisphere it should point due south toward the midday sun. In some cases, for example, on existing buildings, this may not be possible, but any deviation from south should preferably not exceed about 30°.

Second, the array should be tilted down from the horizontal so that the sun's rays at solar noon are normal to its surface. Since the sun's noon elevation varies continuously through the year, a choice has to be made about when to meet this condition. Very often the two equinoxes (on or around March 22 and September 22) are selected, giving the PV tilt shown in Figure 3.11.

On these two dates the array points "perfectly" at the midday sun but is somewhat too low in the summer and too high in the winter—normally a good compromise. This is shown in Figure 3.13.

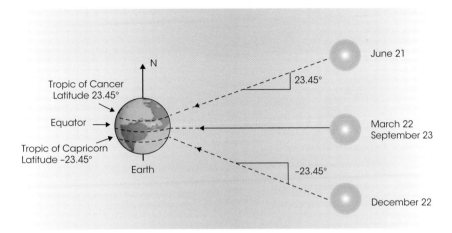

Figure 3.12 An alternative view of the solar trajectories.

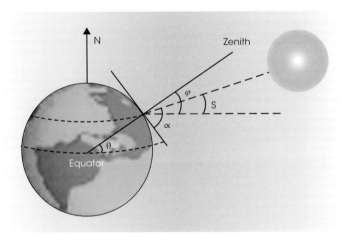

Figure 3.13 The daily maximum altitude angle of the sun as a function of latitude.

In the previous section we noted that the sun's noon elevation at the equinoxes (denoted by angle *a* in the figure) equals 90° minus the latitude of the site. It follows that its *declination* (δ) is equal to the latitude and that the array must be tilted down by this amount (Figures 3.14 and 3.15). For example, in Madrid, latitude 40°N, an array must be tilted down 40° to meet the aforementioned condition. It will then point too high by angle δ (23.45°) at the winter solstice and the same amount too low at the summer solstice.

The aforementioned "equinox criterion" for tilting a PV array is widely adopted, but, as we shall see, it is not essential. Minor variations of tilt angle have very little effect on an array's annual yield, and in any case there are some situations where a different choice of tilt may prove beneficial. A good example is a stand-alone PV system in high latitude, required to provide a steady supply of electricity throughout the year. The winter months are the most difficult and will determine the size of array required, for if the

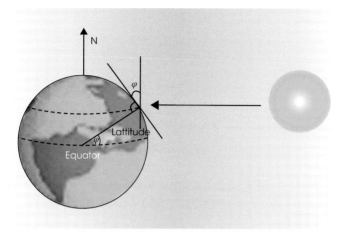

Figure 3.14 A south-facing PV array tilted at an angle equal to the latitude of the location.

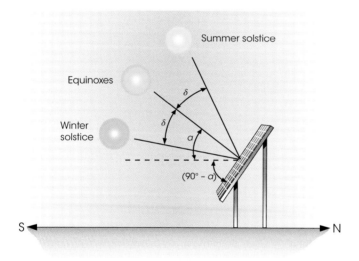

Figure 3.15 Aligning a PV array.

system is able to cope in the winter, it will certainly do so in the summer. So the down-ward tilt of the array is often increased to make the most of winter sunshine. Unusual climatic conditions may also favor different amounts of tilt, for example, in parts of South Asia typified by hot humid summers with overcast skies followed by clear cool winters with plenty of sunshine. If it is required to maximize the annual electricity yield of a grid-connected system, a larger downward tilt may well be helpful. Conversely, in a system required to optimize electricity yield in the summer months, the PV array may be aligned closer to the horizontal. A good example is the summer holiday home described in Section 5.4.1.

In the case of rooftop arrays, tilt is often predetermined by roof geometry, giving little or no flexibility. However it is worth noting that buildings in high-latitude countries such as Norway and Sweden often have high-tilt roofs to encourage snow to slide off easily, whereas roofs in Morocco or Egypt are much more likely to be flat, or nearly so. In this way vernacular architecture tends to suit the sun's trajectory and the preferences of PV system designers.

So far we have concentrated on capturing as much of the sun's direct radiation as possible. This is certainly important, but, as our discussion of the solar resource in Section 1.2 made clear, there is also diffuse and albedo radiation to consider (see Figure 1.10). What happens when we start considering the actual radiation falling on a PV array, taking into account scattered light?

You may find it helpful to refer back to Figure 1.7 showing the large-scale effects of climate on insolation at the Earth's surface. In temperate regions with plenty of "cloudy-bright" weather, the diffuse component can make a surprisingly large contribution to the annual total; for example, in Western Europe it is often over 50%. One effect is to make array alignment less critical because diffuse light tends to come from all over the sky. Another is to reduce the overall importance of shadows in determining the annual energy yield.

To narrow these general ideas down to a particular PV system, we need more detailed local information. Fortunately, the great surge of interest in solar energy in recent years has spawned data on average sunlight conditions for many cities and locations around the world (one valuable source of information is provided by NASA[1]). The data is often presented in the form of 12 monthly mean values of global (direct and diffuse) daily radiation on a horizontal surface, expressed in kWh/m^2. Albedo radiation is not included since it does not affect horizontal surfaces and anyway is highly site dependent. Sometimes the proportions of direct and diffuse light are found by practical measurements with specialized instruments; sometimes they are inferred from the global figure and a *clearness index* summarizing the amount of light scattering caused by clouds and particles in the local climate. Figure 3.16(a) shows a typical distribution for a West European city such as London or Amsterdam with a temperate climate giving plenty of "sunshine and showers" in summer and cloudy skies in winter. The height of each bar represents global radiation, composed of direct (yellow) and diffuse (orange) components. The daily average over the whole year is about $2.8 kWh/m^2$, giving an annual total of about $1050 kWh/m^2$. Part (b) of the figure is for the Sahara Desert. Here the extremely sunny, hot, and reliable climate produces a daily average of about $6 kWh/m^2$ and an annual total of about $2200 kWh/m^2$ on a horizontal surface. Most of the radiation is direct.

Given such figures it is quite easy to make a rough estimate of the annual output from a PV module or array using the concept of *peak sun hours*. If the energy received throughout the year is compressed into an equivalent duration of standard "bright sunshine" ($1 kW/m^2$), then the number of peak sun hours is the same as the global annual figure. For example, London has about 1050 peak sun hours in a year, so a PV module rated at $200 W_p$ can be expected to produce around $1050 \times 200 = 210\,000\,Wh = 210\,kWh/$ year. However this is for a horizontally mounted module—unlikely in London. It also assumes ideal "bright sunshine" conditions, whereas much of London's sunlight is diffuse. Direct light and diffuse light have different spectral distributions, and solar cells do not generally respond equally to them, nor are most cells equally efficient in bright

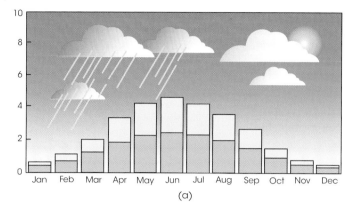

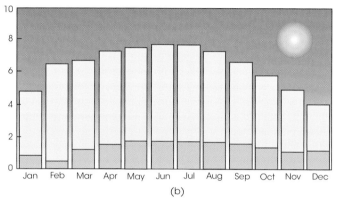

Figure 3.16 Average daily solar radiation in kWh/m² on a horizontal surface: in (a) London or Amsterdam; (b) in the Sahara Desert.

and low-level light. So estimates of array output based on peak sun hours are only very approximate, especially for locations with a large proportion of diffuse light. In most cases actual annual yield is considerably lower. For example, our $200\,W_p$ module in London is more likely to produce 150–160 kWh/year.

It must be emphasized that distributions such as those in Figure 3.16 are normally based on data collected over many years. In a given year, especially in unpredictable climates, they may look very different; it is not unusual to see a 10% variation in annual figures or a 30% variation in monthly ones, and this must be taken into account when making predictions or designing a PV system.

We have so far considered sunlight falling on a horizontal surface. What happens when a PV array is tilted downward to take account of the latitude? How are the figures for global, direct, and diffuse radiation affected? Below we discuss these effects; the methodology and equations for determining the irradiation components on tilted surfaces will be presented later in the Appendix 3.A.

Figure 3.17 shows some estimates for south-facing tilted PV panels in London (latitude 52°N) and the Sahara Desert (latitude 24°N). In each case three different tilt angles are illustrated: 0° (horizontal), shown by blue bars; an angle equal to the latitude, shown by red bars; and 90° (vertical), shown green. You may be surprised at the choice

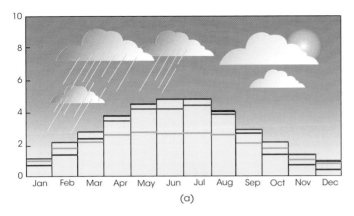

(a)

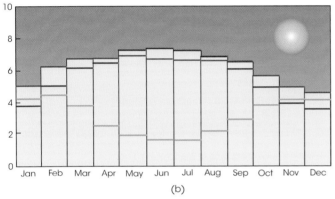

(b)

Figure 3.17 Daily solar radiation in kWh/m² on south-facing inclined PV arrays in (a) London and (b) the Sahara Desert. In each case three values of tilt are illustrated: 0° (blue), the latitude angle (red), and 90° (green).

of 0° and 90°, but actually results for angles closer to the latitude are often almost indistinguishable. And remember that horizontal PV arrays may be installed on flat roofs and vertical ones on building façades (Figure 3.18). The results illustrate several interesting points:

- London: The results for 0° and 52° tilt are quite similar, mainly due to the large proportion of diffuse sunlight, but 0° receives slightly more radiation in the summer and less in the winter; 90° tilt (as on a vertical building façade) receives much less irradiation in the summer months.
- Sahara Desert: The results for 0° and 24° tilt are very similar since we are much closer to the equator. The really big effect is the reduction in radiation for a 90° tilt in summer; when the sun is high in the sky, the radiation is almost all direct and the surface receives very little of it.

Of course such graphs are estimates that cannot take account of fine variations in local climate—for example, different amounts of cloud and shade. As PV enters the multi-gigawatt era, with systems of all shapes and sizes installed around the world, system designers will no doubt have access to ever more performance data collected from working systems.

Figure 3.18 Vertical and in diffuse light: a large PV façade in Manchester, England (*Source:* Reproduced with permission of IEA-PVPS).

3.3.2   Sunshine and Shadow

When positioning a PV array, it is very important to avoid shadows as far as possible for two main reasons. Shading can greatly reduce the output of the modules, and in severe cases it runs the risk of hot-spot formation. What may be termed "occasional shadows" caused by bird droppings, dust layers, or snow on PV modules can obviously be reduced by proper maintenance and cleaning. The situation becomes more complicated if shadows are cast by nearby obstructions. For example, PV roofs may be partially shaded by dormer windows, satellite dishes, chimneys, or ventilation pipes; a small, ill-positioned pipe at 2 m can cause more trouble than a skyscraper at 2 km! Of course small local obstructions should be easier to control and perhaps eliminate. A newly designed roof should always take special care to avoid them.

The degree of shading at different times of year depends upon the sun's trajectory and may be assessed by recasting Figure 3.11 in two-dimensional form, called *sun path chart*, and adding the outlines of buildings, trees, and high terrain that cast shadows over the PV array. Thus a sun path diagram is a graphical representation of the sun's altitude and azimuth angles during the solstice and the equinox dates; often it also shows average monthly sun paths. Figure 3.19 shows such a diagram for latitude 40°N, relevant to world cities including Madrid, New York, and Beijing, and shading obstructions on a particular site. It shows that there are detrimental shading effects in the winter, but shading is not significant during the rest of the year. Sun paths, thus

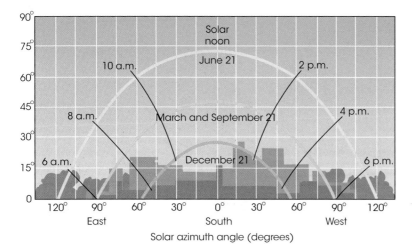

Figure 3.19 A Sun path chart showing shading effects at the solstice and equinox dates. Solar azimuth is the angular displacement from south of the projection of beam radiation on the horizontal plane.

diagrams of the solar altitude α, can be created from the latitude φ, the sun declination δ, and the hour angle ω according to the equation

$$\sin\alpha = \sin\varphi \times \sin\delta + \cos\varphi \times \cos\delta \times \cos\omega \tag{3.2}$$

In practice, predicting shading effects requires placing a panoramic image of obstructions on a solar path diagram correctly aligned for the location, day, and time of taking the image. There are a number of devices on the market for predicting potential shading problems, ranging from handheld viewers to photographic instruments supported by computer software.

The simplest device is the Solar Pathfinder, which has a translucent and highly polished convex plastic dome set on top of a solid base and a compass. The user looks down onto the dome to see panoramic reflection of the obstructions restraining the sunlight at the location. A sun path paper diagram representing the sun path for the location is placed underneath the dome. The sun path diagram then is rotated to its local declination and is pointed to the magnetic south. Slots in the side of the dome allow the user to trace the outline of the reflected obstructions onto the diagram; alternatively a photograph can be taken and superimposed on the sun path diagram. This way we can determine the obstructions that will shade the selected installation site and the times when shading may occur. Figure 3.20 shows the top of the instrument and the outline of the traced obstructions; shading occurs above the white line of the shade tracing shown in the figure on the right.

When shading is unavoidable, it may be possible to reduce its effects by careful planning of module interconnections in the PV array. As we noted in the previous section, a single "bad" or shaded cell in a series-connected string affects all the other cells and can seriously reduce the string's output. The same applies to strings of modules. So it is important to try and prevent one module in a string from becoming shaded at the expense of the others. And if a shadow is large enough to fall on several modules at the same time, it is best if all are members of the same string.

Figure 3.20 Top view of Solar Pathfinder instrument and side shade tracing on a monthly sun path diagram (*Source:* Reproduced with permission of SolarPathfinder).

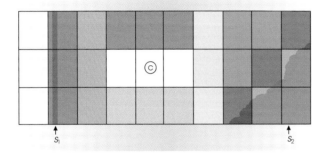

Figure 3.21 Arranging module strings to reduce the effects of shading.

These ideas are illustrated in Figure 3.21 for a PV roof containing 30 modules arranged as 10 parallel strings of three modules each. The various strings, indicated by different green tints, have been arranged to take account of two shadows, S_1 and S_2. The first of these is a narrow stripe formed by a nearby electricity pole, and its effects are reduced by a vertical arrangement of modules. As the shadow moves laterally in line with the sun, it is mainly confined to a single string and affects three modules equally. The second shadow, cast by a neighboring tree, is roughly triangular in shape and falls on the lower right-hand corner of the array. Assuming trees and neighbors are to remain undisturbed, modules may be connected in triangular strings. Once again this minimizes the number of strings affected by the shadow as it moves onto the corner of the roof. And finally there is an unfortunate chimney pipe, labeled c, near the middle of the roof that cannot be moved. Its nuisance value is reduced by a string of "dummy" modules, shown white, which preserves the array's appearance but avoids using expensive real modules that would produce little electricity. In practice it would probably be economical to connect all the unshaded modules in one or two longer strings, a matter we shall return to later.

Although such "array design" is only partly effective, it is virtually cost-free—an important benefit since recurring shadows can degrade an array's output over its entire working life. Finally, it is worth noting that a number of manufacturers now offer module-level power equipment (MLPE) technologies as an alternative to conventional "string" inverters/power optimizers. These technologies aim to reduce shading and mismatch losses by allowing each module in a string to operate at its MPP, regardless of what the others are doing. In principle such devices should be able to overcome many of the shading problems discussed previously. There are essentially two MLPE options that you may come across if you are shopping for a solar system: microinverters (separate inverters for each module) and power optimizers. Both solutions offer similar benefits: better performance when your solar panels are in shade for part of the day. At the same time, solar systems that use these technologies tend to be more expensive than those that use string inverters.

3.4   One-Axis Tracking

Recently, with the evolution of high-efficiency modules, utilities started preferring one-axis tracking systems for ground-mount installations in sunny regions. One-axis tracking systems follow the sun throughout the day to increase energy output and are expected to generate up to 25% more energy than fixed latitude tilt systems. Furthermore, they provide a much flatter power output and can satisfy more of the afternoon peak load than fixed-tilt modules. It appears that they are cost-effective in many high irradiation locations. The only disadvantage is a larger land profile as they need wider spacing between rows and more maintenance than fixed tilt as they have moving parts.

Two common types of one-axis tracking are illustrated in Figure 3.22. In part (a) a PV array rotates from east to west on a horizontal axis, oriented either north or south. In Figure 3.22(b) tracking takes place about a polar axis aligned with the Earth's axis of rotation; this configuration is called polar or tilted single-axis tracking. This limits the sun's offset angle to a maximum of 23.45° from the plane of illumination, giving more efficient overall energy collection than with a horizontal axis (for an explanation of

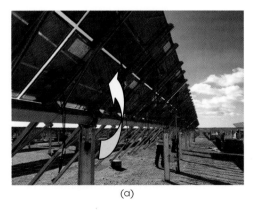
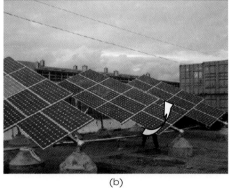

(a)   (b)

Figure 3.22 One-axis tracking: (a) horizontal axis (First Solar); (b) tilted (polar) axis (SunPower).

the angle, you may like to refer back to Figure 3.10 and Section 3.3.1). However the high profile means that more ground area is required to avoid shading of adjacent trackers, wind loading tends to be more serious, and the mounting is more awkward.

The horizontal axis also allows for mounting more PV modules on a given tracking system. The most obvious is to make the tracker larger. A single axis could be made longer or the PV could be stacked in double rows to make each row contain more PV. For a tilted axis tracker, making the rows longer or increasing the width of the PV on a row becomes impractical rather quickly. This is because one end of the tilted axis is sloped up into the air; the impacts of longer rows not only complicates the support structure required to elevate the row but also increases the wind loads, which increase exponentially with height above the ground. As a result, increasing row length as a way to increase array size is not an option for polar axis trackers; thus SunPower also uses long one-axis tracking systems (like Figure 3.20(a)) in their large utility-scale solar farms.

By virtue of having moving machinery and requiring a less dense configuration than fixed-tilt systems, trackers virtually always come at an added cost relative to fixed systems. In order for a tracker to make economic sense, the increased energy harvest must exceed the added cost of installing and maintaining trackers over the lifetime of the system. An additional factor to be considered in the decision to use trackers or fixed systems is land use; tracking systems tend to use additional land because they must be spaced out in order to avoid shading one another as they track the sun. Also fixed racking systems offer more field adjustability than single-axis tracking systems. Fixed systems can generally accommodate up to 20% slopes in the E/W direction, while tracking systems typically offer less of a slope accommodation unless the ground is tilted in the N/S direction facing the equator.

It is also worth noting that a very basic form of tracking can be achieved manually. By adjusting the orientation of a flat-plate module just three or four times a day, in line with the sun's trajectory, over 90% of the electricity yielded by a fully automatic tracker may be obtained. This is an interesting possibility for small systems with just one or two modules, for example, SHS in developing countries where family members are generally on-site and can easily make the adjustments.

Of course the most accurate tracking of the sun requires rotation on two axes to handle both the hourly during a day and the daily during seasonal variations. Such systems are considerably more expensive and are deserved only for concentrating systems where accurate sun tracking is essential.

3.5 Concentration and Two-Axis Tracking

Ever since the dawn of the modern PV age, the PV community has pondered the attractions of concentrated sunlight. After all, if the price of solar cells is very high (and it certainly was in the early days) and is closely related to their surface area, it should make sense to focus the sun's light onto cells of very small area using materials that are less expensive than the solar cells. Furthermore, specialized cells designed to work under concentrated sunlight can achieve considerably higher conversion efficiencies than conventional cells. For example, in our discussion of gallium arsenide cells in Section 2.5.1, we noted that efficiencies around 40% make them suitable candidates for high-concentration PV systems. But the approach is only viable if efficiency improvements and cost savings on the cells more than offset the additional costs of lenses or mirrors plus, in most cases, equipment to track the sun on its daily journey across the sky. Unsurprisingly, there are plenty of skeptics, not least because the cost of conventional solar cells and modules continues to fall. But the jury is still out, and it will be fascinating to see how the market develops in the coming decade.

To summarize, such systems offer two main attractions:

- The area of the solar cells can be greatly reduced.
- Cells designed for high-intensity concentrated sunlight can achieve better conversion efficiencies than standard cells.

However the disadvantages and challenges appear rather onerous:

- Lenses or mirrors must be used to concentrate the light.
- Above a certain level of concentration, it becomes essential to accurately track the sun across the sky, keeping the focused light exactly aligned on the solar cells.
- High concentration is effective for the direct component of sunlight, but not the diffuse and albedo components.
- Focusing and tracking equipment must be robust and properly maintained to match the expected lifetime—say, 25 years—of solar cells and modules.
- Tracking systems are unsuitable for building-integrated PV, including rooftop arrays.

We may therefore expect to see high-concentration tracking systems largely restricted to power plants in areas with a high percentage of direct sunlight. So how is sunlight concentrated and focused onto small-area solar cells? There are two basic approaches: using transparent lenses or reflective mirrors. The first of these is illustrated in Figure 3.23. Part (a) shows a circular lens, normally made of plastic, which concentrates the direct sunlight onto a small solar cell. Simple refractive lenses become very thick if their diameters exceed about 10 cm, so a special form known as the *Fresnel lens* is widely used. Rather than allowing the lens to get thicker and thicker toward its center, the convex surface is collapsed back to a thinner profile in a series of steps. A family of these lenses, each focusing sunlight onto a single solar cell, can be built up as a parquet to make a large flat PV module.

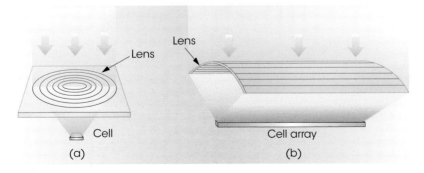

Figure 3.23 Concentrating sunlight onto solar cells using lenses: (a) a circular Fresnel lens with point focus; (b) a linear Fresnel lens with line focus.

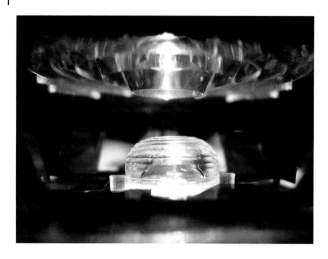

Figure 3.24 Two-stage focusing of light to achieve high concentration (*Source:* Reproduced with permission of EPIA/Isofoton).

Whereas the circular Fresnel lens is *point focus*, the linear domed form of Fresnel lens shown in part (b) of the figure produces a *line focus* onto a long array of cells. Once again the lens profile is collapsed in a series of steps, keeping its thickness reasonably constant around the curve. The curvature increases mechanical strength and avoids optical problems that can arise with more flexible, flat lenses.

In high-concentration systems the necessary focus is sometimes achieved in two stages (Figure 3.24). The main lens performs an initial concentration of the light, directing it onto a *secondary optical element* for further concentration. This also offers an opportunity to ensure that the intensity of light striking the active area of the solar cell is as uniform as possible.

Reflective mirrors provide an alternative to lenses. You are probably aware that a parabolic dish mirror receiving light parallel to its axis brings the light to a point at its focus. This is shown in Figure 3.25(a) with a solar cell mounted at the focus. Another effective configuration is the linear parabolic trough shown in part (b) that focuses the incoming light onto a linear array of cells.

The degree of concentration achieved by a lens or mirror is commonly expressed in *suns*. This is the ratio between the intensity of the incoming sunlight, normally taken as the standard insolation of $1000\,\text{W}/\text{m}^2$, and the average intensity of the light focused

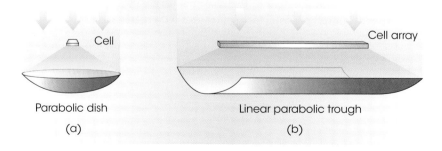

Parabolic dish

(a)

Linear parabolic trough

(b)

Cell

Cell array

Figure 3.25 Concentrating sunlight with reflective mirrors.

onto the active area of the solar cell, or cells. For example, a concentration of 100 suns produces a nominal $100\,kW/m^2$ or $10\,W/cm^2$ at the cell surface. Note, however, that in practice the insolation of $1000\,W/m^2$ is not all direct sunlight, even under clear skies. Typically 85% is direct, and the other 15% diffuse. So a nominal 100-sun concentrator would more likely produce about $8.5\,W/cm^2$ at the cell surface in strong sunlight—and systems are sometimes rated on this basis. The amount of concentration in practical systems varies from as little as 2 or 3 suns in static systems that do not need to track the sun up to 1000 suns in high-concentration tracking systems, some of which now employ multi-junction gallium arsenide solar cells.

Unlike solar cells used in conventional PV modules that must be illuminated over their entire area for efficient performance, small concentrator cells are often designed with an "active area" surrounded by a non-illuminated edge carrying bus bars and connections. This means that the reduction in cell area, one of the main advantages of a concentrator, is less than its number of suns. Furthermore, when many small cells are cut from a semiconductor wafer, there is quite a lot of wastage at the edges. The net reduction in wafer usage is substantially less than might be expected from a simple consideration of the concentration ratio.

Two more aspects of concentration optics should be mentioned briefly. The first of these is *acceptance angle*, the angular range over which a concentrator can accept light from the sun. Clearly, in the case of a tracking system, the greater the acceptance angle, the better, because it minimizes the tracking accuracy required and hence the complexity and cost of the tracking equipment. But perhaps not surprisingly there is a fundamental trade-off between acceptance angle and concentration: increasing the acceptance angle reduces the amount of concentration attainable, and vice versa. An engineering compromise is required.

The second aspect is *nonimaging optics.* As any student who has played with a magnifying glass knows, light from distant objects can be brought to a focus on a sheet of white paper, producing an inverted image of the scene. If the magnifying glass is used to focus the sun's rays onto the paper in an attempt to set it alight, the brilliant circular dot of light is an image of the sun. These are examples of classical *imaging optics.* But in the case of PV concentrators, there is no particular virtue in obtaining an image of the sun. It is more important to illuminate the active area of the solar cell, or cells, as uniformly as possible using a lens or mirror system with as large an acceptance angle as possible. Such design considerations have led to big advances in nonimaging optics applied to PV systems.[2]

High-concentration optics generally demand tracking about two axes (to follow both the azimuth and the altitude of the sun's trajectory during the day) so that the focused light always falls accurately on the solar cells. Two schemes are illustrated in Figure 3.26. Part (a) shows the widely used pedestal form of tracker with rotation about a horizontal (elevation) axis and also a vertical (azimuth) axis. This scheme is simple to install but tends to suffer from high wind loads, producing large torques on the drive system. Large trackers with surface areas up to $250\,m^2$ or more (as the system shown in Figures 3.27 and 3.28) normally adopt a horizontal position in very high winds. Part (b) of Figure 3.26 shows a less common form, known as the roll-and-tilt tracker. Wind loading is generally less serious, but more bearings and supports are needed.

The development of efficient two-axis concentrator PV systems is a very interesting challenge as it requires optimization of four dimensions: the cell efficiency, the module cost, the optics cost and accuracy, and the tracking accuracy and reliability (Figure 3.29).

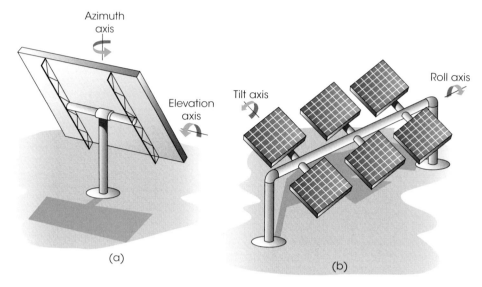

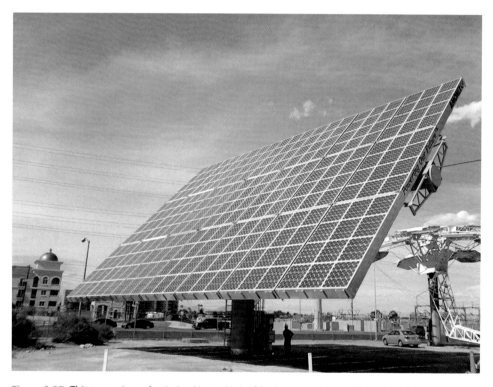

Figure 3.26 Two-axis tracking.

Figure 3.27 This two-axis tracker in Las Vegas, United States, supports multiple point-focus concentrator modules housing multi-junction GaAs solar cells and is rated at 53 kW$_p$ (*Source:* Reproduced with permission of Arzon Solar).

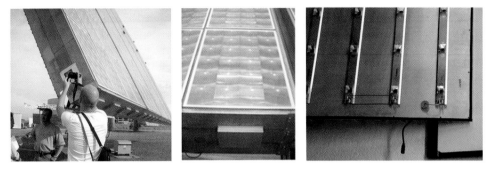

Figure 3.28 The frame, the optics, and the cells within the frame in the Amonix system (*Source:* Reproduced with permission of Arzon Solar).

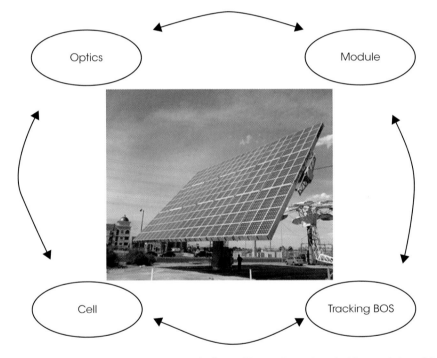

Figure 3.29 CPV system optimization challenge (*Source:* Reproduced with permission of Arzon Solar).

Appendix 3.A

3.A.1 Converting Global Horizontal Irradiation Data to Tilted and Sun-Tracking Surfaces

Predicting solar radiation on an inclined (tilted) south-facing PV array involves the following steps:

1) Solar data collection.
2) Calculation of extraterrestrial radiation.
3) Calculation of clearness index.

4) Determining the beam and diffuse components to the global irradiation on the horizontal surface.
5) Where appropriate, estimate the albedo contribution using reflectivity values for typical ground surfaces. In many cases the albedo is small or insignificant, but snow cover can be particularly relevant.
6) Using a model to calculate the energy incident on the inclined surface.

Typically hourly values of solar data are available and steps 2–6 are conducted on hourly time increments.

3.A.1.1 Solar Data Collection

The main source for 45-year historical horizontal global irradiation data in the United Sates is the National Solar Radiation Database (NSRDB). The NSRDB is a serially complete collection of hourly values of the three most common measurements of solar radiation (global horizontal, direct normal, and diffuse horizontal) over a period of time adequate to establish means and extremes and at a sufficient number of locations to represent regional solar radiation climates. It also includes meteorological data. The database provides data for 1454 stations covering the entire United States for the years 1991–2010 and data for about 300 stations for the previous 25 years (http://rredc.nrel.gov/solar/old_data/nsrdb/1991-2010).

The following table shows the type of data listed in the NSRDB.

Global horizontal radiation in Wh/m^2	Atmospheric pressure in millibars
Direct normal radiation in Wh/m^2	Wind direction in increments of 10°
Diffuse horizontal radiation in Wh/m^2	Wind speed in m/s
Extraterrestrial radiation (ETR) in Wh/m^2	Horizontal visibility in km
Direct normal ETR in Wh/m^2	Ceiling height in decameters
Total sky cover in tenths	Present weather
Opaque sky cover in tenths	Total precipitable water in mm
Dry-bulb temperature in °C	Aerosol optical depth
Dew-point temperature in °C	Snow depth in cm
Relative humidity in percent	Number of days since last snowfall

Solar data for Europe, Africa, and Asia can be found in the PVGIS database provided by the European Commission, Joint Research Centre, Ispra, Italy (http://re.jrc.ec.europa.eu/pvgis).

3.A.1.2 Calculation of Extraterrestrial Radiation

The extraterrestrial radiation is reported in the NSRDB. In any case it is instructive to derive it from the trigonometric relationships describing the movement of the Earth. For this we need to refer to the solar declination angle discussed earlier and the angles shown in Figure 3.A.1.

Solar declination δ, approximately given in radians, is

$$\delta = \pi \frac{23.45}{180} \sin\left(2\pi \frac{284+n}{365}\right) \tag{3.A.1}$$

where n = number of the day of the year ($1 \le n \le 365$).

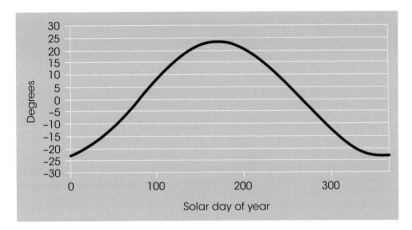

Figure 3.A.1 The geometric and trigonometric parameters used in the following equations.

As shown in Figure 3.12, the declination angle is zero at the equinoxes vernal (spring, about 22 March) and autumnal (22 September). On these days the sun rises exactly in the east and sets exactly in the west. At the summer solstice (~22 June), $\delta = 23.45°$ and at the winter solstice (~22 December), $\delta = -23.45°$ (Figure 3.A.2):

We define as I_o and H_o the hourly and daily, respectively, extraterrestrial radiation on a horizontal surface at the top of the atmosphere; these are given as follows:

$$I_o\left(\frac{\text{kWh}}{\text{m}^2}\right) = 1.367 \cdot \left(1 + 0.033 \cdot \cos\frac{360n}{365}\right) \cdot \cos\theta_Z \tag{3.A.2}$$

$$H_o\left(\frac{\text{kWh}}{\text{m}^2}\right) = \frac{24}{\pi} \cdot 1.367 \cdot \left(1 + 0.033 \cdot \cos\frac{360n}{365}\right) \\ \cdot \left(\cos\varphi \cdot \cos\delta \cdot \sin\omega_s + \omega_s \sin\varphi \cdot \sin\delta\right) \tag{3.A.3}$$

where

θ: angle of incidence
φ: latitude
δ: declination
ω: the hour angle

It is the angular displacement of the sun east or west of the local meridian due to rotation of the Earth on its axis at 15°/h: morning negative, afternoon positive,

$$\omega = \left(\frac{\pi}{12}\right)(t - 12), \tag{3.A.4}$$

where t is the decimal, 24 hours time, for example, 8:45 p.m. is 20.75 and the angles are expressed in radians.

n: number of the day in the year (1 on 1st of January and 365 on 31st December)
ω_s: the sunset angle ($\cos\omega_s = -\tan\varphi. \tan\delta$)
θ_z: zenith angle, the angle between the vertical and the line to the sun, that is, the angle of incidence of beam radiation on a horizontal plane

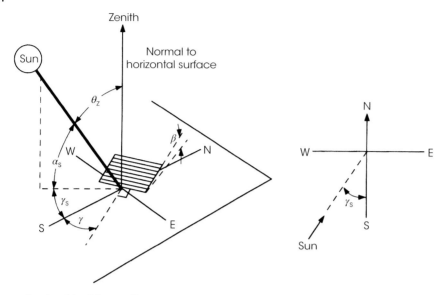

θ : **Angle of incidence**, the angle of the beam radiation and the normal on a surface
θ_z: **Zenith angle**, the angle between the vertical and the line to the sun,
γ: **Surface azimuth angle**, the deviation of the projection on a horizontal plane
of the normal to the surface from the local meridian
γ_s: **Solar azimuth angle**, the angular displacement from South of the projection of beam
radiation on the horizontal plane
β: **Slope** of the plane
α_s: **Solar altitude angle**, the angle between the horizontal and the line to the sun

Figure 3.A.2 The variation of solar declination over the year.

The angle of incidence θ and the zenith angle θ_Z can be determined from the latitude and the sun declination as follows:

$$\cos\theta = \cos(\varphi - \beta)\cdot\cos\delta \cdot\cos\omega + \sin(\varphi - \beta)\cdot\sin\delta \qquad (3.A.5)$$

$$\cos\theta_z = \cos\varphi\cdot\cos\delta \cdot\cos\omega + \sin\varphi\cdot\sin\delta \qquad (3.A.6)$$

3.A.1.3 Determining the Diffuse and the Direct Components of the Global Horizontal Irradiation

We use the top of the atmosphere (extraterrestrial) hourly data (I_o) and hourly data of the global horizontal irradiation (GHI) (I) on the surface of the Earth, and from their ratio we determine what is called hourly **clearness index (k_T)**, a measure of the losses in the atmosphere of sunlight intensity:

$$k_T = \frac{I}{I_o} \qquad (3.A.7)$$

k_T can be correlated with the ratio of diffuse radiation (I_d) to the global horizontal one (I) using one of the three following empirical equations:

$$\text{For } k_T \text{ in the interval } 0 \le k_T \le 0.3, \text{then } \frac{I_d}{I} = 1.020 - 0.248 k_T$$

$$\text{For } k_T \text{ in the interval } 0.3 \le k_T \le 0.78, \text{then } \frac{I_d}{I} = 1.45 - 1.67 k_T \qquad (3.A.8)$$

$$\text{For } k_T \text{ in the interval } 0.78 \le k_T, \text{ then } \frac{I_d}{I} = 0.147$$

Typically data are divided into small ranges of k_T and then are averaged to produce single points. A k_T of 0.5 may be produced by skies with thin cloud cover, resulting in a high diffuse fraction, or by skies that are clear for part of the hour and heavily clouded for part of the hour, leading to a low diffuse fraction.

The beam radiation (I_b) can be calculated if we subtract the diffuse radiation (I_d) from the global; thus $I_b = I - I_d$.

3.A.1.4 Using a Model to Calculate the Energy Incident on the Inclined Surface per Time Increment

The simplest and most conservative model for calculating solar irradiation on a tilted surface is the **isotropic** model, which assumes that the diffuse radiation from the whole sky is of the same strength. In reality the sky around the sun is much brighter than in the whole horizon, and this is described with **anisotropic** models, which are more complicated, yet more accurate.

A common anisotropic model is called HDKR from the names of its developers (Hay–Davies–Klucher–Reindl). More recently, an anisotropic model developed by Richard Perez is found to be even more accurate in high irradiation conditions. The anisotropic models assume that diffuse radiation is composed of three parts: (i) an isotropic part received uniformly from the entire sky dome, (ii) a circumsolar diffuse part resulting from forward scattering of solar radiation and concentrated in the part of the sky around the sun, and (iii) the horizon brightening part (it is concentrated near the horizon and is most pronounced in clear skies).

- **Isotropic model:**

$$I_T = I_b R_b + I_d \left[\frac{1 + \cos\beta}{2} \right] + I\rho_g \left[\frac{1 - \cos\beta}{2} \right] \quad \leftarrow \quad \text{Ground reflected component}$$

$\uparrow$ Beam component $\uparrow$ Diffuse component

where

I_T: total hourly radiation incident on a surface tilted at slope β
I_b: beam radiation

$$R_b = \frac{I_{b,T}}{I_b} = \frac{\cos\theta}{\cos\theta_Z}$$

θ: angle of incidence

θ_Z: zenith angle

ρ_g: albedo of the ground

I: hourly radiation incident on a horizontal surface

$I_{b,T}$: hourly beam radiation on the tilted surface

I_d: diffuse radiation

It is calculated from any of the I_d/I versus k_T correlations.

- **Anisotropic models:**

HDKR model:

$$I_T = I_b R_b + I_{\rho_g}\left[\frac{1-\cos\beta}{2}\right] + I_d\left[(1-Ai)\left(\frac{1+\cos\beta}{2}\right)\left(1+f\sin^3\left(\frac{\beta}{2}\right)\right)\right]$$

where

Ai: anisotrophy index I_b/I_o. It is a function of the transmittance of the atmosphere for beam radiation.

f: modulating factor. $f = \sqrt{I_b/I}$

Perez model:

$$I_{d,T} = I_b R_b + I_{\rho_g}\left[\frac{1-\cos\beta}{2}\right] + I_d\left[\frac{(1-F_1)(1+\cos\beta)}{2} + \frac{F_1 a}{b} + F_2\sin\beta\right]$$

where F_1 and F_2 are coefficient describing circumsolar and horizon anisotropy, respectively; $a = \max(0, \cos\theta)$, and $b = \max(0.087, \cos\theta_Z)$.

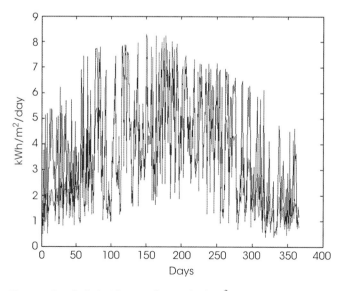

Figure 3.A.3 Daily incident irradiation (kWh/m²) on Massena airport, New York, during year 2000; red shows GHI; blue shows global irradiation on a south facing tilted at latitude plane (44.97°).

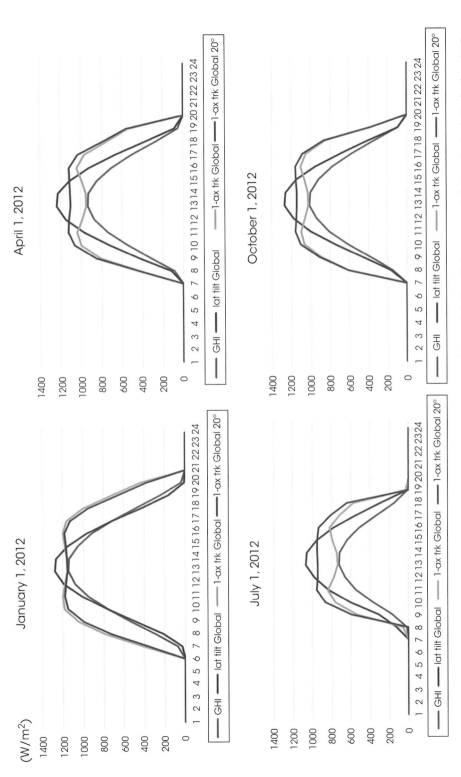

Figure 3.A.4 Daily incident irradiation on the Atacama Desert, Chile. Blue shows GHI; red show global irradiation on a north facing tilted at latitude plane (27°); green shows global irradiation of a one-axis tracker with a horizontal axis; and violet shows the same but with the tracker axis tilted 20° toward the north.

3.A.1.5 Comparisons of Different Configurations

Let us now see the impact of tilting the surface of PV modules and of putting them on one-axis trackers. Figure 3.A.3 shows this for a location in New York with moderate irradiation. The red lines show the daily range over the course of a year of the GHI, and blue shows the global irradiation on a plane tilted at latitude tilt. The later configuration receives a lot more sunlight in the winter and the fall, whereas the irradiation in the summer is somewhat lower. Over the course of the year, the latitude tilt configuration receives significantly more sunlight than the horizontal plane.

Figure 3.A.4 shows the irradiation over different angles for a high solar irradiation site in northern Chile. Note that the location is in the south hemisphere, so January is summer and July is winter. When we installed the modules at a 27° facing north, that is, latitude tilt in this location, we get a noticeable improvement in the irradiation that can be harvested especially during the winter. One-axis tracking extends the duration of the power output and slightly tilting its axis toward the north and optimizes tracking operation for this location.

Self-Assessment Questions

Q3.1 Explain the phenomenon of hot-spot formation in a PV module and how it may be prevented.

Q3.2 Figure 3.5 shows an array of six PV modules. If each module has a voltage 20 V and current 5 A at the maximum power point (MPP), (a) what is the array voltage and current at the MPP; (b) how are these values altered if the six modules are reconnected in series?

Q3.3 A monocrystalline silicon PV module has dimensions 1640 mm × 1000 mm and is **rated** at 300 W_p. It contains 60 cells connected in series.
a) Calculate the module efficiency.
b) Approximately what module voltage and current do you expect at the maximum power point?
c) If the open-circuit voltage is 39 V in standard conditions and the temperature coefficient is –0.3% per °C, what is the voltage at a temperature of 0°C?

Q3.4 A CdTe PV module has dimensions 1860 mm × 1200 mm and is rated at 360 W_p. Its voltage at the maximum power point (MPP) is 177 V.
a) Estimate the module efficiency and the current at the MPP.
b) If the power temperature coefficient is –0.34% per °C, what is the maximum power output at −10°C?

Q3.5 The latitude of New York City is 40.71°N. What is the elevation of the sun above the horizon at solar noon on dates corresponding to (a) the equinoxes, (b) the summer solstice, and (c) the winter solstice?

Q3.6 Define the *Arctic Circle*. What is its latitude, and why?

Q3.7 The average daily solar irradiation on a horizontal surface in Las Vegas, Nevada, is 5.7 kWh/m². Estimate (a) the peak sun hours in a year and (b) the approximate annual output of a horizontally mounted PV module rated at 250 W$_p$.

Q3.8 Why is a PV array often pointed toward the south (or north in the southern hemisphere) and tilted down from horizontal by an angle equal to the latitude of the site? What tilt angle would you expect for an array located in (a) Rio de Janeiro, Brazil; (b) Lisbon, Portugal; and (c) Berlin, Germany?

Q3.9 A PV array in London, England, consists of 20 modules rated at 180 W$_p$ each, facing south and tilted at an angle equal to London's latitude. (a) Use Figure 3.17(a) to estimate the approximate annual solar radiation received by the array; (b) using the peak sun hours concept, estimate the approximate annual output of the array in megawatt-hours (MWh). Why is the latter figure likely to be overoptimistic?

Q3.10 Place the following six city locations in order of their suitability for high-concentration PV tracking systems: (a) Palm Springs, California; (b) Palo Alto, California; (c) Cape Town, South Africa; (d) Durban, South Africa; (e) Crucero, Chile; and (f) Santiago, Chile.

Q3.11 What is "solar declination" and what is its range?

Q3.12 What are the three components of solar irradiance on a terrestrial plane?

Q3.13 What is the average solar irradiance in an early afternoon on a clear day?

Q3.14 Why does concentration of sunlight improve efficiency of a solar cell? Why is there a practical upper limit?

Q3.15 What is the difference between irradiance and irradiation?

Q3.16 Give approximate ranges of energies for different regions of the electromagnetic spectrum. Specifically, give ranges for ultraviolet, visible, near infrared, mid infrared, and far infrared in terms of commonly used units of energy (eV), wavelength (nm and μm), and frequency (GHz and THz).

Problems

Solar irradiation

3.1 The sun is at an altitude of 35° from the horizontal. What is the corresponding air mass?

3.2 Calculate the sun's altitude at solar noon on 21 June in New York (latitude 40.71°N); in Athens, Greece (latitude 37.98°N); and in Antofagasta, Chile (latitude 23.65°S).

3.3   Estimate the zenith angles needed to produce AM1.5, AM2.0, and AM3.0 (hint: AM1.0 occurs at 0°).

3.4   Calculate the zenith angle at solar noon in New York (latitude 40.72°) on July 1 (121th day of the year).

3.5   Use a sun path (Figure 3.19 or 3.20 or a different one) and make a table of shaded times for each month of the year.

3.6   A PV system comprises four rows of south facing modules tilted so that the top of the modules is 3 ft off the ground. If the site is at a latitude of 41°N, determine the spacing between the rows needed to prevent shading from one row to the other. How does this change in a latitude of 30°?

3.7   What are the maximum and the minimum solar elevations at a site with latitude 40.71 and when do they occur?

3.8   The monthly mean daily solar irradiation on a horizontal plane at New York (latitude 40.7°N) and Chania, Greece (latitude 35.5°N), are listed in the table in units of $Wh/m^2/day$:

	Jan	Feb	Mar	April	May	June	July	Aug	Sept	Oct	Nov	Dec
New York	1.9	2.7	3.8	4.9	5.7	6.1	6.0	5.4	4.3	3.2	2.0	1.6
Chania	2.4	3.10	5.0	6.2	7.3	8.4	8.3	7.6	5.8	4.1	2.7	2.0

a)   Determine the mean daily solar radiation outside the Earth atmosphere for each month of the year and the corresponding clearness index k_T.
b)   Determine the daily global radiation for angles of incidence in the range of 0–90°, assuming a ground reflectivity of 0.2 (create a table showing monthly values).
c)   What angle of inclination would you recommend in each location to maximize input to the PV system during the summer months and to obtain the maximum energy during the whole year?

3.9   Assume that we want maximum PV output in the summer in New York between the hours of 2 and 5 P.M. Calculate the orientation and tilt of PV modules that will produce this maximum on July 15.

3.10   Do the same calculation as in Problem 3.9 for Phoenix, Arizona, and Seattle, Washington, and compare the results.

PV performance

3.11   Calculate the daily horizontal extraterrestrial irradiation at New York City and Los Angeles on 21 January by using the formulas listed in Appendix 3.A. (NYC latitude: 40.71°; LA latitude: 34.05°.)

3.12   How many kW (DC rated) would be needed on a rooftop to deliver 6000 kWh/year to a home in at Long Island, NY, when the derated factor (performance ratio) is 0.78? Assume (a) horizontal roof and (b) south-facing rooftop with an inclination of latitude at 15°. The global horizontal irradiation (GHI) at the location (latitude is 40.789°N) is 1630 kWh/m^2/year, and the average insolation on a 25.789° south-facing plane is 1790 kWh/m^2/year.

3.13   Go to the NSRDB website (http://rredc.nrel.gov/solar/old_data/nsrdb/1991-2010/) and download the .csv file with hourly solar data for station with USAF number 723860 (Las Vegas International Airport) for year 1991. You need to choose the column containing global horizontal irradiation (GHI) data produced by the METSTAT model.

   a) What is the horizontal extraterrestrial energy (kWhr) for each of the 24 hours of the 18th day of June? Use the hourly formula given by Equation 3.A.2 to perform your calculations. Sum up the hourly calculations to figure out the daily radiation incident on the surface. Do the same with the extraterrestrial data from your .csv file. Compare the results. You can use Excel or an equivalent spreadsheet program.
   Note 1: You will get same negative values for some hours. That means night-time. Equal all negative values to zero.
   Note 2: Evaluate ω at the midpoint of each hour.

   b) Repeat (a) performing only one calculation using the daily formula given by Equation 3.A.3. The result has to be given in kWh.

   c) What day of the year a horizontal surface located on the equator and at the top of the atmosphere would receive the maximum solar radiation?

   d) The owner of a Las Vegas casino has decided to install a PV system on the roof of his hotel to reduce his annual electricity bill and what to decide between putting the PV horizontally (flat) on his roof or putting them facing south tilted at latitude angle. Use the HDKR model given in class to break global radiation into its beam and diffuse components for the same date as in (b). Calculate hour by hour the energy incident per square meter on the tilted PV panels using the isotropic sky model. Sum up the hourly values to get the daily energy for the same date as in (b) and compare with the daily energy received by a horizontal surface. Explain why there is a difference and what will be the best choice for the casino owner.

3.14   Repeat the aforementioned calculation by using data from National Solar Radiation Database (NSRDB) (http://rredc.nrel.gov/solar/old_data/nsrdb/1991-2010/) and (a) the isotropic model, (b) the HDKR anisotropic model, and (c) the Perez anisotropic model (formulas in Appendix 3.A).

3.15   You are asked to size a PV system for an AC load appliance in a location of 23.45°. Assume that the air condition only operates from May to October. You don't want to oversize the system and so you decide to size for month June rather than August (the hottest month of the year). Explain what do you think that the optimum tilt of your PV system would be.

Ohm's law

3.16 A lamp is designed to consume 60 W when it is connected to a 12 V power source. What is the resistance of the lamp, and what current will flow through and what energy will consume over the course of a month if it operates for 5 hours a day?

3.17 Estimate the total resistance of the network shown below:

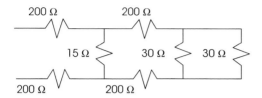

3.18 Suppose that a battery is modeled as a 12 V source in series with a 0.12 Ω internal resistance. (a) What would be the output voltage when the battery delivers a current of 10A? (b) What would be the output voltage when the battery is connected to a 2 Ω load?

3.19 A 12 V battery is connected with a 100 W lightbulb at a distance of 50 ft. If no. 14 ($d = 0.163$ cm) copper wire is used, what would be the power loss on the wire?

Copper Wire Resistance and Amperage Ratings

Wire Gauge Number	Diameter (inches)	Resistance (Ohm/100 ft)	Max Current (A)
14	0.0641	0.2525	15
12	0.0808	0.1588	20
10	0.1019	0.0999	39
8	0.1285	0.0628	40

3.20 A system is delivering 15 A of current through 12-gauge wire to a load 100 ft away. a) find the voltage drop in the wires; b) what percent of the power generated by the PV is lost in the connected wire?

3.21 Based on the wire specifications shown in the Table above, what wire gauge would you have chosen to minimize losses?

3.22 Use a 120 V to connect over a 50 ft distance a 100 W lightbulb with the same wire as above. What would be the power loss on the wire?

3.23 A PV module comprises 72 solar cells 1.5 V each wired in series. Draw the I–V curve when (a) all cells are exposed to the sun and (b) when two cells are shaded; assume i) no bypass diodes and ii) bypass diodes for every cell.

3.24 For the same conditions, estimate the current, voltage, and power in the PV module above connected to a 10 Ω load.

Answers to Questions

Q3.1   Explain the phenomenon of hot-spot formation in a PV module and how it may be prevented.

Q3.2   (a) 60 V, 10 A (b) 120 V, 5 A

Q3.3   (a) 18.3% (b) about 30 V and 10 A (c) 41.9 V

Q3.4   (a) 16.1%, 2.03 A (b) 428 W

Q3.5   (a) 49.29° (b) 72.74° (c) 25.84°

Q3.6   About 66°N

Q3.7   (a) 2080 hours (b) 520 kWh/year

Q3.8   (a) 23° (b) 39° (c) 53°

Q3.9   (a) 1200 kWh/m^2 (b) 4.3 MWh

Q3.10   (e), (a), (c), (b), (f), (d)

Q3.11   It is the angle between the equator plane and the line connecting the centers of the Earth and the sun. It is measured in degrees.

Q3.12   Direct, diffuse, and albedo

Q3.13   1 kW/m^2

Q3.14   Concentration increases photon density making them more energetic; however it also increases the temperature in the cell and after a certain concentration limit cooling is not practical and the thermal losses overcome the gains from concentration.

Q3.15   Irradiance is a power flux (W/m^2), whereas irradiation is an energy flux (kWh/m^2) over a course of time, typically a day or a year.

Q3.16   **Answer:**

Spectrum	Wavelength	Frequency (Hz)	Energy (eV)
Ultraviolet	10 nm–380 nm	8×10^{14}–3.0×10^{16}	124–3.1
Visible	380 nm–750 nm	4.0×10^{14}–8×10^{14}	3.1–1.77
Near Infrared	750 nm–10 μm	3.0×10^{13}–4.0×10^{14}	1.77–0.124
Mid Infrared	10 μm–100 μm	3.0×10^{12}–3.0×10^{13}	0.124–0.0124
Far Infrared	100 μm–1 mm	3.0×10^{11}–3.0×10^{12}	0.0124–0.00124

References

1 NASA. *Surface meteorology and Solar Energy Tables* (2010). eosweb.larc.nasa.gov/sse (Accessed on August 28, 2017).

2 A. Luque. *Solar Cells and Optics for Photovoltaic Concentration,* Adam Hilger: Bristol, Philadelphia (1989).

4

Grid-Connected PV Systems

4.1 Introduction

Photovoltaic (PV) systems are generally divided into two major categories: *grid-connected* (also known as *grid-tied*) systems that are interfaced to an electricity grid and *stand-alone* systems that are self-contained. Over the years it has been customary for books on PV to describe stand-alone systems first, probably because they are seen as "pure PV." Also we should remember that stand-alone systems, including those launched into space and the solar home systems (SHSs) that supply electricity to individual families in developing countries, accounted for much of the PV industry in its early days. But since the 1990s the market has shifted decisively toward PV power plants and installations on buildings connected to an electricity grid. In 2000 grid-connected PV had overtaken stand-alone systems in global market share, and in 2016 more than 98% of solar cell production was being deployed in grid-connected systems. In many ways such systems are simpler to design and describe than their stand-alone cousins. For both these reasons our own story begins with grid-connected PV.

Since most people have seen PV arrays mounted on the roofs of homes, this seems a good place to start. Figure 4.1 shows the elements of a domestic PV installation, typically with an array power between 3 and $8\,kW_p$, interfaced to the local electricity grid. The major advantage of this arrangement is that the output from the PV array is fed into the grid when not required in the home; conversely, when the home needs power that cannot be provided by the PV (especially at night!), it is imported from the grid. In other words the PV system and grid act in harmony and there is an automatic seamless back-and-forth flow of electricity according to sunlight conditions and the electricity demand.

In more detail the various items numbered 1–8 in the figure have the following functions:

1) *PV combiner unit.* This acts as a junction box connecting the modules in the desired configuration.
2) *Protection unit.* This unit houses a DC switch to isolate the PV array and anti-surge devices to protect against lightning. Also an AC switch is placed after the inverter.
3) *Inverter.* At the heart of the grid-connected system, the inverter extracts as much DC power as possible from the PV array and converts it into AC power at the right voltage and frequency for feeding into the grid or supplying domestic loads.

Electricity from Sunlight: Photovoltaic-Systems Integration and Sustainability, Second Edition.
Vasilis Fthenakis and Paul A Lynn.

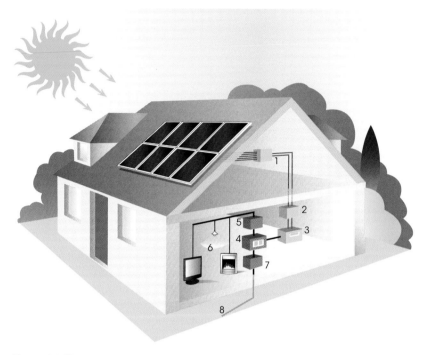

Figure 4.1 Home PV system connected to the grid.

4) *Energy-flow metering.* Kilowatt-hour (kWh) meters are used to record the flow of electricity to and from the grid.
5) *Fuse box.* This is the normal type of fuse box provided with a domestic electricity supply.
6) *Electrical loads.* Domestic electrical loads include lighting, TVs, and heaters.
7 & 8) A junction box connects the home to the utility supply cable.

The adoption of domestic rooftop installations is mushrooming in developed countries in response to the falling prices of PV modules, the support of governments, and the enthusiasm of citizens to do something positive about global warming. Larger grid-connected systems, for example, those installed in schools, offices, public buildings, and factories, extend the power scale from hundreds of kilowatts up to megawatts (MW). All have the advantage of generating solar electricity where it is needed, reducing the losses associated with lengthy transmission lines and cables. And at the top of the grid-connected power scale come multi-megawatt power plants, generally remote from individual consumers, which send all their power to the grid.

4.2 From DC to AC

The *inverter* is the key item of equipment for converting DC electricity produced by a PV array into AC suitable for feeding into a power grid. Inverters use advanced electronics to produce AC power at the right frequency and voltage to match the grid supply.

Figure 4.2 Raising the power level: a 17.6 kW$_p$ grid-connected roof installation on the Oslo Innovation Centre, Norway (*Source:* Reproduced with permission of IEA-PVPS).

While a single inverter may well be sufficient for a domestic installation such as that illustrated in Figure 4.1, multiple units become the norm as we advance up the power scale and their efficiency, reliability, and safety are major concerns of the system designer.

Inverters must obviously be able to handle the power output of a PV array over a wide range of sunlight conditions. Normally they do this using *maximum power point tracking (MPPT)* to optimize the energy yield. DC-to-AC conversion efficiencies up to 98% can be achieved over much of the range, although efficiency tends to fall off if an inverter is operated below about 25% of its maximum power rating. In larger systems with multiple inverters, it can make sense to switch all the power into one unit at sunrise and then, as the sun rises in the sky and the array power increases, bring other inverters successively into play, keeping all working optimally. The switching sequence is reversed toward sunset. Overall, inverter system design is quite a challenge, especially with high-capacity units; few electronic systems are expected to maintain high efficiency over such a wide power range.

From the technical point of view, there are two main classes of inverter: *self-commutated*, where the inverter's intrinsic electronics lock its output to the grid, and *line-commutated*, where the grid signal is sensed and used to achieve synchronization. Inverters are also classified according to their mode of use, with four main types:

1) *Central.* The complete output of an array is converted to AC and fed to the grid. The largest central inverters can exceed 1 MW$_p$ capacity.
2) *String.* This type of inverter is connected to a single string of modules with a typical power range of 1–3 kW$_p$.

Figure 4.3 This Korean power plant uses four 250 kW$_p$ inverters to connect 1 MW$_p$ of PV arrays to the grid. The modules are mounted on horizontal single-axis trackers (*Source:* Reproduced with permission of IEA-PVPS).

3) *Multi-string.* These inverters can accept power from a number of module strings with different peak powers, orientations, and perhaps shading, allowing each string to operate at its own maximum power point (MPP).

4) *Individual.* An increasing number of manufacturers offer PV modules with inverters attached, making each module its own AC power source.

Several factors influencing the choice of inverters for small- and medium-size systems can be explained by referring to Figure 4.4. For simplicity we have shown arrays with just a few modules although most systems contain more—and some a great many more. The array in part (a) consists of two strings of three modules each. In this case all the modules are assumed to be of the same type and rating, with the same orientation and without shading, so the strings are paralleled in the combiner/protection units (1/2) and fed to a single central inverter (3). The inverter is presented with an input voltage equal to three times the individual module voltage and an input current equal to twice the individual module currents. Since the modules are well matched, the MPP selected by the inverter for the whole array ensures that all modules work at, or close to, their maximum output.

In part (b) of the figure, the two strings are dissimilar. They may have different numbers of modules (as shown), or different module types or orientations, or one string may suffer partial shading. For whatever reason they do not produce similar outputs and cannot be efficiently characterized by a single MPP, so each string has its own inverter

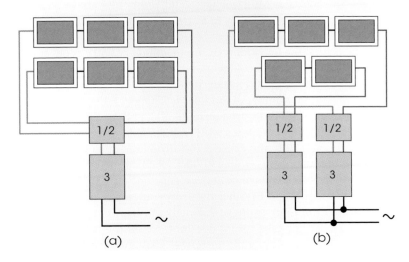

Figure 4.4 PV arrays served by: (a) a single central inverter; (b) two individual string inverters.

and is operated at its own MPP. An alternative is to use a single multi-string inverter. And as we have already pointed out in the previous chapter, manufacturers are now offering power optimizers, one to be connected to each module in a string, allowing every module to work at its own MPP. There are various options for extracting the maximum amount of power from strings and arrays.

To put our discussion in the context of a practical system, suppose we need to specify an inverter for a PV array of about $5\,kW_p$ on the roof of a suburban house. In a sunny climate an array of this size may well generate, over a complete year, electricity equal to the annual requirements of the household. In the summer months the PV will be a net exporter to the grid; in the winter months the solar deficit will be made up from the grid. We will assume that monocrystalline silicon modules rated at $180\,W_p$ have been selected, so 28 will be needed, yielding $5.04\,kW_p$ (the module specification given in Section 3.2.2 will be used in the calculations later). Fortunately, they can all be mounted on the roof at the same tilt angle, and there is no shading, so we may specify a central inverter. Since an array rarely generates its nominal peak power, an inverter rated at slightly less than $5.04\,kW_p$ should be adequate as long as its maximum input voltage and current are never exceeded. We will therefore investigate the suitability of a $5\,kW_p$ central inverter with the following manufacturer's ratings:

Nominal DC input power	5.0 kW
Peak instantaneous input power	6.0 kW
Maximum DC input voltage	750 V
Voltage range for MPPT	250–650 V
Maximum DC current	20 A

We first need to estimate how many modules can be connected in a series string. The maximum number is given by the MPPT voltage of 650 V divided by the MPP voltage of an individual module. The latter is 35.8 V at 25°C, but increases by 0.33% for every degree drop in temperature. Therefore if we allow for sunny winter days with temperatures down to −5°C, the MPP voltage could reach 10% above 35.8 V, that is, 39.4 V. The maximum number of modules in a string for effective tracking is therefore 650/39.4 = 16.5, say, 16.

We should also check that the maximum DC input voltage of 750 V is never exceeded. Once again, the danger condition is a cold winter day with bright sunshine. The module open-circuit voltage of 43.8 V at 25°C rises by 10% to 48.2 V at −5°C. So to keep within the 750 V limit, the maximum number of modules is 750/48.2 = 15.6, say, 15.

The minimum number of modules in a string is dictated by the need to keep the MPPT voltage above 250 V. The module's MPP voltage falls with rising module temperature, which could reach 70°C and cause a 15% drop in MPP voltage to 30.4 V. The minimum number of modules is therefore 250/30.4 = 8.22, say, 9.

To keep within the inverter's voltage limits, we conclude that strings may have any number of modules between 9 and 15. Since the array contains 28 modules, two strings of 14 are acceptable, but not four strings of 7. Finally the array current supplied to the inverter should be checked to make sure it does not exceed the permitted maximum. In this case the peak short-circuit module current is 5.5 A and is little affected by temperature. So two parallel strings of 14 modules will give a peak DC current of 11 A, well below the permitted maximum of 20 A. The inverter is therefore suitable for the job.

Back in Section 3.3.1 we discussed the problem of shading and suggested reducing the effects of recurrent shadows by confining them to as few strings as possible. Where shading is unavoidable, it may be appropriate to use a number of string inverters rather than a single central inverter, giving flexibility to connect the modules in a favorable configuration, perhaps with strings of different lengths.

We have already mentioned the need for inverters to operate efficiently over a wide power range. Some inverters include transformers, and these reduce efficiency slightly. High efficiency is not purely a question of economics; it also relates to keeping inverters cool. For example, if a 5 kW_p inverter is working at full stretch and converting 96% of its input power to AC, the other 4% (200 W) must be dissipated as heat. It is hardly surprising if the manufacturer recommends mounting the unit on an outside, north-facing wall with plenty of air circulation! The cooling issue becomes more and more significant as inverter power-handling capacity increases.

If the electricity grid is turned off for maintenance purposes, or due to a fault, it is very important for an inverter to disconnect itself automatically to avoid putting a voltage on the grid. Otherwise it can endanger personnel working on the grid and may deceive other local inverters into believing that the grid is still operating normally. Sophisticated electronics are included to prevent this potentially hazardous situation, which is referred to as *islanding*.[1,2]

As we gaze at a domestic rooftop system rated at a few kW_p, it is hard to appreciate the engineering challenges posed by scaling up inverters for multi-megawatt power plants. There are major issues of technical performance to be considered including lightning and surge protection, safety, reliability, inverter sequencing, and the mode of connection of tens or hundreds of thousands of PV modules into strings and arrays.[1] The waveform purity and power factor of the inverter output must be satisfactory to the grid operator. Grid-connected inverters are sensitive to fluctuations in grid voltage, frequency, and impedance and will shut down automatically if these parameters stray outside the agreed specification. Islanding, which could be disastrous in a large installation, must be avoided. High-power inverters provide a major challenge to today's electrical and electronic engineers, but these challenges have been largely resolved as discussed in Section 4.8.2.

Figure 4.5 Scaling up: this 1.6 MW$_p$ inverter weighs over 20 tons (*Source:* Reproduced with permission of Padcon GmbH).

Figure 4.6 The Moura power plant in Portugal, rated at 45.6 MW$_p$ (*Source:* Reproduced with permission of IEA-PVPS).

4.3   Completing the System

Various items are required to complete a grid-connected PV system. They may be less eye catching than solar cells and PV modules, but they are essential to a properly engineered installation. Costs, long-term reliability, ease of maintenance, and sometimes appearance are important considerations. They are generally referred to as *balance-of-system* (*BOS*) components.

As the prices of solar cells and modules continue to fall and PV manufacturers achieve the cherished long-term objective of "less than one US dollar per watt," the cost of BOS components can, unless carefully controlled, seriously inflate total system costs. In the past a figure of about 50% has often been quoted for BOS, including inverters. One of the main problems has been a proliferation of components supplied by many manufacturers in small quantities, lacking the economic benefits of scale. As the PV industry continues to grow, there is perhaps a better chance that volume production will drive costs down.

We mentioned and illustrated various BOS components for a domestic PV installation in Section 4.1. It is now time to give a more complete list and add further comments:

- *Module and array mounting structures* (Figure 4.7). Modules and arrays need secure mounting whether on the ground, flat roofs, inclined roofs, or building facades. A great variety of static mounting structures is available, in aluminum, stainless, or galvanized steel. Some allow variable tilt. Generally there should be space left at the back of modules to allow free air circulation.
- *Cabling.* Special double-insulated cables that are UV and water resistant are generally used for the DC wiring from modules to inverters. They must be sized to give low voltage drops, typically less than 2%. Since cable power losses are proportional to the square of the current carried, there is an advantage in reducing current levels by specifying long module strings and high system voltages.
- *PV combiner unit* (Figure 4.8). This acts as a junction box for the various module strings, which are normally connected in parallel. Fuses are provided for each string. The combiner box may include surge protection against lightning and house the main DC isolator switch—providing it is easily accessible—allowing the PV array to be disconnected from the inverter.
- *Protection unit* (Figure 4.8). If the combiner unit does not include a DC isolator, this must be provided separately and be easily accessible. Since a PV array always produces a voltage in sunlight, it must be possible to disconnect it from the inverter for maintenance or testing. The isolator switch is rated for the maximum DC voltage and current of the array. Other safety features, including earthing, vary from country to country and between continents, although regulations are tending to harmonize as the PV industry extends its global reach.
- *Energy-flow metering.* In Figure 4.1 we showed twin kilowatt-hour meters recording the flow of electricity to and from the grid. An alternative is to use a single bidirectional meter to indicate the net amount of electricity taken from the grid. This approach, referred to as *net metering*, implies that electricity exported to the grid by the PV array achieves the same price as imported electricity, regardless of when it is generated. Net metering is, in this sense, beneficial to the homeowner; but it is not suitable for feed-in tariffs (FITs) that offer an attractive price for exported kilowatt-hours or differential tariffs that price electricity according to the time of day or night.

Figure 4.7 Array mountings at the Kings Canyon power plant in Australia (*Source:* Reproduced with permission of IEA-PVPS).

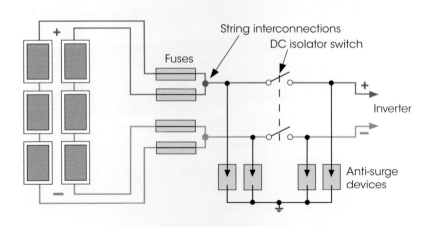

Figure 4.8 Key functions of the combiner/protection units in a domestic PV system.

In any case, most homeowners wish to know how much electricity is being generated by their solar arrays and often choose to have a visual display fitted in a living area of the house. It forms a good talking point with visitors. In addition, most inverters incorporate data-logging facilities allowing the owner to monitor performance using a laptop, and some include wireless data transmission.

As we move up the power scale toward larger grid-connected systems, the importance of accurate performance monitoring grows. Large PV power plants have a full range of instrumentation typical of modern high-tech industrial facilities. And their full complement of BOS subsystems and components account for a substantial part of overall costs.

4.4   Building-Integrated Photovoltaics (BIPV)

4.4.1   Engineering and Architecture

We have seen how a grid-connected system is built up using PV modules, inverters, and BOS components. In previous chapters we included several photographs showing PV roofs and vertical facades. So what exactly is implied by the term *building-integrated photovoltaics* (*BIPV*), and what more is there to be said about giving buildings a "face to the sun?"

PV technology is unique among the renewable energies in its interaction with the built environment. Future generations will find it entirely natural to see PV arrays on roofs and facades, in gardens and parks, on bus shelters and car ports, and as electricity-generating windows and screens inside homes, schools, offices, and public buildings. Most will be grid-connected. And hopefully they will bear testament to the trouble our generation has taken to blend them visually and aesthetically into their surroundings. PV will become increasingly a part of the urban experience.

By contrast, most wind power is generated in wild open country or offshore, and whatever one thinks of the visual impact of large turbines, they rarely impinge on the urban and suburban scene. Wave and tidal power do not affect the daily visual experience of office workers or families—unless they go on trips to marvel at large-scale renewable energy in action. Large PV power plants may be impressive and even beautiful in their own way, but they are not generally noticed by city dwellers.

BIPV is different. It proclaims a message about our care for the environment, it can be anywhere and everywhere, and it matters what it looks like and how people feel about it. Public enthusiasm and support are vital, not least to the PV industry. Architects will, or should, be involved with engineers in the design of solar buildings so that PV is integrated into the fabric in ways that marry technical function with aesthetics. A modern factory producing solar cells, or an exhibition center for renewable technologies, offers an ideal opportunity to create a striking building that makes a highly visible statement about our technological future; a family living in a low-energy timber-framed eco-home may see their PV modules as a symbol of sustainability, an alternative lifestyle. In their very different ways, all wish to proclaim a message about renewable energy that can only be successfully communicated by high-quality BIPV.

Of course there are difficulties. Countries including England, France, Italy, and Spain have a huge stock of old and historic buildings. It would be difficult or impossible to modify most of them to accept PV modules in aesthetically pleasing ways. PV arrays tacked on to existing roofs hardly ever increase their visual attraction. We may like to see them because of what they represent—the owner's commitment to renewable electricity—but our enthusiasm stems from what the PV does, not how it looks.

The main opportunity for successful BIPV, as opposed to PV that is simply superimposed on existing roofs and structures, lies in the creation of new buildings that, from

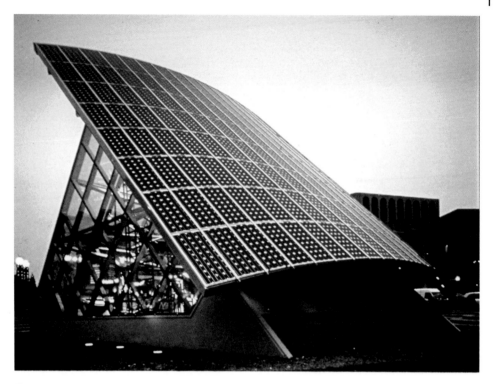

Figure 4.9 Proclaiming a message: the *Solar Showcase* in Birmingham, England (*Source:* Reproduced with permission of EPIA/BP Solar).

the very start, treat PV as an integral part of the design, full of exciting possibilities. It is very encouraging to see architects in some countries—Germany and the Netherlands are good examples—realizing that PV offers far more than a way of producing electricity. While appreciating its technical possibilities and limitations, their primary goal as architects is to ensure that PV enhances the human environment. For them PV is neither add-on nor afterthought, but an important part of the building and a pointer to its function and personality. It must inspire as well as serve its utilitarian purpose. As old housing and building stock is gradually replaced, we may expect PV to exert a growing influence on architectural design, opening up hitherto undreamed-of possibilities.

Apart from aesthetics BIPV has several important economic advantages:

- In most cases the necessary PV support structures, mainly rooftops and building facades, are there anyway. If a roof or façade is made entirely of PV modules, then its cost can be offset against the cost of the building materials it replaces. The same benefit applies to embodied energy and greenhouse gas emission reductions; this is discussed in Chapter 7.
- Both rooftop PV and BIPV do not require additional land—a very important consideration in urban environments and in countries with high population densities where rural land for PV power plants is expensive and in short supply.
- Renewable electricity is generated and mainly used on-site, reducing cable transmission losses.

Figure 4.10 Proclaiming a message: an eco-home in Denmark (*Source:* Reproduced with permission of IEA-PVPS).

So how well do photographs included in earlier chapters (Figures 2.2, 2.25, and 3.18) square up to the expectations of successful BIPV? You may like to refer back and make your own judgments. From the purely technical perspective, it is clear that all these PV installations are integrated on to and into the buildings. But from the architectural point of view, overall appearance is key and a PV array should be a harmonious part of the overall design. If these examples underline the difficulty of defining and agreeing architectural aesthetics, this certainly does not absolve us from trying!

4.4.2 PV Outside, PV Inside

The aesthetics of successful BIPV may be hard to define and judgments are inevitably subjective—yet most of us know instinctively when a solar building feels right for its setting and context. In this section we consider a number of examples to illustrate the wide range of recent international BIPV. Since a picture "is worth a thousand words" in the field of visual impressions, our focus is on photographs accompanied by short explanatory captions.

All PV installations are "outside" in the sense that they must receive sunlight. Building façades and sloping roofs are often highly visible to the public; flat roofs are more likely to be hidden. Any PV array on public display should appeal to passersby and bystanders as well as users and owners of a building. Its environmental statement is offered to the world at large.

Although many PV installations are visible only from the outside, some are also "inside" in the sense that people within the buildings are highly aware of them, and if well designed they can both inspire and delight. Modules may be interspersed with glass windows or arranged as louvers to provide internal shade and ventilation. Some crystalline silicon modules have glass at front and back, allowing light to enter through the gaps between wafers. Thin-film modules can be semitransparent, producing partial shade and generating electricity at the same time. Modules on rooftops that are invisible from the outside may be highly visible on the inside—indeed, this is usually the architect's intention. The advent of tinted and flexible thin-film products means that architects can be increasingly bold and imaginative about incorporating PV into their designs.

It is clear that aesthetic judgments should depend to a considerable extent on whether PV is on the "outside" or "inside." Outside, it interacts with the neighboring buildings and the local landscape and affects a great many people, some of whom are probably skeptics. Inside, it is more self-contained and speaks only to the users of the building who, in most cases, are enthusiastic supporters of renewable energy. It may be helpful to bear these points in mind when assessing the following photographs. They are arranged in two groups labeled *PV outside* and *PV inside*. The selection is designed to show a good international range of solar buildings with different personalities, acknowledging the efforts that many architects are making to enhance the built environment by incorporating PV imaginatively into their designs.

PV Outside
The PV on these buildings and installations is highly visible from the outside.

Figure 4.11 This building in Tübingen, Germany, proudly proclaims its solar identity (*Source:* Reproduced with permission of EPIA/BP Solar).

Figure 4.12 Traditional stone and PV in harmony: a building at the Technical University of Catalonia, Spain (*Source:* Reproduced with permission of EPIA/BP Solar).

Figure 4.13 Architects in countries with a tradition of social housing can spread their influence widely. This example is in Amersfoort, the Netherlands (*Source:* Reproduced with permission of IEA-PVPS).

Figure 4.14 A huge solar pergola at the World Forum of Culture in Barcelona, Spain, supports a 4000 m² PV array (*Source:* Reproduced with permission of EPIA/Isofoton).

Figure 4.15 The Sydney Olympic Games brought PV to the attention of millions with solar-powered lighting and more than 600 1 kW$_p$ arrays on athletes' houses (*Source:* Reproduced with permission of EPIA/BP Solar).

Figure 4.16 This eco-home in Oxford, England, uses PV modules, water-heating panels, and passive solar design to reduce its external energy requirements almost to zero (*Source:* Reproduced with permission of EPIA/BP Solar).

Figure 4.17 PV louvers replace standard glass shading to provide a dual function (*Source:* Reproduced with permission of EPIA/BP Solar).

Figure 4.18 A PV-covered walkway at an exhibition center in Japan (*Source:* Reproduced with permission of IEA-PVPS).

Figure 4.19 A 1.6 km PV array gives added purpose to a highway sound barrier in Germany (*Source:* Reproduced with permission of EPIA/Isofoton).

PV Inside

The PV systems on these buildings add a great aesthetic value in addition to energy savings.

Figure 4.20 Sunlight and shadow: a striking interior at the Energy Research Centre of the Netherlands (*Source:* Reproduced with permission of EPIA/ECN).

Figure 4.21 In harmony with nature: 30 kW$_p$ of glass/glass modules at the National Marine Aquarium, Plymouth, England (*Source:* Reproduced with permission of IEA-PVPS).

Figure 4.22 Thin-film semitransparent modules allow dappled light into this building in Germany (*Source:* Reproduced with permission of EPIA/ Schott Solar).

Figure 4.23 Customer satisfaction: a shop in Tours, France (*Source:* Reproduced with permission of EPIA/Total Energie).

These photos show that PV has aesthetic value in addition to its apparent environmental and economic value and the great service that provides to the people in developing countries that lack access to reliable electricity. However, cost reductions in grid-integrated systems, mostly in developed countries, have been the major driver for a phenomenal growth for PV.[3] At this point it makes sense to examine the growth of the markets before we discuss large-scale grid-integrated PV.

4.5   The Growth of Global PV Markets

Over the last 10 years, the market for photovoltaics as measured by their cumulative installed capacity has been growing by about 45% per year. Between 2005 and 2015, global solar PV capacity increased from approximately 5 to 220 GW, and it was estimated that it would exceed 320 GW at the end of 2016 (Figure 4.24). This strong growth can be attributed to cost reductions, induced by innovations in a market environment that was created by national and state subsidy programs. Early on most of the market growth happened in Germany catalyzed by a renewable portfolio standard (RPS) that provided feed-in tariffs (FITs) to solar and wind installations in the country.

The scale and profitability of the German market enabled efficient scaling up of PV manufacturing worldwide. Italy started a strong incentives program in 2010 and China and the United States introduced FITs a year later. As of 2016, China has both the largest annual deployment and the largest PV manufacturing capacity in the world (Figures 4.25 and 4.28 correspondingly).

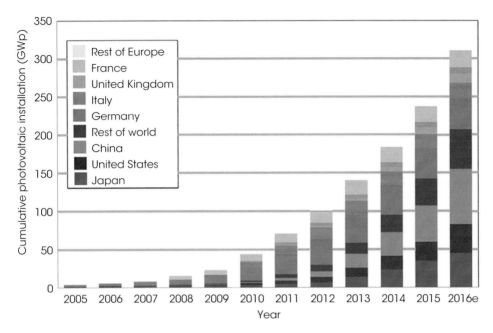

Figure 4.24 Global growth of PV deployment (*Source:* A. Jäger-Waldau; PV Status Report 2016; EUR 28159 EN).

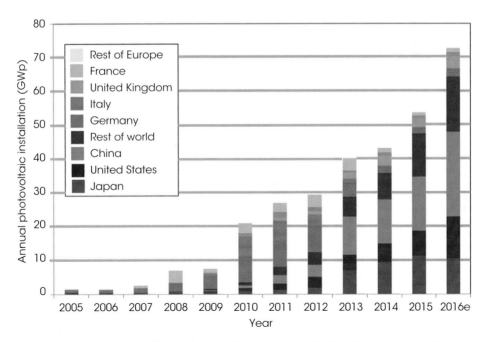

Figure 4.25 Annual growth of PV deployment (*Source:* A. Jäger-Waldau; PV Status Report 2016; EUR 28159 EN).

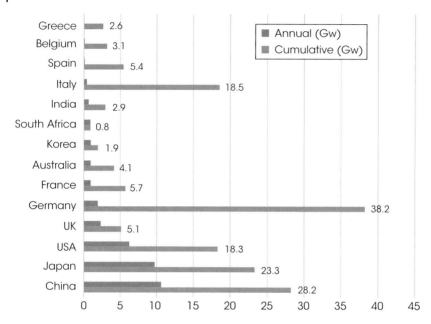

Figure 4.26 Top 14 countries in terms of cumulative capacity as of 2014; shown in order of increasing annual capacity from top to bottom (*Source:* Data from Report IEA PVPS T1-26:2015).

The decrease in rooftop installed costs can almost entirely be attributed to the drop in module prices, which fell from an average of $5/W in 1998 to approximately $0.6/W in 2014, whereas utility-scale system cost reductions were also enabled by increased efficiencies in installation. Further cost reductions are targeted on marketing, administrative, and permitting costs (called "soft costs"), which make up approximately 50% of residential and commercial installations in the United States but only 10% in Germany. Investment in solar PV installations was encouraged recently by substantial fall in the costs of solar PV that resulted largely from its widespread deployment and lately by substantial overcapacity/oversupply (Figure 4.26).

As shown in Figure 4.30 the initially booming markets in Greece, Belgium, and Spain almost disappeared, and the market in Germany, which has been the greatest consumer of PV during 2003–2012, has slowed down a lot; currently the strongest markets are in China, Japan, and the United States. On the positive side, Italy, Greece, and Germany have now enough installed capacity to produce 7–8% of their annual electricity demand with PV.

So far, the increase in solar PV installations has been supported by FIT and purchase power agreements (PPA) that reduce a project's risk as long-term returns are guaranteed, typically for 10–20 years. Such incentives and financing mechanisms are discussed in Chapter 7. Beyond Europe, the largest PV markets are in China, the United States, Japan, and India. The market in China rose to the top in 2013 and remained there, largely in response to the introduction of a national FIT. In 2015, PV installed capacity in China exceeded 38 GW, among which 32 GW is PV power plants and 6.2 GW is distributed PV systems.

In the United States, falling prices combined with state incentives and the extension of federal investment tax credits (ITC) doubled the market, bringing the total operating capacity to more than 40 GW at the end of 2016. The highest growth in the US market has been on utility-scale systems as those are the least costly and can compete with

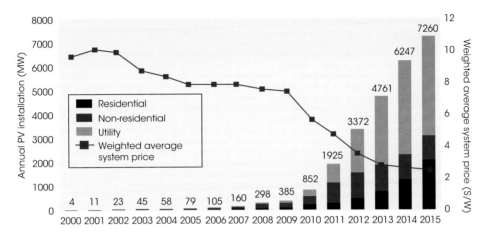

Figure 4.27 Trends in deployment increase and price decrease in the United States (*Source:* Adapted from Solar Energy Industries Association).

conventional power costing (Figure 4.27). California remains the nation's largest market, followed by New Jersey and Arizona. One would expect this high growth for the sunny south California and Arizona; in New Jersey the growth was enabled with efficient state policies. The US PV installed capacity increased from 2 to 40 MW in just 6 years (2010–2016). This impressive growth was catalyzed by the extension of the federal ITC and a rapid decrease of the utility-scale installed cost, as shown in Figure 4.31. An increase by another order of magnitude by 2030 is feasible provided that large-scale storage at costs of approximately 2–4 cents/kWh will be available.[3,4]

Japan continues to rank third globally for total operating capacity. On the other hand, most of the growth in Japan has been in residential systems reflecting the difference in land and solar resource availability between the two countries. India has been another recent solar power success story; the installed capacity grew to 5 GW within 5 years and is projected to grow to 50 GW or more by 2024. The Indian case is especially interesting for the large populations in developing countries worldwide who do not have access to reliable electricity. The PV market in India started with small solar systems replacing diesel generators in remote areas, and when the reliability of PV was established, it grew to large-scale PV deployment across the country. Chile is another fast-growing market, having installed about 0.4 GW in 2014; the north of Chile has the richest solar resources of the world, and there is a vision prevailed among academics that solar electricity from the Atacama Desert there can serve, in addition to Chile, large loads in Rio de Janeiro, San Paolo, and Buenos Aires.

How can this growth be maintained and even accelerated? Let us look at the United States, which is a difficult market because of low electricity prices. If the United States averages 15 GW of new solar per year from 2016 to 2030, the 2030 target of the Grand Solar Plan will be materialized. Catalysts for further growth include:

a) Maintenance of financial investments and stable regulatory environment.
b) The Clean Power Plan, which is currently challenged by the court, could open up new state markets to solar.
c) The cost of PV would continue to fall although at a lesser rate than recent cost reductions. System costs would be reduced as modules are made more efficient and "soft costs" are reduced. These elements will be further discussed in Chapter 7.

However, increased penetration of solar makes it more difficult to compete with conventional generation, as it would have to become dispatchable to displace base power. The solutions for making PV dispatchable (thus available on demand) include (i) investing in high-voltage direct current (HVDC) transmission to transfer solar from the SW to the rest of the country[4]; (ii) increasing the size of the grid balancing areas to harvest the benefit of regional geographical diversity in supply and demand; (iii) combining solar and wind resources; (iv) increasing the flexibility of the grid, with, for example, natural gas turbines displacing coal and nuclear power plants; (v) improving forecasting and thus certainty of PV output; (vi) adding energy storage; and (vii) implementing demand management.

While *energy storage* serves its purpose on the supply side of the equation, *demand management* can provide additional support on the demand side. As more devices in buildings become controllable, smart, and networked, customers and service providers will gain the ability to shift load away from traditional peak periods to periods of higher solar and/or wind production.

4.6   Current Status of the PV Industry

PV manufacturing capacity has grown rapidly in response to booming global demand, initially in Europe, Japan, and the United States; thereafter, leadership in production shifted to China, which expanded its manufacturing capacity massively to meet growing international solar PV demand (Figure 4.28). In recent years manufacturing capacity expanded much more quickly than demand for PV panels; fortunately for the Chinese manufacturers, a big domestic market was quickly developed there and it absorbed the overcapacity. Annual worldwide solar cell production reached 40 GW in 2014 with about 80% of it being multicrystalline and monocrystalline Si-based PV. China and Taiwan accounted for approximately 75% of global cell production based entirely on Si. In the United States approximately 32% of the module production was thin film, mostly CdTe PV.

2011 was a transition year for the PV industry; due mainly to an oversupply of modules from China, module prices fell more than 40% during the year. Installers of solar PV systems and electricity consumers greatly benefited from falling solar PV prices, but solar PV manufacturers around the world, and particularly those in the United States and Europe, experienced financial losses or shrinking profits with long-lasting impacts. Cell, module, and polysilicon manufacturers struggled to make profits or even survive amid excess inventory and falling prices, declining government support, and slower market growth for much of these years. During this period, there was significant industry consolidation worldwide to lower costs and become more competitive; several large companies became bankrupt. Among the US and European manufacturing firms that survived, several shifted their production to Asia where incentives are given and labor is cheaper. Trade tensions have arisen between the United States, Europe, and China, resulting in the imposition of import tariffs by the United States and the EU on solar panels from China. Many solar PV manufacturing firms continued their vertical integration by expanding into project development to remain competitive. Large companies developed new business models and partnered with electric utilities, real estate developers, sports teams, and retailers.

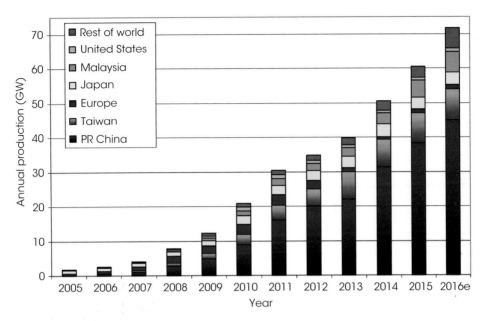

Figure 4.28 Annual PV global production (*Source:* A. Jäger-Waldau; PV Status Report 2016; EUR 28159 EN).

4.7   Large PV Power Plants

We next discuss the large PV systems that have brought prices down to "grid parity" in several regions of the world. Until a few years ago, the idea of a PV plant generating megawatts seemed unlikely to most people, but in 2008 there were around 1000 plants worldwide rated at $1 MW_p$ and above and in 2015 many plants in 100–500 MW range were added. The great driver of this revolution has been the generous financing of PV electricity in certain countries, most notably Germany, Spain, and the United States. Germany and the United States had seen steady increases in capacity for many years; then, in 2007–2008, a remarkable surge took place in Spain due to its government's introduction of a highly attractive tariff of 0.44 euro cents/kWh. In 2008 alone Spain installed $2.7 GW_p$ of PV, including some $700 MW_p$ of power plants rated above $10 MW_p$. The fast evolution to greater plant sizes is shown in Figures 4.29 and 4.30. Spain's achievement in a single year was remarkable. It must be added, however, that the Spanish government reduced the power plant tariff substantially toward the end of 2008, slanting the future more toward roofs and facades, and placed a cap of $500 MW_p$ on annual PV installation for the following few years. Even though the immediate boom was over, Spain's experience, although short-lived, changed international perceptions of what is possible and provided a massive boost to the PV industry.

Other countries active in PV power plant installation are pushing global cumulative capacity into the multi-gigawatt era. Germany and the United States are especially prominent, but Japan, Italy, Portugal, France, Greece, Korea, and as of 2015 Chile and India all deserve mention.[5] About three-quarters of plants have static arrays; the rest use single- or double-axis tracking, the great majority without sunlight concentration.

Figure 4.29 A 7.2 kW system in Marchal, Spain (*Source:* Reproduced with permission of First Solar).

Figure 4.30 The 23 MW$_p$ La solar farm in Magascona, Spain (*Source:* Reproduced with permission of IEA-PVPS).

In the rest of this chapter we will discuss the economic drivers and the system integration challenges of two growing segments of large PV deployment, namely, commercial and utility installations.

4.7.1 Commercial and Industrial Installations

Installations on public and commercial roofs proliferate in Australia (Figures 4.31 and 4.32) and recently started taking off in the Unites States and other countries, catalyzed by Investment Tax Credits (ITC) and demand charge reductions in other countries. Utility tariffs for commercial and industrial customers are commonly divided into demand and consumption. Many utilities in the United States charge demand based on the peak 15–30 minutes of load each month, and these charges apply to the whole month. Thus substantial demand charges can be incurred due to load variability. Demand charge reduction can be achieved by storage alone, PV alone, and even more effectively with a combination of the two. Peak demand reductions also benefit the utility by reducing overloads in congested distribution grids. An example of such a system is the "JFK Solar Park" (Figure 4.33) located on roofs of commercial buildings near JFK International Airport, which is operated under New York State's "remote net metering" program; the electricity produced at JFK Solar Park is consumed by Bloomberg LP at its headquarters in Manhattan.

4.7.2 Utility-Scale PV

Over the last 5 years the deployment of large utility-scale scale PV power plants has been impressive. They first made their appearance about 10 years ago in Germany and

Figure 4.31 The 1.2 MW PV rooftop system at the University of Queensland, Australia (*Source:* Reproduced with permission of Trina).

Figure 4.32 The 1.17 MW PV rooftop system at the Adelaide Airport, Australia (*Source:* Reproduced with permission of Trina).

Figure 4.33 Section of the 1.6 MW "JFK Solar Park" located on the roofs of three airfreight logistics buildings near JFK International Airport in New York (*Source:* Reproduced with permission of JFK Solar Enterprises LLC).

Spain, and since 2010 they have become the fastest-growing segment of the PV market in China and the United States.

Large ground-mount PV plants have grown in the United States more than residential applications due to drastic cost reductions and increased familiarity of utilities with the

Figure 4.34 The 1.3 MW PV power plant at Dimbach, Germany (*Source:* Reproduced with permission of First Solar).

Figure 4.35 Section of the 10 MW$_{ac}$ PV power plant at Tibet Sangri, China (*Source:* Reproduced with permission of Trina).

reliability of PV power plants, which, in turn, led to power purchase agreements (PPA) between the plant owner and the utility. China and the United States also have the largest PV plants (a 500 and 520 MW in China, two 550 MW in the United States). As of 2015, other emerging markets for large-scale multi-MW PV plants include Chile and South Africa.

Figure 4.36 The 300 MW$_{ac}$ PV power plant at Yunnan Jianshui, China (*Source:* Reproduced with permission of Trina).

Figure 4.37 Section of the 290 MW$_{ac}$ PV power plant, Agua Caliente, AZ, USA (*Source:* Reproduced with permission of First Solar).

The biggest advantage of large ground-mount PV plants is their easy installation (shown in Figures 4.38, 4.39, and 4.40) and economies of scale that make the total costs much lower than those of rooftop installations. Also these plants can be controlled according to the utility requirements and help the stability of the grid. On the other

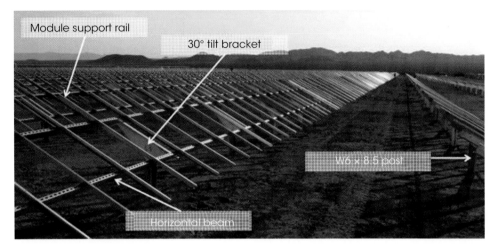

Figure 4.38 Start of PV plant construction; support and mounting structures (*Source:* Reproduced with permission of First Solar).

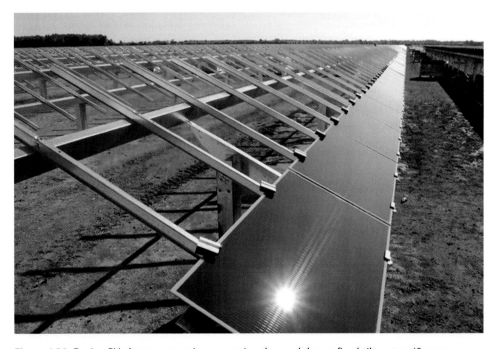

Figure 4.39 During PV plant construction, mounting the modules on fixed-tilt system (*Source:* Reproduced with permission of First Solar).

Figure 4.40 Section of the 52.5 MW$_{ac}$ one-axis tracking PV plant during construction, Shams Ma'an, Jordan (*Source:* Reproduced with permission of First Solar).

hand, smaller distributed PV systems have the advantage of being at the point of demand so that transmission and distribution power losses are eliminated.

At the time of writing the biggest PVPS in the world are the 550 MWac Desert Sunlight, 550 MWac Topaz, and 579 MWac Solar Star plants in south California. The first two were constructed by First Solar using their CdTe PV modules, and the third was constructed by SunPower using their monocrystalline silicon modules. Each of these projects was completed ahead of a 3-year schedule and is operating under PPA with regional utilities (Pacific Gas and Electric (PG&E) and Southern California Edison). It is noted that two of these plants are now owned by a Warren Buffett company; the big capital is behind PV now!

These and other projects have taken the price of PV electricity under US Southwest irradiation conditions down to 6–7 cents/kWh, ahead of the projections made in 2008 by one of the authors[4] and also ahead of the US Department of Energy (US-DOE) projections made in 2010 (Figure 4.45), which materialized into the SunShot Vision Study.[6] Figure 4.45 shows estimated and projected price reductions for utility-scale PV expressed in levelized cost of electricity (LCOE). The LCOE averages all the costs during the operating life of a project; it is used for comparing the cost of energy technologies with different operating conditions, as technologies like solar have higher capital costs (and consequently higher financing costs), whereas fossil fuel-based power technologies have higher operating (and fuel) costs. The LCOE is further discussed in Section 7.1.

Figure 4.41 52.5 MW$_{ac}$ one-axis trackers, almost completed, Shams Ma'an, Jordan (*Source:* Reproduced with permission of First Solar).

Figure 4.42 Section of 550 MW$_{ac}$ Desert Sunlight, California, PV plant during construction stages (*Source:* Reproduced with permission of First Solar).

Figure 4.43 Section of the 550 MW$_{ac}$ Desert Sunlight PV power plant (*Source:* Reproduced with permission of First Solar/NEXTera Energy).

Figure 4.44 Close-up on another session of the 550 MW$_{ac}$ Desert Sunlight PV plant (*Source:* Reproduced with permission of First Solar/NEXTera Energy).

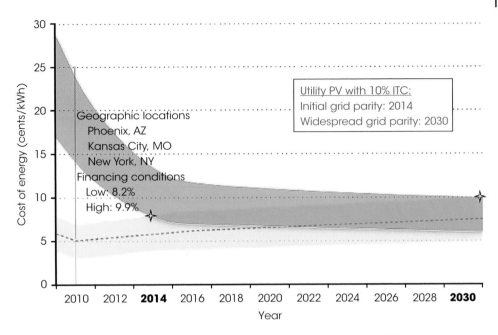

Figure 4.45 Levelized cost of electricity (LCOE) for utility conventional grid and PV power (*Source:* Reproduced with permission of US-DOE Solar Technologies Program).

4.8 PV Grid Connection and Integration

4.8.1 The Electricity Grid

The electricity grid is a complex network of power lines, designed to transport energy from suppliers to loads. In its normal configuration, power plants feed electricity into the high-voltage (>100 kV) transmission grid, which transports the energy to demand centers where the voltage is stepped down to deliver power to individual customers on the distribution grid. Both the European and the US electricity grids are more than a century old. The US electricity grid is a conglomerate of many smaller grid systems that were built in the late 19th century, each regulated by separate utility companies. The grid eventually grew into three major "interconnects" on which all power plants operate synchronously. They are the Eastern Interconnection, Western Interconnection, and Electric Reliability Council of Texas (ERCOT); there are very few links between the three interconnects. Figure 4.46 shows these interconnects, the North American Electric Reliability Corporation (NERC) interconnect subregions, and about one hundred generator, load, and transmission balancing areas. Balancing authorities integrate power resources to meet demand within balancing areas; they are responsible for maintaining interconnection frequency and controlling the flow of power so that overloading of transmission lines is avoided. The US power transmission grid consists of 300 000 km of lines operated by 500 companies.

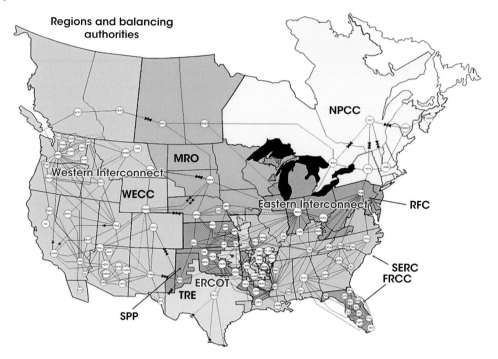

Figure 4.46 The US electric grid: three major interconnects, eight NERC subregions, and about 100 balancing authorities (*Source:* http://www.eesi.org/briefings/view/021617wires. CC BY 4.0).

The largest synchronous (by connected power) electrical grid is that of continental Europe connecting over 400 million customers in 24 countries. Although synchronous, some countries operate in a near island mode, with low connectivity to other countries, and there are plans to increase connectivity and host more renewable energy while enhancing the reliability of the grid. To this effect, the European Electricity Grid Initiative (EEGI) has the following objectives: (i) to transmit and distribute up to 35% of electricity from dispersed and concentrated renewable sources by 2020 and a completely decarbonized electricity production by 2050 and (ii) to integrate national networks into a market-based, truly pan-European network to guarantee a high-quality of electricity supply to all customers and to engage them as active participants in energy efficiency.

Now let us look at the basics of electricity grid operation to better understand the challenges and benefits of integrating renewable and distributed energy in the grid. Supply and demand of active power on the grid must always be in balance, or the grid frequency will deviate too far from its set point (60 Hz in the United States, 50 Hz in Europe), causing connected appliances to shut down or get damaged. The frequency of the system would vary as load and generation change. If there is more generation than demand, frequency goes up; if there is less generation than demand, frequency goes down. During a severe overload caused by tripping or failure of generators or transmission lines, the power system frequency will decline, due to an imbalance of load versus

---

Grid Electricity Frequency

The utility frequency, or power line frequency, is the frequency of the oscillations of alternating current (AC) in electric power transmitted from a power plant to the end user. In large parts of the world, this is 50 Hz, although in the United States and parts of Asia it is typically 60 Hz. Places that use the 50 Hz frequency tend to use 220–240 V, and those that use 60 Hz tend to use 100–127 V. Both frequencies coexist today (Japan uses both).

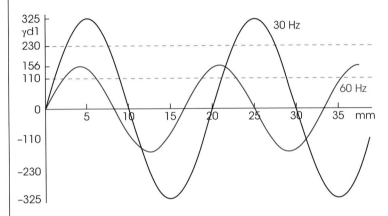

In conventional thermoelectric and hydro systems power is supplied by rotating AC generators. The frequency of each generator is directly proportional to its speed (*N*) and number of electric-magnetic poles (*P*) in the motor according to the equation

F (Hz) $= P \times N/120$, where *F* is in Hertz and *N* is in rotations per minute (RPM).

The RPM for gas turbines (in a 60 HZ system) is typically 3600 and for nuclear steam plants is 1800. Synchronous generators connected to the grid run at various speeds, but they all generate the same frequency by varying the number of poles (*P*).

---

generation. Loss of an interconnection while exporting power will cause system frequency to rise. Also, temporary frequency changes are an unavoidable consequence of changing demand. Automatic generation control (AGC) is used to maintain scheduled frequency and interchange power flows, and the presence of many generators and a large distributed load allows for easy frequency management. Control systems in power plants detect changes in the network-wide frequency and adjust mechanical power input to generators back to their target frequency. Modern PV power plants can also respond to frequency regulation by curtailing or increasing power instantaneously (see Section 4.8.2).

4.8.2 Grid-Friendly PV Power Plants

As PV generation grows to the point of making a significant contribution to the grid, the PV industry is developing large PV power plants that support grid stability and reliability. A modern utility-scale PV power plant is a complex system of large PV arrays and multiple power electronic inverters, and it can contribute to mitigate impacts on grid stability and reliability through sophisticated "grid-friendly" controls (Figure 4.47).[7] Components of a typical multi-MW, utility-scale PV power plant are shown in Figure 4.48 including power conversion and electrical equipment, such as PV panels, inverters, switchgear, grid interconnection, power plant controller (PPC), supervisory control and data acquisition (SCADA), and communication systems.

Figure 4.47 Utility-scale PV operations center in Mesa, Arizona, USA (*Source:* Reproduced with permission of First Solar).

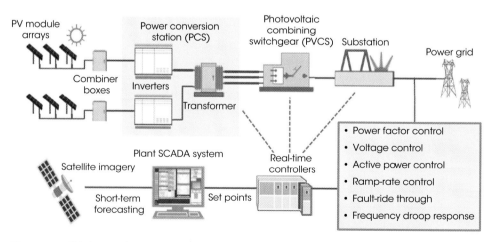

Figure 4.48 PV plant grid integration and control system (*Source:* Reproduced with permission of First Solar).

"Grid-friendly" PV plants help to stabilize the grid by incorporating voltage regulation, active power controls, ramp-rate controls, fault-ride through, and frequency response.[7] These services are outlined in the Box below:

Grid-Friendly PV Capabilities

Fault-ride through is the capability of electric generators to stay connected during short periods of voltage dips in electric distribution networks, which could lead to a widespread loss of generation.

Automatic active power frequency control. The frequency in alternating current grids is kept constant within strict limits—typically very close to 50 or 60 Hz. The frequency drops if more energy is consumed than generated. The opposite occurs if there is an energy surplus—the grid frequency increases.

Static voltage support. In order to protect the connected loads, the voltage must be kept within defined limits—which applies to the distribution grid in particular. PV inverters that have the ability to provide controlled inductive or capacitive reactive power can help guarantee the required voltage quality. They can also be used to compensate for phase shifts caused by transformers, large motors, or long cable sections.

Dynamic grid support. Until recently, PV plants had to disconnect from the grid immediately under fault conditions, thus unbalancing the grid even further. Inverters with dynamic grid support functions act within milliseconds, preventing the grid failure from spreading further. Dynamic grid support ensures that the inverter is ready to feed energy into the grid immediately after a drop in the grid voltage.

A plant-level control system, which controls a large number of individual inverters to affect plant output at the grid connection point, is a key enabler of such features (Figure 4.48). The PPC monitors system-level measurements and determines the desired operating conditions of various plant devices to meet the specified targets. It manages the inverters, ensuring that they are producing the real and reactive power necessary to meet the desired settings at the point of interconnection (POI). For example, when the plant operator sends an active power curtailment command, the controller calculates and distributes active power curtailment to individual inverters. In general, the inverters can be throttled back only to a certain specified level of active power without causing the DC voltage to rise beyond its operating range. Therefore, the plant controller dynamically stops and starts inverters as needed to manage the specified active power output limit.

The actively controlled plant also has the ability to minimize the impact of cloud cover. Accommodation of the reduction of power output due to partial plant shading is done by increasing the output of the inverters from the unaffected sections of the plant (Figure 4.49).

These "grid-friendly" capabilities, essential for increased penetration of large-scale PV plants into the electric grid, are operational and available today for utility-scale PV plants ranging from several megawatts to several hundred megawatts. These advanced plant features enable solar PV plants to behave more like conventional generators and actively contribute to grid reliability and stability, providing significant value to utilities and grid operators. They also use the active power management function to ensure that

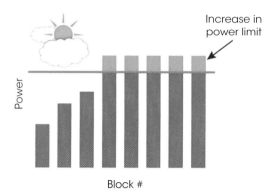

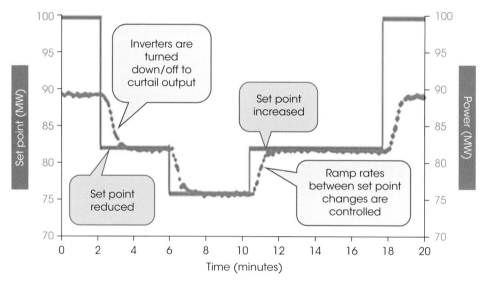

Figure 4.49 Impact of cloud passage in utility PVPS operation; the plant comprises of eight power blocs (*Source*: Reproduced with permission of First Solar).

Figure 4.50 Power curtailment at different levels: this figure shows field data from a PV plant operating at around 90 MW power. The brown lines show the power set points and the blue shows the supplied power (*Source*: Reproduced with permission of First Solar).

the plant output does not exceed the allowed ramp rates, to the extent possible. It cannot, however, always accommodate rapid reduction in irradiance due to cloud cover and storage may be needed in such cases. Figure 4.50 shows field data from a PV plant operating at around 90 MW power.

The brown lines show the power set points and the blue shows the supplied power. The plant controller turns down the inverters (and turns off some of them if required) to achieve the new set point. Note that the turndown of power is gradual to meet the specified ramp-rate limit.

A discussion of electricity markets is necessary to better understand the effects of variable generation on the grid. The National Renewable Energy Laboratory (NREL), AES, the Puerto Rico Electric Power Authority, First Solar, and the ERCOT have conducted a demonstration project on two utility-scale PV plants to test the viability of providing important grid ancillary services from these facilities. This demonstration showed that active power controls can leverage PV's value from being simply a variable energy resource to providing additional ancillary services that range from spinning reserves, load following, ramping, frequency response, variability smoothing, and frequency regulation to power quality. Specifically, the tests conducted included variability smoothing through AGC, frequency regulation for fast response and droop response, and power quality.[8]

4.9 Electricity Markets and Types of Power Generators

Prices in energy markets are set by the dynamics of supply and demand under the geographical constraints imposed by the reach and limitations of the transmission grid. For this reason, large grid systems are split into smaller markets at major nodes or zones, each of them establishing a price for each hour of the day according to supply and demand. This is called *nodal pricing*, or locational-based marginal pricing (LBMP). Imbalances of supply and demand on single nodes can be overcome by importing electricity from other zones, as long as there are not transmission constraints.

In the USA, balancing of supply and demand on the grid is achieved by independent service operators (ISO) using seasonal, week-ahead, day-ahead, and hour-ahead load forecasts. The ISO ranks all generator bids based on their bidding price and fills up the load forecast of each hour of the day starting from the cheapest bidder, taking into account transmission line limits. Any differences between the forecast and actual load are settled, in real time, in the ancillary services market, which consists of regulation and reserve resources. Regulation resources can quickly adjust their output to accommodate changes to the balance of supply and demand, upon receiving a signal from the ISO. To protect against the risk of a plant outage, the ISO also has in-service spinning reserves that can be on full capacity within 10 minutes and non-spinning reserves that can respond within 30 minutes.

The grid operations outlined earlier are important when we consider the impacts of high penetration of variable generation (i.e., solar and wind) into the grid. Let us now discuss the different types of power plants according to the services they provide. Nuclear and large coal-fired and gas-fired power plants that operate 24 hours a day serve a baseload, typically the demand that is required throughout the year (8760 hours). Peaker plant are smaller units, typically natural gas- or diesel-fired power plant that operate only during the high demand days of the year, for example, hot summer afternoons in tropical and subtropical climates (depicted as up to 1000 hours in Figure 4.51). Medium-size natural gas and coal power plants satisfy the balance of daily loads (depicted as intermediate load in the same figure).

Peaker Plants

The market clearing price of electricity is set by the marginal price for the last MW needed to meet the load. During peak load hours, a large percentage of the generator fleet is dispatched, including "peaker plants," which are more expensive, less efficient, and more polluting than conventional generators.

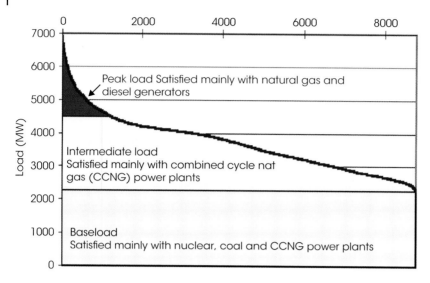

Figure 4.51 Example of a load duration curve.

These plants set a high electricity clearing price for all energy delivered in that time period, and it is therefore economically desirable to prevent their dispatch. Solar generators have the potential to minimize the need for peakers, as peak demand is typically AC driven, and therefore effectively reduce the market clearing price while mitigating emissions.

Steam and Gas Turbines

Nuclear and coal power plant use steam turbines designed according to the Rankine cycle, a closed-loop expansion and condensation of steam generated by the fission or combustion of fuel. Natural gas and diesel-fired turbines employ the Brayton cycle comprising alternative isentropic and isobaric process. There are two types of gas-fired power plants, namely, single-cycle or open-cycle gas turbine (OCGT) and combined-cycle gas turbine (CCGT) plants. OCGT plants consist of a single compressor/gas turbine that is connected to an electricity generator via a shaft. Air is compressed by the compressor and its oxygen is burned with natural gas in the combustion chamber of the gas turbine that drives both the compressor and the electricity generator. Almost two-thirds of the gross power output of the gas turbine is needed to compress air entailing high thermal losses, and the remaining one-third drives the electricity generator. CCGT use both a Brayton and a Rankine cycle as the latter recovers rejected heat from the exhaust of the gas turbines to produce steam that drives a steam turbine and generates additional electric power. The Rankine engines have some flexibility within a range by adjusting the flow of steam into the turbine, but the Brayton engines have even greater flexibility as they have little thermal inertia and are capable to cycle and ramp quickly (10–30 minutes start-ups, ~8% per minute ramping).

Typical thermal efficiency for utility-scale electrical generators operating at design capacity is around 33% for coal and oil-fired plants between 35 and 42% for OCGT and 56–60% for CCGT. Their efficiencies decline when operating at partial capacity.

Intermediate Load-Following Plants

The power output of hydroelectric power plants, gas turbines, and combined-cycle gas turbines (CCGT) can be effectively adjusted, and they are used to follow the variation of demand load throughout the day. Load-following power plants run during the day and early evening when the demand is higher and are either shut down or greatly curtailed output during the night, when the demand for electricity is the lowest. Coal power plants with sliding pressure operation of the steam generator can generate electricity at part-load operation up to 75% of the nameplate capacity and can also be used for load following although their ramping rates are slower than those of gas turbines.

Baseload Plants

Hydroelectric power plants can be efficient for both intermediate load-following and baseload applications. Thermoelectric baseline generators (nuclear, large coal) use the Rankine cycle (discussed in previous text box) and are built to operate at their maximum or near-maximum output 24 hours a day and are expensive to ramp or cycle as this incurs physical wear due to their high thermal inertia. CCGT power plants are also used for serving baseload, and as the price of natural gas is being reduced, they are displacing coal in baseline power plants.

Intermediate and peak load power plants can also provide important voltage and frequency stabilization services in addition to supplying the required loads.

Regulation Reserves

These reserves are normally supplied by generators that have the ability to be dispatched up or down remotely (commonly referred to as automatic generation control).

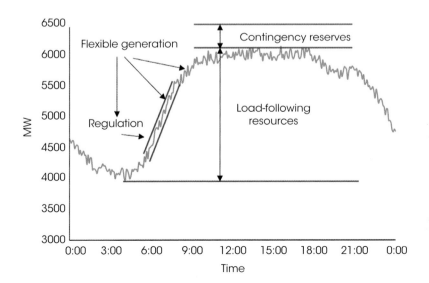

Figure 4.52 Power generators for load balancing and regulation.

Contingency Reserves

These can be supplied by generators that are online (spinning) and by "quick-start" generators that can be started and turned up within 15 minutes. The amount of power that a generator can contribute is limited by its ramp rate and the difference between its current dispatch level and its maximum capability. Contingency can also be supplied by demand response controlled by the system operator.

4.10 The Variability Challenge and Solutions

The electric grid system and its market operations were designed to deal with variability of demand and supply on different timescales, mainly by dispatching controllable generators and, to a lesser degree, by using electricity storage systems. A high penetration of variable solar and wind electricity into the grid creates challenges for the grid operators who need to reliably satisfy load demands every hour of the day. Thus it is important to closely investigate the variability of the solar resource on seasonal, diurnal, and cloud-induced time domains (Figure 4.53).

The seasonal and diurnal variability are precisely described with a set of geometric equations, describing the Earth's rotation and its elliptical movement around the sun. Clear sky irradiation can be predicted with very high accuracy, but stochastic variability due to cloud coverage is cause of concern for grid reliability in high solar penetration scenarios. The following solutions are available for reducing or mitigating such variability: (i) geographical diversity/transmission interconnections, (ii) solar forecasting, and (iii) energy storage. A combination of these solutions would in most cases provide the minimum cost solution as variability is reduced with geographical diversity and controlling it is easier and less expensive when we have accurate forecasting. Also demand management has a significant role in handling the diurnal variability.

In the following we discuss solar forecasting and geographical diversity and high-voltage transmission lines; energy storage is the subject of Section 4.11.

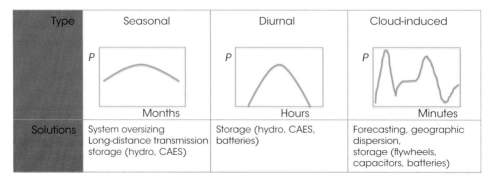

Type	Seasonal	Diurnal	Cloud-induced
	P / Months	P / Hours	P / Minutes
Solutions	System oversizing Long-distance transmission storage (hydro, CAES)	Storage (hydro, CAES, batteries)	Forecasting, geographic dispersion, storage (flywheels, capacitors, batteries)

Figure 4.53 Solar resource fluctuations and options to mitigate them.

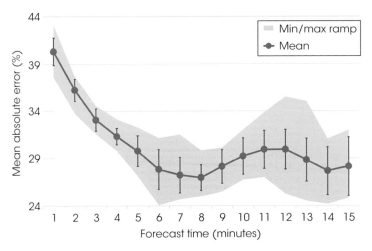

Figure 4.54 Forecast mean average errors (MAE) are reduced by 39–24% when sky imaging is integrated with satellite data[9] (*Source:* Courtesy of Dantong Yu, NJIT).

Forecasting

Irradiation data are available in real time from satellite measurements, but their time and scale resolution is rather coarse. For individual PV plants, this can be improved with ground-based sky-imaging hardware and software that can provide local irradiation information 1–15 minutes ahead by estimating the movement of the clouds and associated shading. An irradiance sensor network, together with multipoint optical imaging systems, can provide three-dimensional spatial information about clouds and their motions. As shown in Figure 4.54, this information leads to greatly improved accuracy and system reliability in forecasts, especially for cloudy and partly cloudy conditions wherein the fluctuation of solar energy is the largest, and the value-added gain in forecasting is the highest. With this information PV operators could foresee a forthcoming rapid ramp-rate change and handle it with providing extra power, gradually curtailing output, or dispatching power from storage. For large ramp rates, use of storage would be required, but this is minimized by considering the changes in total output aggregate of the PVPS or of the aggregate of PV plants in the same balancing area.

Geographical Diversity/Transmission Interconnections

Irradiance measured at a single point can change drastically; however, the aggregate output of geographically dispersed PV systems is much smoother than that of a single point. Such smoothening is also observed within large multi-MW PV, which typically occupy larger areas that are not totally shaded in a partially cloudy day. This effect is shown in the histogram shown in Figure 4.55 of daytime minute-to-minute irradiation changes (called ramp rates) observed at a 5 and 80 MW plants near each other in Ontario, Canada, over the course of a month. The 80 MW plant exhibits relatively less variation compared with the 5 MW plant. Figure 4.55 shows how the aggregate power production from the whole 80 MW plant is smoother than the irradiation profile on one point of the plant.[10] The smoothening of the fluctuations is even greater in the multi-hundred MW PV plants constructed in the United States, China, and other countries.

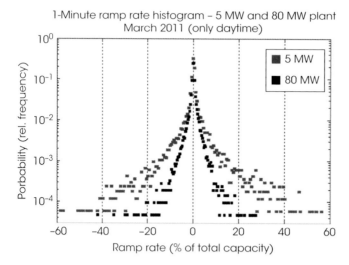

Figure 4.55 One-minute ramp-rate histograms for a 5 and 80 MW plants near each other in Ontario, Canada. The larger the plant, the smaller the ramp rates.

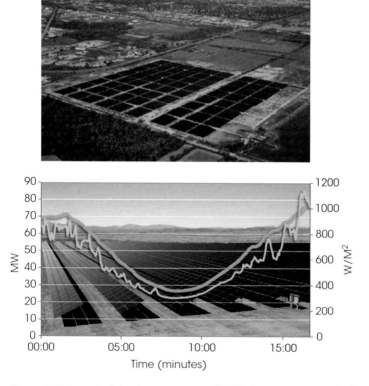

Figure 4.56 Impact of cloud passage on an 80 MW plant power output in Ontario, Canada. The green line shows irradiation measured on one point (W/m²), and the orange line shows the smoothening effect from the aggregate of the inverters (MW) (*Source:* Reproduced with permission of First Solar).

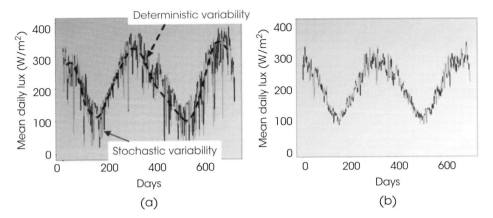

Figure 4.57 Daily variability over the course of 2 years in (a) Los Angeles and (b) an area of $190 \times 190 \, \text{km}^2$ around the city.

Understanding the spatial and temporal characteristics of solar resource variability is important because it helps inform the discussion surrounding the merits of geographic dispersion and subsequent electrical interconnection of photovoltaics as part of a portfolio of future solutions for coping with this variability. Unpredictable resource variability arising from the stochastic nature of meteorological phenomena (from the passage of clouds to the movement of weather systems) is of most concern for achieving high PV penetration because unlike the passage of seasons or the shift from day to night, the uncertainty makes planning a challenge. A detailed discussion of these variability aspects is given by Perez and Fthenakis.[11] The drastic effect of geographical diversity on reducing the stochastic variability of solar irradiation is shown in Figure 4.57.

4.10.1 Long-Distance Transmission Lines

High-Voltage (HV) Transmission Lines

High-voltage (HV) transmission lines can be AC or DC. The HVDC lines have larger interconnection losses than HVAC, but much lower transmission losses.

While HVAC systems merely require transformers that step down the voltage to the next lower level, HVDC systems rely on power electronics to convert between DC and AC networks. Transformers and power electronics are located in two converter stations at each end of the line. Typically the line is bipolar at a voltage of ±800 kV, so the load is carried by two cables in parallel. Each pole consists of six aluminum core steel reinforced (ACSR) conductors.

Geographical diversity can greatly reduce the fluctuations of solar energy. As discussed before, increasing the geographical area of PV, by either connecting an aggregate of dispersed small systems or by constructing large power plants, decreases the cloud-induced fluctuations of their total output. In addition, long interconnects can reduce diurnal intermittency, by taking advantage of different time zones; in theory, if we could connect anti-diametric locations of the globe, we could resolve the diurnal intermittency (e.g., by connecting China with the United States) as well as the seasonal variability (e.g., northern Chile with the Southwest United States). The long distances involved will necessitate HVDC lines, a technology that exists and can be further improved if there is a market. The transmission of electricity over 2000 miles of HVDC lines typically entails a 10% electromagnetic loss versus a 22% or higher loss using high-voltage AC power lines of the same distance. Also construction of HVDC power lines typically requires 37% less land area than constructing high-voltage AC ones. The technology is well established, but there are cost and siting challenges that need to be addressed. Currently, HVDC transmission lines with a capacity of 5 GW are operating in China utilizing 800 kV technology, and in the future a doubling of this capacity is expected.[12] The array converters will be boosting array voltages of around 1 kV DC to a "gathering" voltage of around 50 kV DC. A second DC-to-DC converter will be used to boost the "gathering" voltage to the transmission voltage of 800 kV DC. These converters are already in common use throughout the electric utility industry in power-conditioning devices and HVDC power lines in Europe and the United States. Transmission experts McCoy and Vaninetti reported in 2008 a cost of $0.02/kWh for constructing a network of HVDC lines with 800 kV.[12] The same experts foresee a dramatic improvement in technology and in unit cost with a plan for large deployment and domestic production of the large quantities of heavy cables required. Future developments in superconducting cables would make the vision even more feasible. The biggest challenge regarding long-distance transmission is that would need approvals from a plethora of regional and national jurisdictions. For the United States, Fthenakis and collaborators Zweibel and Mason have defined the needs of HVDC from the SW for solar and the High Plains for wind to the rest of the country.[3,4] Crossing national borders may be more involving, but we have plenty of boundary crossing transmission lines in Europe, United States–Canada, and elsewhere. Developing large global electricity grids is not a far-fetched idea; it could follow the paradigm of global telecommunications networks, which were enabled by fiber-optic technology connecting continents with underwater cables; it is expensive but it is technologically feasible.

4.10.2   Grid Flexibility

The flexibility of a power system, that is, its ability to vary its output to meet the demand, depends on the mix of its generators. Its flexibility is constrained mainly by the baseload generators, usually nuclear and coal power plants. The operation of nuclear power plants is quite inflexible, and that of coal power plants is flexible only within a narrow range.

We discussed earlier the need for adjusting power output for load following and for satisfying peak demand. In addition to these duties, in order to enhance the reliability of the grid, a number of generators are operated at partial load with their surplus megawatts available to ensure reliability under rapid changes on demand (frequency regulation) or an unplanned outage of a unit or a transmission line (contingency reserves). Frequency and contingency reserves operating at partial load are called spinning reserves. There are also units that provide grid stability (voltage, frequency, reactive power). Overall, a number of generators operating at partial load account for

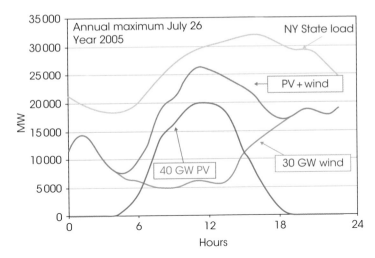

Figure 4.58 Synergy of PV and wind in New York State (*Source:* Nikolakakis and Fthenakis[13]. Reproduced with permission of Elsevier).

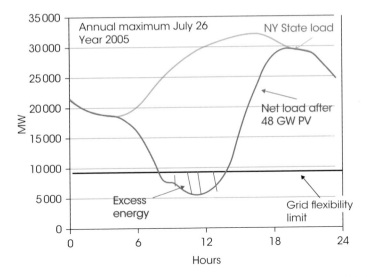

Figure 4.59 Effect of grid flexibility on PV energy delivery (*Source:* Nikolakakis and Fthenakis[13]. Reproduced with permission of Elsevier).

the functions of load following, frequency regulation, and spinning reserves. Solar and wind penetration into a power system displaces generation from conventional units, and often their combination can increase such displacement. Figures 4.58 and 4.59 shows a result from simulations based on hourly wind, solar, and load data in NY State.[12] The figures show the load and PV and wind outputs for a summer day where a peak of the load for that year occurred; it is shown that PV and wind together serve the load better than each of them separately. From a utility's perspective, the main advantages of this penetration are fuel economy, emissions reduction, and reduced need for increasing overall system capacity.

The integration of large amounts of variable renewable energy supply into the grid may necessitate additional means for frequency regulation and increase the requirements for ramping rate of load-following units. Thus, such integration on the grid may translate into additional costs to cover the greater amount of flexibility, ramping capability, and operating reserves needed in the system. However, such costs are minimized when PVs are dispersed geographically over large regions.

The limit of the penetration of renewable energy into an electricity grid depends on the mix of generators of the system. As described previously, there are two main types of generators: inflexible baseload units and the more flexible cycling units. The former units are designed to operate at full output, and they often provide most of the winter demand or 35–40% of the annual peak capacity. The penetration of wind and solar power cannot drive the net load below the limit imposed by the number and the type of baseload generators and the amount and type of reserves. This limit depends on the ability of conventional baseload generators to reduce significantly their output, and on economic and mechanical constraints. For example, coal plants can vary their output from full to half capacity, but if this is done frequently, it would demand costly maintenance. (The flexibility limit that separates the flexible from the inflexible capacity and, hence, the flexibility of a system is defined as the percentage of the annual peak capacity that is flexible.)

High levels of wind and solar energy penetration may stress the system because of its flexibility limit; there will be hours throughout the year where the net load is brought below it. Then, the amount of energy below the flexibility limit cannot be absorbed and must be curtailed. This is more of a problem for incorporating wind power than solar power, since winds are stronger during the night when the load levels are low. The more flexible a power system is, the higher is the penetration achievable, and the less the restraint on renewable electricity. The amount of energy to be cutback can be determined with a cost analysis; often it makes economic sense to curtail small amounts of energy as a trade-off in penetration.

The irregularity of renewable resources and the limit on the grid's flexibility both pose restrictions on the maximum penetration achievable in a system. There are some interesting studies on the maximum renewable penetration that can be realized without storage. It was shown that the maximum annual energy penetration attainable from solar energy alone in the Texas grid is around 10% if no energy is curtailed, but it increases to 22% if we allow for 10% of the PV energy to be curtailed.[13] Studies focusing on the New York yielded almost identical results and also showed a great synergy between wind and solar resources in meeting the NYISO loads.[14]

4.11 Energy Storage

Energy storage can increase the flexibility and reliability of the system; it can offer the following services:

a) *Renewable energy capacity firming*: This process refers to combining a storage technology with sources of wind or solar energy so that the power output is more certain. It offsets the need to purchase additional dispatchable capacity (for offsetting renewable generation ramping), and when local, it relieves congestion and compensates for the need for having transmission and distribution equipment.

Table 4.1 Categories of electricity storage technologies.

Categories	Applications	Operation timescale	Technologies
Power quality	Frequency regulation, voltage stability	Seconds to minutes	Flywheels, capacitors, superconducting magnetic storage, batteries
Bridging power	Contingency reserves, ramping	Minutes to ~1 hour	High-energy-density batteries
Energy management	Load following, transmission/ distribution deferral	Hours to days	CAES, pumped hydro, high-energy batteries

b) *Renewable energy time-shift*: Excess of renewable energy is stored for use at peak demand hours. For example, low-value wind energy generated at night when the demand is low can be stored and sold later through the energy market. Excess solar energy during noon hours can be stored for use in the evening peak hours.

Electric storage technologies are differentiated by various attributes, such as rated power and discharge time. In general, there are three major categories of large-scale energy storage technologies: power quality, bridging power, and energy management. The main difference between them is the timescales over which they operate and their power and energy capacities (Table 4.1).[15]

Power quality refers to the set of parameters (e.g., voltage and frequency regulation, reactive power, fault-ride through) that must be continuously satisfied for electrical systems to operate as expected. The storage technologies that are best suited for ensuring this continuity must provide a large power output on very short timescales, typically seconds. Since power delivery occurs in such a brief period, large storage capacities are not necessary. These technologies include superconducting magnetic energy storage (SMES), electric double-layer capacitors (EDLCs), flywheels, and batteries (Figure 4.60).

On the other hand, technologies that provide power over longer timescales for applications such as load leveling and peak shaving are used for *energy management*. Whereas power-quality applications deal with short-term and unpredictable fluctuations in power output, energy management technologies address variability that largely is predicted by peak and off-peak demand. Emphasis is placed on storage capacity and less so on instantaneous power, and the timescales involved are much longer than those needed for power-quality technologies. The upper region of Figure 4.60 provides a few examples of storage mechanisms used for energy management: pumped hydro, compressed air energy storage (CAES), and flow batteries. Between these two boundaries lie storage technologies used for *bridging power* to ensure continuity when switching from one source of energy to another.

Let us now discuss individual technologies.

4.11.1 Power-Quality Storage Technologies

4.11.1.1 Superconducting Magnetic Energy Storage

Among the most efficient storage technologies are SMES systems. They store energy in the magnetic field created by passing direct current through a superconducting

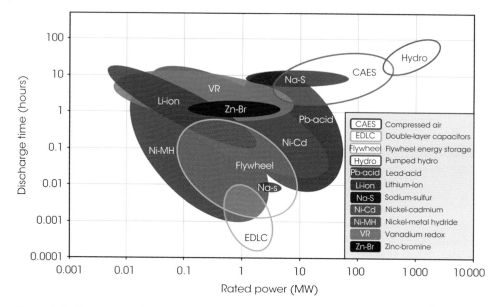

Figure 4.60 Comparison of storage systems in terms of discharge times and rated power.[15]

coil; because the coil is cooled below its superconducting critical temperature, the system experiences virtually no resistive loss. Four components comprise a typical SMES system: the superconducting coil magnet (SCM), the power-conditioning system (PCS), the cryogenic system (CS), and the control unit (CU).

The major disadvantage of SMES is the high cost of refrigeration and the material of the superconducting coil. Research is being made into so called "high-temperature" superconductor (TSC) technology that utilizes liquid nitrogen operating at 65–77°K (about −200°C) rather than the costly liquid hydrogen required for a very-low-temperature superconductor; nevertheless, the costs of HTSC material remain high. SMES systems with large capacities of 5–10 GWh involve large coils, several hundred meters in diameter that must be kept underground, adding to the expense of the system.

However, SMES have very high round-trip efficiencies (e.g., ~95%). In addition, SMES systems can discharge almost all the energy stored in the system with a high power output in a very short time, making them ideal for power-quality applications. SMES systems improve power quality and system stability in several ways. Following an interruption, such as a downed power line or generator, a SMES unit can dampen low-frequency oscillations and mitigate voltage instability by providing both real and reactive power to the power system. On the demand side, an SMES can balance fluctuating loads by releasing or absorbing electricity according to demand. They also can be used as a backup power supply for critical loads that may be sensitive to disturbances in power quality; their fast response time allows them to inject power in less than one power cycle.[14]

4.11.1.2 Electric Double-Layer Capacitors

EDLCs, also known as supercapacitors or ultracapacitors, offer another solution to ensure quality and short-term reliability in power systems. Like a conventional capacitor, electricity in an EDLC is stored in the electrical field between separated plates; the capacitance is a function of the plates' area, the distance between them, and the dielectric constant of the separating medium. However, whereas standard capacitors employ two plates of opposite charge separated by a dielectric, EDLCs consist of two porous electrodes immersed in an electrolyte solution, a structure giving a highly effective surface area and minimal distance between electrodes.

Compared with regular batteries, EDLCs have lower energy densities and higher power densities, making them suitable for power-quality applications. For short-term high-power applications, electricity discharge in an EDLC is not limited by the rates of chemical reaction rates as is the case with batteries. In addition, EDLC–battery hybrids that incorporate the benefits of both technologies often are used for distributed energy storage.

This hybrid storage offers a dynamic solution to problems related to off-grid PV: the EDLC provides the power necessary for large fluctuations in power demand, while the battery remains the source of continuous electricity over long periods. In this arrangement, the battery's size is geared for a constant load rather than for peak current demand, which can be up to 10 times the normal operating current, and may only need to be satisfied for a few seconds at a time. Because the EDLC handles high currents, the battery does not experience deep discharges, and thus its life is extended.

Batteries in off-grid PV hybrid storage systems, when paired with EDLCs, experience less discharge depth, which translates to longer lifetimes and smaller batteries. In addition, the hybrid arrangement increases the reliability of the PV system on both large and small timescales. Increased power quality ensures that fluctuations in load demand will not adversely affect the stand-alone system, making off-grid PV systems a viable option where they might not be without energy storage technology.[14]

4.11.1.3 Flywheels

Energy in a flywheel is stored in the form of rotating kinetic energy, in contrast with batteries and SMES where energy is stored in chemical and electrical form, respectively. Peak power for flywheels depends on the application, ranging from kilowatt to gigawatt scales. One major advantage of flywheels is long life, longer than 20 years and independent of depth of discharge; that is, unlike electrochemical batteries, flywheels operate equally well whether discharges are few and deep or frequent and shallow.

The most mature commercial application for flywheels is providing an uninterruptable power supply, taking advantage of the flywheel's high power density and fast recharge time. Short bursts of power are administered when power line disturbances occur, 80% of which last for less than a second. For some applications, a flywheel can coast for over an hour to zero charge.

4.11.2 Bridging Power

Batteries can be used for both power-quality and bridging power applications. The most common ones are lead-acid batteries.

4.11.2.1 Lead-Acid Batteries

Sealed, valve-regulated lead-acid (VRLA) batteries have been the most common type of batteries in PV residential systems (Figure 4.61). There are two types of VRLA batteries: absorbed glass mat (AGM) and gelled electrolyte. The former store the electrolyte on a glass mat separator composed of woven glass fibers soaked in acid. The latter immobilize the electrolyte in a gel. Hybrid VRLA batteries encompass the power density of AGM design and the improved thermal properties of the gel design. Of all types of lead-acid batteries, the hybrid VRLA proved to be the technology best suited for PV stand-alone lighting systems.

Operational Principles of Pb-Acid Batteries

Discharge:

- PbO_2 is converted to $PbSO_4$ at the positive electrode, absorbing electrons from the electrolyte; the electrons flow out to an electric circuit.
- At the negative electrode, Pb is converted into $PbSO_4$, emitting electrons into the electrolyte.
- Sulfate ions, which are responsible for the conversion, are replenished from the electrolyte.
- Layers of $PbSO_4$ grow on both electrodes.

Recharge:

- $PbSO_4$ is converted into PbO_2 at the positive electrode, emitting electrons into solution.
- Elemental Pb is formed at the negative electrode.
- Electrolyte level (SO_4—) returns to previous levels.

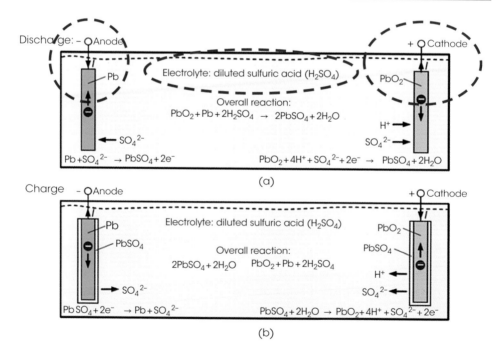

Figure 4.61 Pb-acid battery operational principles; shown anode, cathode, electrolyte, and associated reactions[16] (*Source:* Mertens[16]. Reproduced with permission of Wiley).

PV panels are not ideal sources for charging lead-acid battery because they generate power intermittently. One proposed method to extend the lifetime of the VRLA batteries is combining them with supercapacitors into a hybrid storage system utilizing the high power density, longer life cycle, and fast charge–discharge times of the supercapacitor to supply short bursts of power during times of peak demand and motor starting. The more energy-dense VRLA battery supplies energy continuously over longer periods. Incorporating a supercapacitor allows the battery to be sized according to the demands of normal operating current rather than to that of peak current while avoiding deep discharge, maintaining a high state of charge (SOC), preventing sulfation and stratification, and extending the battery's life. Though supercapacitors are expensive, cost reduction through technology development and market growth may enable such supercapacitor–VRLA battery hybrids affordably to provide more reliable storage for PV systems.

4.11.2.2 Lithium-Ion Batteries

At this point of time (mid-2016) among all battery technologies, lithium-ion (Li-ion) batteries have the largest potential for future development and implementation. Both their energy density and power density are high compared with other battery technologies (Figure 4.62), making them ideal for portable applications and a good candidate for PV-supporting applications as well.

Li-ion batteries have long lifetimes and low self-discharge rates. Electricity can be charged and discharged very quickly with high power output, with no memory effect. Round-trip efficiency is 90% or higher. A Li-ion battery stores electricity when voltage is applied to it causing Li-ions to travel from the metal oxide cathode through the electrolyte separator to the graphitic carbon anode. Electricity is discharged when the ions travel in the opposite direction; Figure 4.63 depicts this mechanism. The disadvantages of this technology include its high cost and sensitivity to extreme conditions.[17]

Despite the large power density of a Li-ion battery, like any battery it deteriorates when exposed to deep discharging and overcharging. In fact, much of the cost of Li-ion

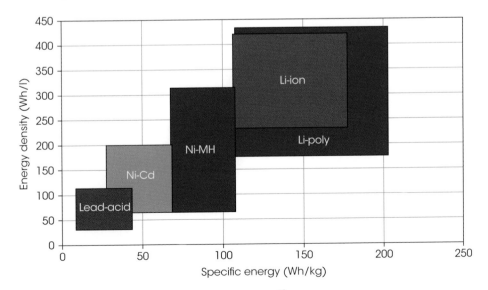

Figure 4.62 Battery types: energy density comparisons.[15]

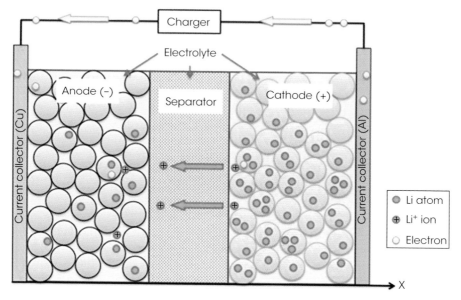

Figure 4.63 The basic working mechanism of a Li-ion battery (Energy and Power Group, University of Oxford).

batteries lies in the overcharge protection units to prevent such events from occurring. High temperatures can further decrease the battery's life.

However, despite its drawbacks, Li-ion battery technology offers an interesting solution to grid-scale energy storage. Due to their high storage capacities, fast charging rates, and relatively small sizes, Li-ion batteries are popular for use in electric vehicles. If implemented on a large scale, fleets of plug-in electric vehicles can offer not only as cleaner modes of transportation but also distributed sources of stored energy for the grid. In time, when their battery capacity is increased, large numbers of electric vehicles powered by batteries would add flexibility and stability to a smart grid and enable further penetration of renewable energy.

Tesla, which has been using Li-ion batteries in their cars, started in 2015 marketing a battery for stationary applications at a cost of $350/kWh for residential (10 kWh systems) down to $250/kWh for commercial (100 kWh) systems. The latter can also be used for community systems that connect several residential ones and take advantage of both the smoothening of the solar and the demand variability and the economies of scale from the larger battery systems.

Another rechargeable Li-based technology with potential automotive applications is the lithium-ion polymer (Li-poly) battery that is like a Li-ion battery, wherein the liquid electrolyte is replaced by a solid polymer electrolyte. The cost of Li-poly batteries currently is prohibitive, but their increased production may lower the cost in future. Hyundai announced plans to use Li-poly batteries in its HEVs, and an Audi A2 powered by Li-poly batteries recently set the record for distance traveled on a single battery charge.

4.11.2.3 Flow Batteries

In flow batteries, unlike conventional batteries, the charged electrolyte is stored in a separate tank and circulated when needed. This configuration essentially decouples the

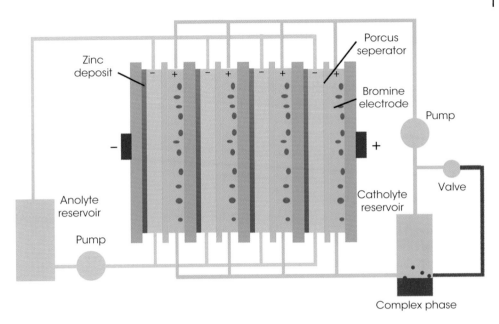

Figure 4.64 Schematic of a flow-assisted battery.

power and energy aspects of the batteries and allows them to be sized independently: the power is determined by the size of the electrochemical cells, and the energy storage is limited only by the size of the external storage tank and volume of electrolyte. It also avoids the problem of self-discharge present in most battery technologies. However, flow batteries typically have lower energy densities than most portable batteries when the storage and reactor tanks are accounted for; they also require using additional components, such as pumps and sensors.

There has been limited deployment of at least two types of flow batteries, vanadium redox and zinc bromine; other types are under development, such as polysulfide bromide ones. Redox flow batteries consist of two half cells, each containing dissolved species in different oxidation states, separated by an ion-exchange membrane. Hybrid flow batteries such as zinc bromine batteries (Figure 4.64) contain metallic species deposited in one of the half cells.

VRB Power (currently Prudent Energy) invented the vanadium redox battery energy storage system (VRB-ESS™) built on a 175 kW modular basis. Large-scale installations include a 1 MWh per 500 kW storage system for the China Electric Power Research Institute (CEPRI) as part of a project that includes 78 MW of wind capacity, 640 kW of PV capacity, and 2.5 MW of energy storage. In addition, the company is building in California vanadium redox battery systems with SunPower's PV systems, in coopera-tion with PG&E, KEMA, and Sandia National Laboratories.

Flow batteries are being developed on the residential PV system scale. HomeFlow is a 30 kWh/10 kW zinc bromine flow battery intended for such usage. The product is being developed by Premium Power, which strives to become the "Dell computers of flow batteries" by bringing the technology directly to homes through modular design and inexpensive manufacturing. Its product line also includes larger zinc bromine batteries with capacities/rated powers of 45 kWh/15 kW, 100 kWh/30 kW, and 2.8 MWh/500 kW that can be installed at the community level to curb peak power demand.

The key characteristic of flow batteries is their independent scalability: reactors can be scaled up in response to increasing power demand, while storage tanks and electrolyte solution can be added for more energy storage capacity to accommodate additional renewable generators. At the utility scale, where capacity and cost are more influential than volume requirements, flow batteries are a promising technology for renewable energy systems.

4.11.3 Energy Management Storage Technologies

4.11.3.1 Pumped Hydro Energy Storage

Pumped hydro energy storage (PHES) is the most widely used large-scale energy storage technology. It utilizes elevation difference between natural (or man-made) reservoirs to increase the potential energy of water by pumping it into the higher reservoir and later produce electricity by reversing the operation of the pump running it as a turbine. The schematics of a pumped storage plant are shown in Figure 4.65. In general, PHES plants have a round-trip efficiency of around 75% and can have discharge capacities of more than 20 hours. Projects may be practically sized up to 4000 MW and operate at about 76–85% efficiency, depending on design. Pumped hydro plants have long lives, on the order of 50–60 years. As a general rule, a reservoir having 1 km in diameter, 25 m deep, and an average head of 200 m would hold enough water to generate 10 000 MWh.

The earliest plant in the United States was built in the late 1920s, and the last pumped storage plant commissioned was in the 1980s, when environmental concerns over water and land use severely limited the ability to build additional pumped hydro capacity. Most of the PHES storage facilities were constructed during the 1970s and the 1980s,

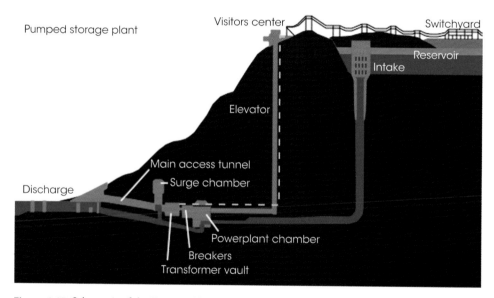

Figure 4.65 Schematic of the Raccoon Mountain Pumped Hydro Plant (Wikipedia).

Table 4.2 Energy storage technology cost estimates (Akhil *et al.*[20]).

Storage type	$/kW	$/kWh	Hours	Total capital ($/kWh)
CAES (100–300 MW underground storage)	590–730	1–2	10	600–750
Pumped hydro (conventional 1000 MW)	1300	80	10	2100
Battery (10 MW)				
Lead-acid, commercial	420–660	330–480	4	1740–2580
Sodium–sulfur (projected)	450–550	350–400	4	1850–2150
Flow battery (projected)	425–1300	280–450	4	1545–3100
Lithium-ion (small cell)	700–1250	450–650	4	2300–3650
Lithium-ion (large cell, projected)	350–500	400–600	4	1950–2900
Flywheel (10 MW)	3360–3920	1340–1570	0.25	3695–4313
Superconducting magnetic storage	200–250	650000–860000	1 second	380–489
Supercapacitors (projected)	250–350	20000–30000	10 seconds	300–450

Source: Ibrahim *et al.*[17]. Reproduced with permission of Elsevier.

and their capacity just in the United States grew to 20 GW. The oil crisis of 1973–1976 and associated fuel price increases catalyzed this growth as PHES were planned to use nighttime power from nuclear plants to satisfy day peak loads. Even though the capital cost of PHES was always higher than conventional generation, the difference was small till late 1980s. PHS started being less competitive since the 1980s as nuclear power deployment became standstill and CCGT capital costs and the price of natural gas were reduced. At this time, the capital cost of PHES is almost twice that of CCGT (the overnight capital cost of CCGT in 2006 was around $800–1100 compared with the estimated cost of $2100 of a 10 hours conventional PHES plant in 2009). An order of magnitude estimation of energy storage costs can be found in a handbook produced by the Electric Power Research Institute[18]; these are summarized in Table 4.2.

The reader is advised to look at the source for details of those estimates and keep in mind that cost may vary with the price of commodity materials and the location of a project.

4.11.3.2 Compressed Air Energy Storage

Compressed Air Energy Storage (CAES) converts grid electricity to mechanical energy in the form of compressed air stored in underground (or surface) reservoirs. The source of input energy can be excess off-peak electricity or renewable electricity coming from wind or solar farms. To convert stored energy back to electricity, the compressed air is released through a piping system into a turbine generator system after having been heated. When compression and expansion are rapid, the processes are near adiabatic; heat is generated during compression, and cooling occurs during expansion. The first is associated with large energy losses as compression to 70 atm can produce temperatures of about 1000°C, so necessitating cooling.

For large CAES plants, a large storage volume is required and underground reservoirs are the most economically viable solution. Such reservoirs can be a salt formation, an aquifer, or depleted natural gas field. When the volume confining the air is constant, pressure fluctuates throughout the compression cycle. Constant pressure operation in hard rock mined caverns is achievable by using a head of water applied by an above-ground reservoir. For smaller CAES plants (e.g., <5 MW), air can be stored in above-ground metallic tanks or large on-site pipes, such as those designed for carrying natural gas under high pressure.

CAES Compressor Thermodynamic Operation

It is required to compress air from P_1 to P_4. Ideal isothermal compression is shown by *a-b-j-h*. Single-stage compression is shown by *a-b-c-k-h*.

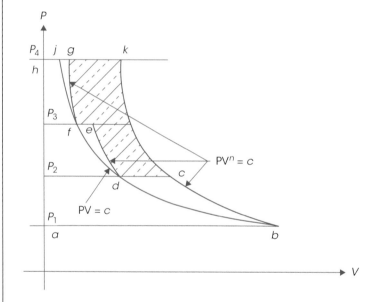

Assume a three-stage compressor process is used. The air is compressed from P_1 to P_2 ($a \rightarrow c$) and is transferred into a receiver and cooled to its original temperature ($c \rightarrow d$); then the air is transferred from the receiver to a second cylinder and compressed to P_3 ($d \rightarrow e$). The air is then transferred to a second receiver and cooled back to its original temperature ($e \rightarrow f$) and transferred again to a third cylinder and compressed to P_4 ($f \rightarrow g$).

The overall process is represented by curve *a-b-c-d-e-f-g-h*. The cooling brings the process closer toward the ideal isothermal (constant temperature) curve. The saving in work done per cycle is identified by the shaded area.

A typical CAES power plant comprises a compression and a generation train connected through a motor/generator device. During the compression mode, electricity runs dynamic compressors that compress air at pressures of 70 bars or more. Because of the high pressure ratio required, compression takes place in a series of stages separated by cooling periods. Cooling the air is necessary to reduce power consumption and meet the cavern's volume requirements. The higher the number of stages, the greater is the efficiency attained; however, this increases the cost of the system. During the expansion mode, motor operation stops and clutches engage the generator drive. Air is released to run the expanders after having first being heated in properly designed combustors. Heating the air assures high efficiency and avoids damaging of the turbomachinery due to low temperatures resulting from the rapid expansion of air and the Joule–Thomson effect. A recuperator sited after the exit from the expanders recovers some of the energy of the heated air before it is released to the atmosphere. Even though fuel is needed to run a CAES power plant, the input for a certain power capacity is around 65% less than the amount required to run a Gas Turbine (GT) because around two-thirds of the energy produced by a GT is used to run its compressor. Thus, when the compressors are fed by renewable electricity, the emissions of a CAES power plant are 35% of those produced by a GT of the same capacity.

CAES Efficiency (McIntosh Plant)

$$\text{Heat rate}, HR = \frac{Q_{in}}{E_{out}} = 4550 \text{ kJ/kWh} \left(4800 \text{ Btu/kWh}\right)$$

$$\text{Electricity ratio} = \frac{E_{in}}{E_{out}} = 0.8$$

$$\text{Gross efficiency} = \frac{E_{out}}{E_{in} + Q_{in}} = 50\%$$

$$\text{Round trip efficiency} = \frac{E_{out}}{E_{in} + n_{NG}Q_{in}} = 78\%$$

where

E_{in} = electrical energy to compressor
n_{NG} = natural gas thermal to electricity efficiency approximately 48%
Q_{in} = thermal energy in fuel
E_{out} = generated energy in turbine

Currently, two CAES power plants are operating. The world's first facility is the Huntorf CAES plant that has operated since 1978 in Bremen, Germany. It is a 290 MW facility, designed to provide black-start services to nuclear power plants located nearby, along with spinning reserves and VAR support as well as cheap off-peak electricity. It stores air up to 1000 psi (68 atm) in two depleted salt caverns located 2100 and 2600 ft under the ground; it offers up to 4 hours of power generation. The second CAES plant is a 110 MW power plant operating in McIntosh, Alabama, since 1991 (Figure 4.66). It pressurizes air to 1100 psi (75 atm) and has electricity generation cycle of up to 26 hours between full charges. The McIntosh plant also has a heat recuperator in the expansion train that reduces fuel consumption by 25% compared with the Huntorf plant that does not include recuperation.

Deregulation and the current structure of electricity markets now allow storage technologies to participate in the electricity market and profit from their operation. As an example, the NYISO includes markets for installed capacity, energy, and ancillary services and for preventing transmission congestion. CAES power plants have the

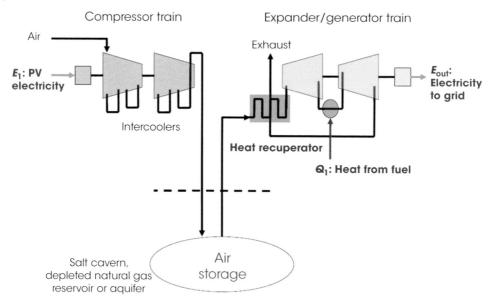

Figure 4.66 Schematic of a CAES plant source (*Source:* Compressed Air Energy Storage (CAES) Conference[19]).

following attributes that make them suitable for large-scale, diurnal, multiday, and seasonal energy storage:

1) CAES and pumped hydro are the only storage technologies that offer the high capacities (>100 MW) for long periods.
2) CAES has an approximately flat heat rate at part-load conditions.
3) CAES has the lowest annual estimated cost for an 8 hours discharge system that includes the cost for O&M, electricity used during the charging cycle, fuel requirements (if nonadiabatic CAES systems are used), and capital carrying charges.
4) CAES has the potential for the lowest levelized annual cost of storing and delivering power, due to inexpensive air storage and greater operational efficiency than other systems.
5) A CAES power plant can participate in the capacity market, provide load-following services and energy arbitrage, operate as an ancillary reserve, and offer VAR control or energy shift for renewable applications. More than one of these services are possible at the same time.

Self-Assessment Questions

Q4.1 What is the cost "learning curve" for cumulative PV production?

Q4.2 What is currently the installed system cost for large solar farms ($/kW), and what is the corresponding electricity cost ($/kWh) for production in south California and other southwest regions of the USA?

Q4.3   Which country has been the biggest customer for PV during 2003–2013, and how was this accomplished?

Q4.4   How can variability of PV output due to cloud coverage be smoothened (reduced)?

Q4.5   What is the effect of the size of a large solar farm on the variability of the total power output from the plant?

Q4.6   What options exist for managing cloud-induced fluctuations in PV power output?

Q4.7   What are the general characteristics of "grid-friendly" PV power plants?

Q4.8   What is the function of a controller in a PV stand-alone system?

Q4.9   What is the temperature effect in Pb-acid batteries?

Q4.10   How does compressed air energy storage work?

Q4.11   Indicate next to each of the seven following types of storage listed below if they are typically used for providing storage for: A. second/minute fluctuations; B. up to 4 hours; C. diurnal; D. seasonal.
1) Capacitors
2) Li-ion batteries
3) Pb-acid batteries
4) Flow batteries
5) Flywheels
6) CAES
7) Pumped hydro

Problems

4.1   A PV system consists of 48 panels (modules) supplying 15 V and 4 A each. How do you recommend that they are connected to make use of two inverters? What inverter specification would you recommend?

4.2   An energy efficient single family house in New York is to be fitted with a rooftop PV array that will displace (on an annual basis) all the electricity that the home uses. How many kW (dc, rated) will be required and what roof area will be needed if the modules are 17% efficient? Insolation on latitude tilt plane is about 4.5 kWh/m^2/day.

4.3   Develop an estimate of the performance of the system above using the PVWatts Calculator (accessed at: http://pvwatts.nrel.gov/) and typical meteorological year (TMY) data at your location.

4.4   Develop an estimate of the performance of the system above using NREL's System Advisor Model (SAM) (https://sam.nrel.gov/), hourly irradiation data from the

National Solar Radiation Database (NSRD) and load date for a full year for your location and residence.

4.5 What size (in watts) of a PV system do you need to produce 1000 kWh of DC electricity a year at a remote location that has an average of 5 peak sun hours per day?

4.6 Find the necessary number of modules to supply 800 W power accounting for: 95% charge controller efficiency; 2% wiring loss; 10% battery charge–discharge loss; 4% soiling loss. (Assume 12 V 100 W STC modules.)

4.7 At a remote area people who lack electricity want to take advantage of solar power to pump underground water. Size the necessary battery bank (how many batteries needed?). Assume a PV system can be sized accordingly.

Necessary power 1080 W.
Operating time 18 h/day.
Desired storage time is 1 day.
Depth of discharge is 80%.
One battery gives 12 V and 225 Ah.

4.8 A pump is used to draw water from a water reservoir 100 m deep in the ground. Assume water temperature of 15°C, density of 1 l/kg, and pump efficiency of 95% (5% in thermal losses). (a) Determine the power required for a water flow of $1 \, m^3/$ min through a pipe of a 12 cm diameter. (b) Determine the exit velocity if the water exits through a 2 cm nozzle at the end of the pipe.

4.9 A compressor operating at steady state draws air at atmospheric pressure and 20°C and compresses it to 10 bar. At the exit the temperature is 520 K and air velocity is 3 m/s. The rate of heat transfer from the compressor to its surroundings is 180 kJ/min. (a) Assuming ideal gas law, calculate the power input to the compressor and (b) repeat the calculation using a suitable compressibility factor and provide the reference.

Answers to Questions

Q4.1 A relationship of cost decline with every doubling of installed capacity; it captures economies of scale and efficiency improvements as the market grows.

Q4.2 As of mid-2017, installed cost has been around $1000/kW corresponding to electricity cost of 5–6 cents/kWh in the US-SW.

Q4.3 Germany, because it enacted a law giving financial incentives to renewable energy.

Q4.4 By geographical diversity, creating large grid balancing areas.

Q4.5 The greater the size of the PV plant, the greater the land occupied and this smoothens the shading effect of the clouds.

Q4.6 Other than geographical diversity, batteries and other storage systems can help.

Q4.7 They can provide voltage support, active power control, ramp-rate control, fault-ride through, and frequency response.

Q4.8 Optimizing PV performance and providing optimal battery charging while protecting the batteries from overcharging.

Q4.9 Pb-acid batteries function best at room temperature (~20°C). Operating a battery at elevated temperatures may improve performance but prolonged exposure will shorten life. If, for example, a battery operates at 30°C instead of 20°C, the cycle life is reduced by 20%. At 40°C, the loss of life expectancy can be 40%. On the other hand, the performance of batteries drops drastically at low temperatures. At −20°C most batteries stop functioning.

Q4.10 It uses electric energy to pump air into caverns (compression) and releases pressurized air into turbines (expansion) to generate electricity.

Q4.11 1A; 2B; 3B; 5A; 6B–C–D; 7B–C–D

References

1 S.R. Wenham *et al. Applied Photovoltaics*, 2nd edition, UNSW Centre for Photovoltaic Engineering: Sydney (2009).

2 F. Antony *et al. Photovoltaics for Professionals*, Earthscan: London (2007).

3 K. Zweibel *et al.* A solar grand plan, *Scientific American*, 298(1), 64–73 (2008).

4 V. Fthenakis *et al.* The technical, geographical and economic feasibility for solar energy to supply the energy needs of the United States, *Energy Policy*, 37, 387–399 (2009).

5 PVTECH, *SolarServer*. PVresources.com (Accessed on August 23, 2017).

6 U.S. Department of Energy, *SunShot Vision Study*, (February 2012). http://energy.gov/eere/sunshot/sunshot-vision-study (Accessed on August 23, 2017).

7 M. Morjaria *et al.* A grid-friendly plant the role of utility-scale photovoltaic plants in grid stability and reliability, *IEEE Power and Energy Magazine*, 12(3), 87–95 (2014).

8 V. Gevorgian and B. O'Neill. *Advanced grid-friendly controls demonstration project for utility-scale PV power plants*, NREL/TP-5D00-65368. (2016). http://www.nrel.gov/docs/fy16osti/65368.pdf (Accessed on August 23, 2017).

9 Z. Peng *et al.* 3D cloud detecting and tracking system for solar forecast using multiple sky imagers, *Solar Energy*, 118, 496–519 (2015).

10 R. van Haaren *et al.* Empirical assessment of short-term variability from utility scale solar-PV plants, *Progress in Photovoltaics: Research and Applications*, 22(5), 548–559 (2014).

11 M.J. Perez *et al.* On the spatial decorrelation of stochastic solar resource variability at longer timescales, *Solar Energy*, 117, 46–58 (2015).

12 P. McCoy and J. Vaninetti. It's doable, *EnergyBiz*, 5(2), 40 (2008).

13 T. Nikolakakis and V. Fthenakis. The optimum mix of electricity from wind- and solar-sources in conventional power systems: evaluating the case for New York state, *Energy Policy*, 39(11), 6972–6980 (2011).

14 P. Delholm *et al.* Bright future, solar power as a major contributor to the U.S. grid, *IEEE Power and Energy Magazine*, 11(2), 22–32 (2013).

15 V. Fthenakis and T. Nikolakakis. Storage needs and options for solar renewable energy, *Comprehensive Renewable Energy*, 1, 199–211 (2012).

16 K. Mertens. *Photovoltaics, Fundamentals, Technology and Practice*, John Wiley & Sons, Ltd., West Sussex (2014).

17 H. Ibrahim *et al.* Energy storage systems: characteristics and comparisons. *Renewable and Sustainable Energy Reviews*, 12, 1221–1250 (2008).

18 A.A. Akhill *et al. DOE/EPRI 2013 Electricity Storage Handbook in Collaboration with NRECA, SAND2013-5131*, Sandia National Laboratories: Albuquerque (2013).

19 Integrating Wind-Solar-CAES. *2nd Compressed Air Energy Storage (CAES) Conference and Workshop*, Center for Life Cycle Analysis, Columbia University, October 20–21 (2010). http://www.clca.columbia.edu/CAES2workshop_proceedings.pdf (Accessed on August 24, 2017).

20 A.A. Akhil *et al. DOE/EPRI Electricity Storage Handbook in Collaboration with NRECA*, Sandia National Laboratories: Albuquerque (2015). http://prod.sandia.gov/techlib/access-control.cgi/2015/151002.pdf (Accessed on September 19, 2017).

5

Stand-Alone PV Systems

5.1 Remote and Independent

Imagine living in a remote farmhouse, supplied with electricity by an elderly diesel generator and a long way from the nearest electrical grid. The generator needs replacing—but you dislike polluting fumes—the cost of diesel fuel always seems to be rising, and the local electricity utility has just quoted a large sum to connect you to the grid network. How about photovoltaics (PV) as an alternative? What are the possibilities and pitfalls if you decide on a completely independent stand-alone system?

Figure 5.1 shows a possible scheme. The farmhouse roof faces east–west, making it unsuitable for mounting a PV array, so the modules (1) are placed on an adjacent field, south-facing and tilted at an optimum angle. They are interconnected at the array and the DC electricity flows via an underground cable into the farmhouse. The site is windy and exposed so it is decided to include a wind generator (2) in the system. The PV array and wind generator have separate *charge controllers* (3) to regulate the flow of current into a battery bank (4) that acts as an energy store. This is essential because the energy generated by wind and PV is variable and does not coincide with household demand (especially at night in the case of PV!). The battery bank voltage is normally 12 or 24 V DC, but may be higher in a large system. An inverter (5), connected to the battery bank, produces AC at the national supply voltage and frequency (e.g., 220 V at 50 Hz in Europe, 110 V at 60 Hz in North America and Japan) and supplies the household loads via a fuse box (6), allowing you to use standard AC appliances (7). Note that there is no electric fire in the scheme; generally speaking renewable electricity is too precious to be used for space heating, and an alternative such as a wood-burning stove is more suitable.

You may like to contrast this scheme with the grid-connected home illustrated in Figure 4.1. Apart from the wind generator, the major difference between the two systems is the replacement of the grid by a battery bank. Grid connection is relatively straightforward. The PV array in Figure 4.1 is not required to supply all of the household's needs; indeed in most cases it supplies considerably less, and the homeowner pays the electricity company for the shortfall. We may think of the grid as an infinite "source and sink," able to supply or accept any amount of electricity on demand, at any time of day or night. But our stand-alone system enjoys no such luxury. The battery bank is a strictly finite "source and sink," and its capacity needs careful consideration. Too little capacity, and the electricity supply is unreliable; too much,

Electricity from Sunlight: Photovoltaic-Systems Integration and Sustainability, Second Edition.
Vasilis Fthenakis and Paul A Lynn.
© 2018 John Wiley & Sons Ltd. Published 2018 by John Wiley & Sons Ltd.
Companion website: www.wiley.com/go/fthenakis/electricityfromsunlight

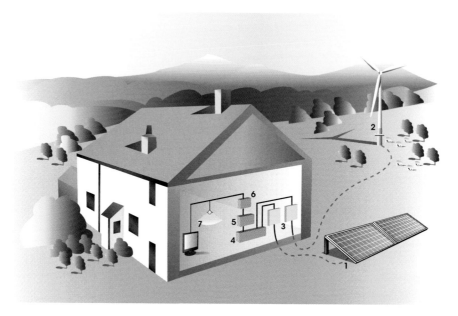

Figure 5.1 Remote and independent: a stand-alone system for a farmhouse.

and the capital cost of batteries becomes excessive. Being autonomous has its problems! As we shall see, the "sizing" of a PV generator and battery bank to provide an acceptable balance between reliability and cost is a major challenge to the designer of a stand-alone PV system.

As far as the PV modules are concerned, a few points should be added to the account given in Chapter 3. Historically, most PV modules were designed to be suitable for battery charging, and some still are. Typically a crystalline silicon module containing 36 cells connected in series gives an open-circuit voltage of about 20 V and a maximum power point (MPP) at 17 V under bright sunlight. This is well suited to direct charging of a 12 V lead-acid battery—the most common type—that reaches about 14.5 V as it approaches full charge. The surplus module voltage is needed to overcome small voltage drops across the blocking diode and charge controller and to ensure effective charging in reduced sunlight or at high module operating temperatures.

PV modules that are suitable for 12 V battery charging may also be used in grid-connected systems. A good example is the Swiss PV array composed of 36-cell modules previously shown in Figure 3.1. And Figure 2.1 illustrated 72-cell modules that would be suitable for charging a 24 V battery bank. Of course, grid connection favors higher system voltages with strings of series-connected modules, whereas battery charging requires modules connected in parallel or series–parallel. In recent years the increasing dominance of grid-connected systems has led manufacturers to offer a wider choice of module sizes and voltages, including many that are not suitable for direct battery charging—a point to be borne in mind when selecting modules for a stand-alone system.

The system shown in Figure 5.1 is fairly sophisticated, involving two sources of renewable energy, battery storage, and an inverter to provide continuous AC power to the household. Various other stand-alone PV schemes are possible, depending on the application. Starting with the simplest, they are the following:

- *Without battery storage or inverter.* A PV module can supply a DC load directly. A simple example is the type of small solar fountain that floats on a garden pond: the PV sends its current directly to a DC motor driving a pump. The fountain plays only when the sun shines. A more serious application is water pumping for village water supply, irrigation, or livestock watering, where a PV array supplies a DC motor driving a pump that delivers water to a holding tank whenever the sunlight is sufficiently strong.
- *With battery storage, without inverter.* Low-power consumer products such as solar calculators and watches belong to this category. So do solar-powered garden lights. Moving up the power scale, a variety of electrical loads, including low-energy lights and a small TV, may be run directly from DC batteries. Many of the solar home systems (SHSs) used in developing countries to supply a small amount of PV electricity to individual families are of this type. A typical SHS comprises a battery, a charge controller, and a single PV module (see Figure 1.15). Other examples are DC systems for remote telecoms, security systems, and medical refrigeration.
- *With inverter, without battery storage.* This type of system produces AC power from a PV module or array and is appropriate when AC electricity is useful at any time of day. For example, AC motors are sometimes used for pumping schemes in preference to DC motors because of their rugged reliability and cheapness (although this must be set against the cost of inverters).

In conclusion, stand-alone PV systems encompass a wide variety of applications with power levels from the miniscule up to a hundred kilowatts or more. Until the 1990s they were the bread-and-butter business of most PV companies, but the recent huge increase in grid-connected systems means that today they account for less than 5% of annual PV module production. However this is not to diminish their huge importance for the families, communities, and businesses that rely on them, including millions of people in developing countries (Figure 5.2). In countries such as the United States and Australia, there are robust markets for systems in the $1-20\,kW_p$ range, installed on remote farms and in holiday homes that formerly relied on diesel generators (and may still use them for backup supply). So-called *hybrid systems* integrating PV with wind, hydro, biofuel, or diesel generators are attractive where there are large seasonal variations in sunlight levels. And at the top end of the power scale come independent mini-grids in remote mainland areas or on islands, often supplied by several power sources including PV, which provide electricity to whole communities. We shall return to these and other applications later in this chapter. But before venturing out into the wider world, we need to discuss batteries, charge controllers, and inverters and explain how a stand-alone system is designed for cost-effectiveness and reliability.

5.2 System Components

5.2.1 Batteries

Reliable energy storage is crucial to most stand-alone PV systems. Without it operation of the system is confined to daylight hours when the sunlight is sufficiently strong; with it the user becomes independent of the vagaries of sunlight and can expect electricity by

Figure 5.2 Off the grid: PV water pumping for a Moroccan village (*Source:* Reproduced with permission of EPIA/Isofoton).

night and day. Many new types of storage battery have come on the market in recent years, including nickel–cadmium, nickel–metal–hydride, and lithium-ion, but since the great majority of present stand-alone PV systems use the more traditional lead-acid type, we shall concentrate on it in this section.

You are probably familiar with 12 V vehicle batteries, and at first sight they might seem suitable for storing the output of a PV array. But there are important differences between the duty cycle of a standard vehicle battery and a PV storage battery. A vehicle battery's most arduous duty is to supply large currents, typically hundreds of amps, for a very short time to the engine's starter motor. The battery is not supposed to be substantially discharged, except on rare occasions. But a PV battery delivers smaller currents for much longer and must routinely withstand *cycling*, in other words going through many hundreds or even thousands of charge–discharge cycles without damage. Its duty is rather similar to that of a "leisure" battery used for running electrical appliances in caravans and boats. But the particular requirements of PV systems in terms of efficiency, reliability, and durability have led manufacturers to develop specialized deep-cycle batteries for the PV market. It should be added that the cheapness and universal availability of standard vehicle batteries means that in practice they are often used in low-power SHSs in developing countries, and with reasonable success due to the low current levels involved.

High-quality lead-acid batteries for stand-alone PV systems must have long working lives under frequent conditions of charge and discharge. Since PV electricity is precious, especially during long cloudy periods or in winter, the batteries must also display low self-discharge rates and high efficiency. Self-discharge rates of around 3% per month are fairly typical. Efficiency is assessed in three ways:

- *Coulombic* or *charge efficiency*, the percentage of charge put into a battery that may be retrieved from it, typically 85% for lead-acid.
- *Voltage efficiency*, reflecting the fact that the voltage when discharging is less than when charging, typically 90%.
- *Energy efficiency*, the product of coulombic and voltage efficiencies, typically 75%. (Unfortunately some manufacturers quote the coulombic efficiency as "battery efficiency," which can be misleading.)

We see from these figures, and especially the one for energy efficiency, that even high-quality lead-acid batteries cause substantial energy losses in a stand-alone system. Not that all the energy produced by a PV array has to go through the battery charge–discharge process: during periods of strong sunlight, the batteries may be fully charged much of the time, and the PV electricity can be passed straight to the loads.

So far we have talked about 12 V batteries. But as you are probably aware, a 12 V battery is made up of six electrochemical cells connected in series, each with a nominal voltage of 2 V. A 6 V battery contains three such cells. High-capacity cells may also be purchased individually and connected in series. Each has positive and negative electrodes made of lead alloy, in an electrolyte of dilute sulfuric acid. Two main categories of cells (and batteries) may be identified:

- *Flooded* or *wet*, using a liquid electrolyte that must be regularly topped up with distilled water. Adequate ventilation must be provided for hydrogen given off during charging.
- *Sealed* or *valve regulated*, sealed with a gastight valve, only allowing gas to escape in the event of overpressure. In normal operation the comparatively small amounts of hydrogen and oxygen produced during charging are recombined to form water, so no topping up is required. An alternative type of sealed battery uses gel electrolyte. In general, sealed batteries require a strict charging regime, but need very little maintenance.

Batteries recommended for multiple cycling in PV systems often have special electrodes in the form of tubular plates. If not discharged with more than 30%, they typically survive several thousand charge–discharge cycles; if regularly discharged by 80%, about a thousand cycles.

The capacity of a cell or battery is normally quoted in *ampere hours* (Ah), which is the product of the current supplied and the time for which it flows. For example, if a fully charged 12 V battery can provide 20 A for 10 hours, its capacity is 200 A h (unfortunately many people refer to such a battery as "200 amp"). And since its voltage is 12 V, the total energy stored is $200 \times 12 = 2400$ W h or 2.4 kWh.

However, it is important to realize that the capacity and energy efficiency of a battery depend on the rate at which it is discharged. The faster the discharge, the lower the capacity. Therefore, when a manufacturer quotes a battery capacity as 200 A h, this refers to a particular discharge time such as 10 hours, and this should be specified. The capacity is said to be 200 A h *at the 10 hour rate*. In general we get the most energy from a battery by discharging it as slowly as possible. A 100 hour rate is often considered

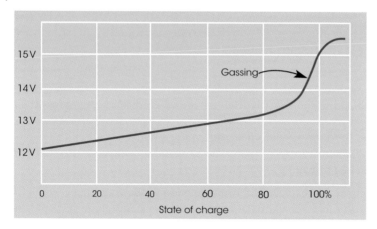

Figure 5.3 Typical charging characteristic of a 12V lead-acid battery.

more relevant to PV applications. Battery capacity also depends significantly on temperature. The rated capacity normally applies to 20°C and reduces by about 1% for every degree drop in temperature.

We now consider how the voltage of a lead-acid battery varies during charge and discharge. This is very important because, as we shall see in the next section, the charge controller used to regulate current flow from a PV array into a battery (or battery bank) uses voltage as a "control signal" to protect it from damage and prolong its working life. Once again we can use our 12 V battery as an example.

When a battery is put on charge at constant current, its voltage varies in the manner shown in Figure 5.3. Initially close to 12 V, it rises steadily as the *state of charge (SOC)* increases. In the final phase it increases more rapidly, reaching over 14 V as full charge (SOC = 100%) is approached. If the battery is of the flooded type, this last phase is accompanied by *gassing* in the liquid electrolyte, producing free hydrogen and oxygen. Excessive gassing can occur if charging is continued and may cause damage to the plates; it is very important to provide adequate ventilation to avoid the risk of explosion. However occasional, controlled, overcharging known as *equalization charging* is helpful as the gassing tends to stir up the electrolyte and prevent stratification into different levels of acid concentration. Note that overcharging must always be avoided in sealed batteries, and equalization is not relevant.

A good charging scheme, which helps keep the battery in top condition, is to provide an initial *boost charge* using all the available current, then, as the SOC approaches 100%, an *absorption charge* at constant voltage and low current; and finally a *float charge* to keep the battery gently topped up. Of course, in a PV system dependent on variable sunlight, with none at night, we cannot expect an optimal charging regime. We return to this point a little later.

We next consider what happens during discharge. Figure 5.4 shows typical voltage characteristics when our 12 V, 200 Ah battery is discharged at constant current. The curve labeled *10 hours* is for discharge at 20 A for 10 hours, which reduces the voltage from its starting value down to about 11 V, the point at which the manufacturer recommends disconnection of the load to prevent damage. Note that, at this point, the amount of charge used is 100%—the battery's full nominal capacity. But if we discharge

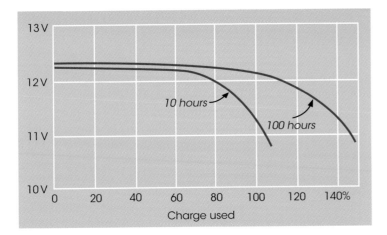

Figure 5.4 Typical discharge characteristics of a 12V lead-acid battery.

it at the slower rate of 2A for 100 hours, we get the curve labeled *100 hours*. The voltage holds up better, and the total available charge is substantially increased, emphasizing once again that the usable capacity of a battery depends significantly on the rate of discharge. It is also very important to note that severe overdischarge of a lead-acid battery, or allowing it to remain at a low SOC for lengthy periods, should be avoided whenever possible. In a wet battery the main danger is *sulfation*, the formation of large lead sulfate crystals on the plates, leading to damage and loss of capacity.

In a practical PV system, we cannot expect charging and discharging to occur at constant current or in regular cycles of constant depth. The situation is far more complicated and depends on the availability of sunlight compared with the user's demands for electricity. In general we may identify *daily* fluctuations of sunlight and demand and *seasonal* fluctuations. In sunny summer weather the battery bank is likely to spend more of its time close to full charge (SOC = 100%), with relatively small daily reductions due to demand; but under overcast conditions, or in the winter months, the electricity consumption pattern may lead to periods of low SOC with the risk of supply cutoff. Annual records of charge–discharge cycles in a PV system often appear somewhat random and irregular. Nevertheless, the main points outlined previously are useful guides to the performance, care, and maintenance of lead-acid batteries, pointing to the ways in which they may be protected by suitable charge control strategies.

5.2.2 Charge Controllers

A charge controller is used to regulate the flow of current from the PV array into the battery bank and from the battery bank to the various electrical loads. It must prevent overcharging when the solar electricity supply exceeds demand and overdischarging when demand exceeds supply. Various subsidiary control and display functions, depending on the price and sophistication of the unit, are included to protect the batteries from damage and to ensure an operating regime that maximizes their performance and length of life. Batteries are an expensive part of most stand-alone systems, especially those required to provide a highly reliable electricity supply day and night, so the relatively modest cost of a good charge controller is money well spent.

In the previous section we saw how the voltage of a battery changes during charging and discharging and noted that it is used as a control signal to regulate current flow. The two paramount tasks of the charge controller are prevention of battery overcharging and excess discharging. Overcharging is avoided by disconnecting the PV input whenever the battery voltage reaches an *upper set point*, normally preset at 14.0 V for float charging, 14.4 V for boost charging, and 14.7 V for equalization charging of a flooded (wet) 12 V battery. Excess discharge is prevented by disconnecting the load and/or giving a warning by light or sound whenever the voltage falls to a *lower set point*, normally about 11 V. Between these extremes charging and discharging continue in accordance with the amount of sunlight falling on the PV array and the demands of the load.

Ideally, a charge controller continually estimates the battery SOC and uses it to regulate the current accepted from the PV array. Actually, this is more difficult than it sounds because SOC is not simply related to instantaneous battery voltage, but depends on past history. For example if a battery has been supplying load current for some time and its voltage has fallen, then on disconnection it slowly recovers, even without further charging. Conversely if it has been on charge for some time and the voltage has risen, when charging ceases, it slowly falls back to a lower level. In other words the voltage signal detected by the charge controller is not a straightforward indicator of SOC. Effective controller algorithms must take past history as well as present voltage into account in assessing SOC and select boost, float, or equalization charging accordingly.

A closely related issue is that of *hysteresis*. When the upper set point is reached and the PV array is disconnected to prevent overcharging, the battery voltage immediately starts to fall back, even if no load is connected. How far should it be allowed to fall before reconnection? Too much, and there will be long interruptions to charging; too little, and there will be frequent on/off oscillations. So the gap, or hysteresis, between disconnect and reconnect voltages is a compromise that must be chosen carefully. A similar situation arises at the lower set point. After disconnection at 11 V, the voltage must be allowed to recover by a reasonable amount before automatic reconnection.

Charge controllers come in many shapes and sizes. In the case of a low-power SHS based on a single PV module and 12 V battery supplying a few low-energy DC lights and a small TV, a simple unit to control a few amps of current at 12 V is all that is required. Figure 5.5 shows the external circuit connections for such a unit. Typically, there is a

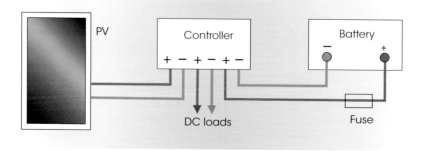

Figure 5.5 A simple scheme for a low-power solar home system (SHS).

row of six terminals, one pair each for the PV, DC loads, and battery, making installation very straightforward. Note that a fuse has been included close to the positive battery terminal, generally a wise safety precaution in case of a short circuit.

Moving up the power scale, suppose we have a $1\,kW_p$ PV array feeding a 24 V battery bank with a peak solar current of about 30 A. A suitable controller is likely to offer a number of features such as

- Choice of flooded or sealed lead-acid batteries
- Protection against reverse polarity connection of PV modules or batteries
- Automatic selection between boost, float, and equalization charging regimes, depending on the estimated SOC of the battery bank
- Protection against battery overcharging and deep discharging, excessive load currents, and accidental short circuits
- Prevention of reverse current at night
- Display of such parameters as battery voltage and/or estimated SOC, PV, and load currents and warning of impending load disconnection

The cost of the unit will clearly depend on how many features are included, and as we move towards the top of the power range, protection and monitoring functions become ever more important and sophisticated.

How do charge controllers perform their central task and regulation of current flow into and out of a battery bank? There are three basic designs on the market: *series* controllers, *shunt* controllers, and *maximum power point tracking* (*MPPT*) controllers. Once again we will illustrate ideas and values using a 12 V system, but they apply equally to higher system voltages if voltage values are scaled up in proportion.

The main functional elements of a series controller are illustrated in Figure 5.6. It includes an electronic switch known as the *low-voltage disconnect* (*LVD*) to prevent battery damage if the voltage falls below some critical value, normally chosen at 11 V. The diode is included to ensure that reverse current cannot flow back into the PV at night. And to the left of the figure, a second electronic switch (*S*), usually a MOSFET, has the vital task of overseeing the charging of the battery. When *S* is closed the PV current is sent to the battery; when *S* is open charging is interrupted. In most modern designs the required switching sequence is achieved by a subtle process known as *pulse width modulation* (*PWM*). Current is released to the battery in rapid pulses of

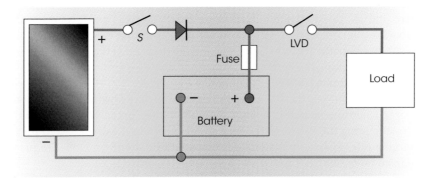

Figure 5.6 Series charge control.

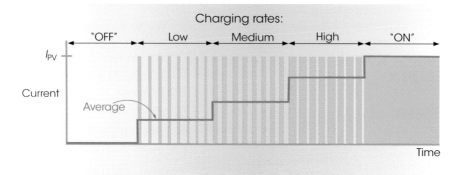

Figure 5.7 Battery charging with pulse width modulation (PWM).

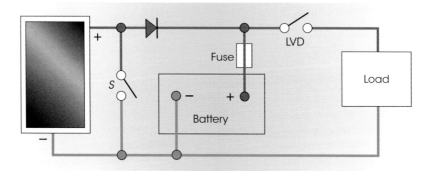

Figure 5.8 Shunt charge control.

variable width so that the *average* current, which determines the charging rate, is constantly adapted to take account of the battery's SOC. This is explained in Figure 5.7. The charging rate can be varied continuously between "OFF" when the battery is fully charged and "ON" when the available solar current is all sent to the battery. In the "OFF" condition the pulse width is zero (in effect, no pulses); in the "ON" condition it is maximum (pulses contiguous). Three intermediate pulse widths are shown as examples of different charging rates (low, medium, and high). Note that the current switches between zero and I_{PV} the available output from the PV array. The clever part of the PWM approach is, of course, to design a control algorithm that continuously changes the pulse width in sympathy with the SOC, making the best use of the PV's output while at the same time protecting the battery.

The operation of a shunt controller is illustrated by Figure 5.8. Here, the electronic switch S is connected across the PV array rather than in series with it, so the battery receives charge when the switch is open. Charging is interrupted when the switch is closed, short-circuiting the PV. Most modern shunt controllers also use PWM to regulate the charging rate. Inclusion of a diode is essential in a shunt controller to prevent the battery from being short-circuited when switch S is closed.

Supporters of the shunt concept often claim that it offers better charging efficiency than the series alternative. Switching losses (which should be small anyway) only occur

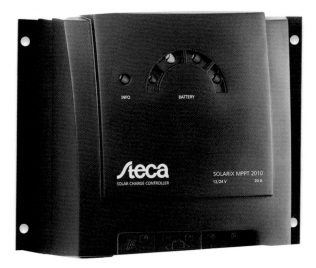

Figure 5.9 This MPPT controller can control a 12 or 24 V system with PV array power up to 500 W$_p$ and MPP voltages up to 100 V. With dimensions 19 × 15 × 7 cm, it weighs 900 g (*Source:* Reproduced with permission of Steca Elektronik GmbH).

when the solar current is being rejected; whereas in a series controller switching losses detract from power being sent to the battery. But there are two offsetting disadvantages. First, the shunt controller's switch, normally a MOSFET, needs a larger heat sink and must carry the full short-circuit current of the PV array, possibly in strong sunlight and for long periods. And second, although PV modules do not generally object to being short-circuited, there may be a risk of hotspot formation due to a "bad" cell or severe shading (as discussed in Section 3.2.1). In practice there are far more series than shunt controllers on the market, together with a few that combine the two design approaches to produce series/shunt hybrids.

We now move to the third basic type of design—the MPPT controller. Until fairly recently its more complex electronics and higher costs made it something of a niche market product, mainly reserved for the larger stand-alone systems. But as with many aspects of PV, technological advances coupled with volume production have reduced the cost sufficiently to make MPPT controllers attractive in a wide range of systems, even down to a few hundred peak watts. The potential advantage is clear: by working a PV array at its MPP, rather than at a voltage determined by the system's batteries, it is possible to extract considerably more energy and improve system efficiency.

Allowing the PV array to operate at a different voltage from that of the battery bank opens up another important possibility. As we pointed out in Section 3.2.2, the rapid development of grid-connected systems, larger PV modules, and new thin-film technologies has tended to shift manufacture toward higher module voltages unsuited to the direct charging of batteries. Specifying an MPPT controller allows a wide choice of modules that could not be used with a more conventional series or shunt design (Figure 5.9).

The basic scheme of an MPPT charge controller is shown in Figure 5.10. The key element is a *DC-to-DC converter* that allows the PV module or array to operate at a

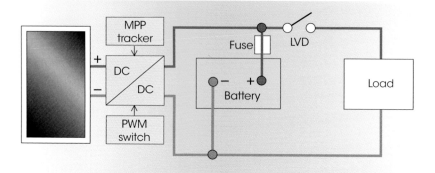

Figure 5.10 Extracting the most from a PV array: the MPPT charge controller.

different DC voltage from that of the battery or battery bank. Designs fall into two main categories: *boost converters* that raise the input voltage to a higher level and *buck converters* that reduce it. Buck converters are more common in PV applications, reducing the relatively high voltages of modern PV modules (or series-connected modules) to the lower voltages of battery banks. DC-to-DC converters have undergone extensive development in recent years, and their ability to "transform DC" finds many applications in electronic engineering. In the case of an MPPT charge controller, the really innovative part is not the voltage change, but rather the ability to sense the MPP of the PV array as sunlight levels change, at different times of day and in variable weather. Typically this is achieved with an algorithm that performs continuous electronic tracking of the array's MPP together with periodic sweeps along its *I/V* characteristic to confirm that the true MPP is being detected rather than a local power maximum. The voltage on the input side of the converter is automatically adjusted to the MPP voltage.

One further feature is required, this time on the output side of the DC-to-DC converter. As in more conventional charge controllers of the series or shunt type, the available output from the PV array, at its new voltage level, is presented to the batteries as a train of rapid current pulses of variable width using a PWM switch. Once again, an advanced control algorithm ensures that charging rate is continuously adapted to the estimated SOC of the batteries.

5.2.3 Inverters

We described a stand-alone PV system for a remote farmhouse at the start of this chapter (see Figure 5.1). It includes an inverter, connected to the battery bank, for supplying AC to the various household electrical appliances. Of course, inverters are not required in systems having only DC loads, but when they are used, it is important to understand the special features required in independent, stand-alone units. As the stand-alone PV market develops, customers are increasingly opting for AC systems because they like the flexibility offered by a wide range of consumer electronics, household appliances, electric tools, and even washing machines. AC systems are also used by hospitals and remote telecommunications sites and for running machinery in small factories. Well over 50% of newly installed stand-alone systems are AC.

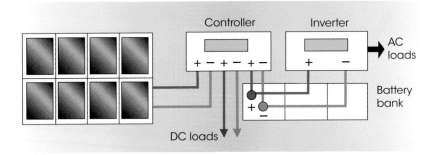

Figure 5.11 Typical connections for a mid-range stand-alone system.

The grid-connected inverters described in Section 4.2 are not suitable for stand-alone systems. An important difference is that whereas a grid-connected inverter must generate AC at precisely the right frequency and phase to match the grid supply, a stand-alone unit is not so constrained. It generates its own AC without any need to lock into a grid and is necessarily self-commutated. Although the generated waveform must suit the various AC loads, it need not satisfy an electric utility. And whereas a grid-connected inverter is supplied directly from a PV array and often performs MPP tracking, its stand-alone cousin is fed from storage batteries at a more or less constant DC voltage, leaving the task of extracting energy from the PV array to a charge controller that may itself work on the MPPT principle.

Figure 5.11 shows the connections for a typical mid-range stand-alone system with a PV power between, say, 1 and $2\,kW_p$. It is similar in many ways to the much lower-power, DC only, SHS in Figure 5.5, with an inverter added; however it is redrawn to emphasize several features of a larger, more sophisticated, system:

- The PV is an array rather than a single module.
- A battery bank replaces a single battery, giving more storage capacity.
- The charge controller has an electronic display (or a set of colored LEDs) indicating parameters such as battery voltage, SOC, PV current, and load current.
- The inverter, connected directly to the battery bank, also indicates its operating conditions.

We have not shown fuses because adequate fusing is normally included in the charge controller and inverter. Note also that many controllers supply 12 V loads; if these are not required, the system voltage may be set to a higher value such as 24 or 48 V. Some controllers offer dual voltage operation, typically 12 or 24 V, but limited to the same maximum current, so they can regulate a more powerful system, provided the higher voltage is selected. Since the inverter is connected directly to the battery bank, it should include a disconnect function if the battery voltage falls below the lower set point.

In Section 4.2 we considered ways in which the choice between various types of inverter is influenced by module connection schemes and the problem of unavoidable shading. Although similar considerations apply in principle to the stand-alone case, in practice the situation tends to be simpler. Shading is rarely a problem because PV arrays may often be sited on open ground, with all modules facing the same direction, rather than confined to potentially awkward urban rooftops; this, together with the

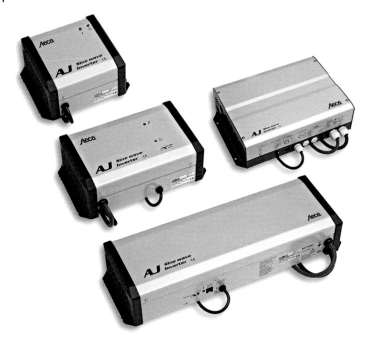

Figure 5.12 This family of inverters covers the power range 200W to 2kW (continuous), with system voltages of 12, 24, and 48V (*Source:* Reproduced with permission of Steca Elektronik GmbH).

moderate size of most stand-alone systems, means that a single central inverter is generally suitable.

The user of a stand-alone inverter should look for the following technical features:

- A power rating sufficient for all loads that may be connected simultaneously
- Accurate control of output voltage and frequency, with a waveform close to sinusoidal (low harmonic distortion), making the AC supply suitable for a wide range of appliances designed to run off a conventional electricity grid
- High efficiency at low loads, and low standby power draw (possibly with automatic shutdown when all loads are turned off), to avoid unnecessary drain on batteries
- Ability to absorb or supply reactive power in the case of reactive loads
- Tolerance of short-term overloads, particularly caused by motor start-up

Inverter efficiency is especially important in a stand-alone system that must obtain all its energy from precious sunlight without grid backup. Maximizing efficiency and minimizing standby consumption do not come cheap, but the resulting energy savings may allow the system designer to specify a smaller PV array and battery bank, leading to overall cost savings. Unfortunately some inverter manufacturers only quote maximum efficiency, or efficiency at full rated output, disguising unfavorable performance under low-load conditions. Figure 5.13(a) shows two typical efficiency curves, red for an inverter incorporating a low-frequency transformer and orange for a unit with a high-frequency transformer. Both suffer from severely reduced efficiency when delivering less than about 10% of their rated output. With rising load the efficiency reaches a maximum over 90% and then tails off again. But there are subtle differences between the

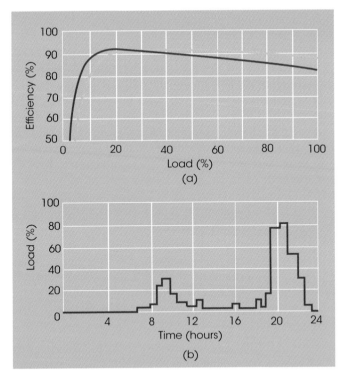

Figure 5.13 (a) Efficiency curves for two types of inverter; (b) a daily load profile for a solar home.

two: the unit with the low-frequency transformer does better at low load, and vice versa (switching losses associated with HF electronics are relatively dominant at low load, whereas magnetic losses in a low-frequency transformer are greater at high load). Such differences can be important when choosing an inverter for a particular duty.

Many stand-alone systems, including those in solar homes, spend much of their time on low load with peaks at certain times of day. Figure 5.13(b) shows a representative daily load profile for a home running a wide range of electrical appliances for lighting, cooking, and household machines. Most of the time the inverter load is less than 20% of its rated output, with peak periods in the morning and evening. This is exactly the sort of situation where careful attention to the inverter's low-load efficiency and standby power requirements is likely to pay dividends.

Like other aspects of PV engineering, the stand-alone inverter scene is advancing rapidly. AC systems are finding favor for small SHSs in developing countries, and individual PV modules with integrated inverters have entered the market. Some manufacturers offer inverters combined with charge controllers in single units; others use a modular design approach so that many inverters can be stacked together to increase power-handling capacity. Toward the top end of the power range, there is ever-increasing sophistication in monitoring, data logging, and intelligent power management. And some of the most powerful units are being used as central inverters for mini-grids of $100\,kW_p$ or more, often integrating PV with other energy sources and providing renewable electricity to remote villages and island communities.

5.3   Hybrid Systems

Stand-alone systems that rely on natural energy flows in the environment must inevitably cope with intermittency. Their main defense against unreliability and loss of service is a battery bank to store incoming energy whenever it is generated and feed it out to the electrical loads on demand. But in many cases system reliability may be enhanced, and the size of the battery bank reduced, by a hybrid system based on two or more energy sources. PV and wind power are often attractively complementary, especially in climatic regions such as western Europe where low levels of winter sunshine tend to coincide with the windiest season of the year (and, of course, wind does not refuse to blow at night!). You may have seen examples of small PV–wind hybrid systems at roadsides, powering traffic control, or telecommunications equipment. We illustrated a larger one for a remote farmhouse in Figure 5.1. Worldwide, many large hybrid systems are based on the valuable partnership between PV and wind.

Stand-alone electrical systems in isolated areas, including those for homes and farmsteads, are often referred to as *remote area power supplies (RAPS)*, a market traditionally satisfied by diesel generators. For those of us who like to champion renewable energy, it may be rather hard to extol the virtues of hybrid systems based on PV and diesel fuel, but they do offer advantages in many practical situations and are widely used. The benefits may be summarized as follows:

- It may be too expensive, in terms of the PV array and battery store, to provide a sufficiently reliable service with PV, especially where solar insolation is highly seasonal. For example, does it make economic sense to install a PV system that can cope with occasional high load demands in winter when sunlight is in short supply? A hybrid system with a backup diesel generator may be a better option.
- Diesel engines are very inefficient when lightly loaded, giving poor fuel economy. Low running temperatures and incomplete combustion tend to produce carbon deposits on cylinder walls (glazing), reducing service lifetimes. It is advisable to run engines above 70–80% of full-rated output whenever possible. But a lone diesel generator that can cope with occasional peak demands is likely to run at low output much of the time. Better to turn it off and use PV and the battery bank when electricity demand is low. The diesel can boost charge the batteries if necessary, at a high charging rate.
- In addition to rising fuel costs, unpleasant fumes, and the noise of diesel engines, it may be difficult to obtain reliable fuel supplies and engine maintenance services in remote locations. PV needs no fuel and, provided the battery bank is looked after properly, should be low maintenance.
- If an existing diesel installation needs upgrading, the addition of PV may be a good solution. Being essentially modular, PV may be added in small stages, raising system power capacity in line with increasing demand.

We see that the combination of PV with diesel can offer distinct environmental and economic benefits compared with a diesel generator on its own. Each energy source is used to best advantage, taking account of its special features. Substantial savings on diesel fuel and maintenance can be realized in those hybrid systems where a diesel generator remains the most realistic option for meeting occasional high load demands and providing security of supply.

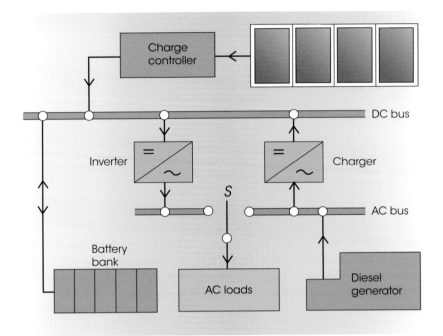

Figure 5.14 A PV–diesel hybrid system.

Figure 5.14 illustrates a common form of PV–diesel hybrid system. The PV array feeds its electricity into a main cable highway known as the *DC bus* (short for bus bar) via a charge controller, and the diesel generator supplies AC electricity to an equivalent *AC bus* that supplies the AC loads. The AC bus and DC bus are connected by an inverter and a battery charger, which may be combined in a single unit. This allows the diesel generator to charge up the battery bank if required and the battery bank to supply AC to the electrical loads. A master switch *S*, operated either manually or automatically, effects changeover between the diesel generator and the battery bank for supplying the AC loads, depending on operating conditions. Intelligent use of this arrangement ensures that the diesel engine is always run fairly hard to satisfy a high load demand or boost charge the battery bank. At other times the PV and battery bank take over.

This system, in which the AC loads are switched between the diesel generator and the battery bank plus inverter, is conceptually simple and quite common in practice. It is straightforward to implement as a system upgrade for an existing diesel installation. An alternative *parallel hybrid* configuration dispenses with the changeover switch and uses automatic control circuits and a more sophisticated inverter–charger to bring in the diesel generator when necessary.[1] Such a system can often meet the load demand in a more optimal way without the need for human supervision. A fuller account of the technical, economic, and environmental aspects of diesel hybrid stand-alone systems is given elsewhere,[2] and we will meet a sophisticated island mini-grid of this type in Section 5.5.2.

We conclude this section by a return to our starting point: the potential of hybrid systems, including those based on PV and wind energy, to raise the reliability and reduce overall costs of renewable electricity in remote areas. In principle it is possible to include several different sources (not necessarily including diesel generators). Care is needed over system integration, for example, in choosing several stand-alone inverters that cannot, in general, be interconnected because of the need to synchronize their AC outputs in frequency and phase. But, once again, modern electronics including power conditioning and control units come to the rescue, with increasingly elegant solutions to the needs of the PV systems engineer.

5.4 System Sizing

5.4.1 Assessing the Problem

In the popular imagination, science provides firm answers to firm questions, leaving little to chance when it comes to technical decision-making. But things are not as simple as that. For example, while almost all experts agree that global warming due to greenhouse gas emissions poses a major threat to life on Earth, there are wide-ranging views on its exact severity and timescale because the supporting evidence is essentially statistical. Scientists and engineers are trained to understand technical uncertainty, but it often confuses the public and offers scope for vested interests to declare the whole idea erroneous or exaggerated.

   In this book most of our discussion is based on "hard science," and we have been able to describe the performance of individual system components such as PV modules, batteries, and inverters with considerable accuracy. But there are two major chapters in the PV story where chance and uncertainty play a key role. Interestingly, but perhaps unsurprisingly, they are at opposite ends of nature's range of operations—one dealing with the miniature and the other with the large scale. The miniature, discussed in Chapter 2, concerns the quantum nature of light and the random way in which solar photons are absorbed or transmitted by solar cell materials. Although we avoided the mathematical details, you may be sure that the underlying theory is replete with probabilities! The second topic, the large-scale one we are about to tackle, concerns system sizing—deciding how much PV power and battery storage is needed for a particular stand-alone system, based on estimates of local insolation patterns, electricity demand, and required reliability of service. A few moments reflection will surely convince you that such estimates must always be hedged about with uncertainty. Indeed, so much so that the "sizing problem" is often considered the most difficult aspect of system design.

   This is primarily a stand-alone rather than a grid-connected problem because independent systems lack the support of a powerful electricity grid acting as a flexible "source and sink." A stand-alone system, especially when powered by PV alone, cannot realistically achieve total reliability. There is inevitably a trade-off between reliability and cost, forcing the system designer (and customer) to face some difficult choices. We can illustrate the dilemma using four stand-alone PV scenarios with very different operational expectations:

- *PV in space.* Launched into space on long missions without any prospect of replacement or repair, the PV arrays on spacecraft are surely the most extreme examples of stand-alone systems. 100% reliability is certainly the aim, probably over many years and at almost any cost, because spacecraft are entirely dependent on their PV power supplies. Fortunately, there is one simplifying factor: insolation in space, beyond the Earth's volatile atmosphere, is highly predictable, removing one major source of design uncertainty.

- *PV-powered refrigeration.* PV is increasingly used to power refrigerators for storing vaccines and medicines in remote hospitals in developing countries. Failure of the electrical supply may be life-threatening as well as highly inconvenient and expensive, so reliability is obviously a major requirement.

- *PV-powered traffic signs.* Also a "professional" application, traffic signals to warn drivers that they are speeding, or that there is an obstruction ahead, should obviously be dependable. But how dependable and over what timescale? What if the PV electricity runs out for a few days, and foggy weather makes an accident more likely? Will the highway authority's budget stretch to units containing more PV and larger batteries?

- *PV for a solar home or farmhouse.* We have already illustrated a stand-alone system for a farmhouse (see Figure 5.1). The size of the PV array and battery bank will obviously depend on the input from the wind turbine, the owner's choice of electrical appliances, and the amount of use. There is plenty of room for flexibility and cost saving here although it may be very difficult to decide such issues at the design stage. Generally speaking, security of supply is judged less important than for the "professional" systems mentioned previously, even though to be without lights and a TV in dead of winter is not an attractive option! In a holiday home used mainly in the summer months, occasional supply failure may be quite tolerable.

It is clear from the previous examples that the designer of a stand-alone PV system is faced with difficult decisions and choices. They can be approached in various ways. Sizing methods based on practical experience and "rules of thumb" are quite often used and may provide sensible, cost-effective solutions without much appreciation of the background science. PV sizing software is also widely available, although there is always a danger of using inaccurate input data or failing to appreciate the underlying assumptions. At the other extreme, analytic methods that attempt to put figures (including probabilities) to the many individual factors and components in a PV system promise more accuracy and scientific insight, yet they are also highly dependent on the robustness of input data and assumptions.[3] Our own approach, similar to one recommended elsewhere,[4] is intermediate in sophistication yet sufficiently detailed to highlight the main technical issues. We will illustrate it with an example based on a holiday home in southern Germany that is mainly used in the summer months.

We start the design process by considering the range of electrical appliances required by the homeowners, the power that each appliance consumes, and the average amount of daily use. This allows us to specify the total amount of electricity required in an average day, which is basic information needed to size the PV system. The table shown in Figure 5.15 includes eight low-energy lights and a TV (often considered priorities for homeowners) plus a number of other appliances reflecting individual needs and preferences. Note that by multiplying the power of each by its estimated average daily use, we arrive at its consumption in watt hours (W h) per day and, at the bottom of the table, the total estimated consumption for the whole system—in this case 2200 W h (2.2 kWh) per day. This is the amount of electrical energy to be supplied by the PV

Appliance		Power (W)	No.	Average hours/day	Average Wh/day
Light		11	8	3	260
TV		60	1	4	240
Computer		60	1	3	180
Refrigerator		80	1	24 (on-off)	500
Kettle		1000	1	0.2	200
Microwave oven		700	1	0.4	280
Food mixer		400	1	0.15	60
Washing machine		800	1	0.6	480
				Total	2200

Figure 5.15 Appliances and energy requirements for a stand-alone system.

system and is fairly typical for an SHS that includes a good range of modern appliances (by contrast, simple SHSs in developing countries based on a single PV module and a battery often provide just 200–300 W h/day). In this case the homeowners wish to use standard AC appliances, so an inverter must be included in the system.

It is extremely important to specify the most energy-efficient appliances available and, wherever possible, to avoid those involving heating. Electric fires for space heating must be considered taboo, PV electricity being far too precious to be used for warming human bodies! The kettle in the previous list might also be thought extravagant; a daily consumption of 200 W h is sufficient to provide about ten cups of coffee or tea, and whether its great convenience is worth, the energy cost is clearly a personal choice. The same applies to the microwave cooker (but note that it is only switched on for very short periods). A washing machine is high on many people's list, but it must not be used to heat the water; running it once or twice a week rather than daily would be very helpful. In short, everything should be done to reduce daily usage, especially of high-consumption units, with the aim of reducing the size and cost of the PV system. We are here confronting a reality that escapes most people living in developed economies: electricity cannot always be taken for granted and used casually but must sometimes be treated as a precious resource.

Having decided on the daily amount of electricity required, we are ready to tackle two key aspects of system design—the power of the holiday home's PV array and the capacity of its battery bank.

5.4.2 PV Arrays and Battery Banks

In the previous section we estimated 2.2 kWh as the average daily electricity requirement for a holiday home in southern Germany, and it is now time to decide on the amount of PV and battery storage needed to meet the specification. In this section we shall often refer to arrays and battery banks, terms appropriate for medium-size and larger systems, but our approach is also valid in principle for small systems containing a single PV module and battery.

The first task is to work out the size of the PV array: how much peak power should it have to satisfy the electricity demand? As it stands, the 2.2 kWh/day applies throughout the year whereas the amount of sunlight falling on the array is bound to be seasonal. So if we size the array to cope with the "worst" month for sunlight—usually December in the northern hemisphere—the owner is likely to be paying a lot of money for an array that is unnecessarily powerful in summer. Since this is a holiday home, it may be more sensible to restrict the 2.2 kWh daily usage to the summer months.

We will assume that the array can be sited on adjacent open ground, facing south, and inclined at a suitable tilt angle. Back in Section 3.3.2 we discussed the amount of daily solar radiation falling on south-facing inclined PV arrays, and Figure 3.17 showed typical data in the form of monthly mean values for London and the Sahara Desert. We also introduced the concept of *peak sun hours* for estimating an array's annual output. This involves compressing the total radiation (direct plus diffuse) received throughout the year into an equivalent duration of standard "bright sunshine" (1kW/m^2). The same concept may be used for daily radiation. For example, if an inclined array receives an average insolation of $3 \text{kWh/m}^2\text{/day}$ in April, this is considered equivalent to 3 peak sun hours, so an array rated at, say, 2kW_p is predicted to yield $3 \times 2 = 6 \text{kWh/day}$. Although it is an approximation that tends to be overoptimistic for arrays receiving a high proportion of diffuse radiation, it offers a very straightforward way to estimate array output in a particular location.

Figure 5.16 shows daily solar radiation levels in the same form as Figure 3.17, but using published data[5] relevant to the holiday home's location in southern Germany. Three representative values of tilt are illustrated: 33°, 48° (the latitude angle), and 63°. As expected, 33° does best over the summer months when the sun is high in the sky (an even smaller tilt would give better results at midsummer but at the expense of other times of year). A tilt of 48° gives good results around the time of the equinoxes in March and September, and 63° is marginally preferable over the winter months. We also see that radiation levels in December are only about one-third of those in midsummer, so a PV array big enough to supply the home's electricity in December would be three times oversized in June. Clearly, choosing an array to cope with the "worst" month of the year would be a very expensive option.

At this stage the system designer must surely discuss alternatives with the homeowners. For example, they might agree to restrict their demand for 2.2 kWh/day to the months March–September, covering the main holiday period, in exchange for a smaller PV system at lower cost. Over this 7-month period the 33° tilt angle is a good choice. The "worst" month is now taken as March, for which the average daily radiation

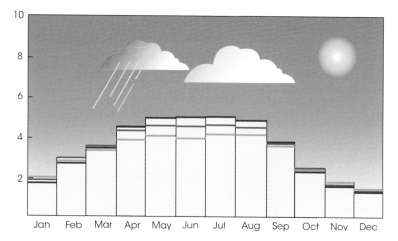

Figure 5.16 Daily solar radiation in kWh/m² on south-facing inclined PV arrays for a location at latitude 48°N in southern Germany. Three values of array tilt are illustrated: 33° (blue); 48° (red); and 63° (green).

is 3.5 kWh/m². This figure can be used for sizing the array. The homeowners will have to make do with considerably less electricity over the winter months, unless the total is boosted by an alternative energy source, or perhaps they will agree to forgo the use of refrigerator, microwave oven, and washing machine and cut down on the drinking of coffee! Unlike the "professional" PV systems mentioned in the previous section, a "leisure" installation should offer plenty of opportunities for energy saving, trading convenience, and reliability against cost.

Using the peak sun hours concept, we may express the average daily amount of electricity available for running the home's appliances, E_D as

$$E_D = P_{PV} S_p \eta \tag{5.1}$$

where P_{PV} is the rated peak power of the PV array, S_p is the number of peak sun hours per day in the month of interest, and η is the overall system efficiency (discussed hereafter). Therefore the peak power of the array is given by

$$P_{PV} = \frac{E_D}{S_p \eta} \tag{5.2}$$

In the case of the holiday home, $E_D = 2.2$ kWh/day, $S_p = 3.5$ hours in March, and we will assume a system efficiency of 60% ($\eta = 0.6$), so that

$$P_{PV} = \frac{2.2}{3.5 \times 0.6} = 1.05 \, kW_p \tag{5.3}$$

We therefore predict that a PV array rated at just over 1 kW$_p$ will supply the daily load requirement of 2.2 kWh during the months March–September.

The overall system efficiency η takes account of various power reductions and losses that prevent the PV array's nominal output from getting through to the household's AC appliances. A figure of 60% may seem disappointing, but is fairly typical of such stand-alone systems. It is derived by multiplying together efficiencies for the various system components, expressed as numbers between 0 and 1 (e.g., an efficiency of 85% is expressed as 0.85). Although it is difficult to give exact figures the following are fairly typical:

- *PV modules* (0.85). Power output is less than the rated value in standard "bright sunshine" ($1\,kW/m^2$), due to such factors as raised cell operating temperatures, dust or dirt on the modules, and aging. Also, modules are not generally operated at or close to their MPP (unless a controller with MPP tracking is used).
- *Battery bank* (0.85). The charge retrieved from the battery bank is substantially less than that put into it (see Section 5.2.1).
- *Charge controller, blocking diodes, and cables* (0.92). There are small losses in all these items.
- *Inverter* (0.9). This is a typical figure for a high-quality inverter, bearing in mind that it must sometimes work at low output power levels (see Section 5.2.3).

The product of all these figures is 0.6, or 60%. If MPP tracking is used and the system is DC only (no inverter), the system efficiency might approach 70%. But in practice it is hard to predict how components will behave in variable sunlight and ambient temperatures or how the system will actually be used by the homeowners as they become familiar with it, so the aforementioned figures should be treated with caution.

In view of all these uncertainties, plus the vagaries of the weather, oversizing a PV array by a reasonable amount—say, 20%—is often recommended. In the previous example $1.2\,kW_p$ would obviously improve reliability of supply, but it is, as ever, a question of cost. An alternative is to regard PV as an essentially modular technology that can easily be upgraded. So it would be possible to install a $1\,kW_p$ array initially and expand it later if required.

The remaining task is to size the battery bank. The biggest decision is how many "days" of battery storage are required. Too few, and a spell of unusually dull or wet weather may cause a serious loss of electricity supply. Too many, and the battery bank becomes unnecessarily large and expensive. Five days of usable battery storage (in the previous example, equal to $5 \times 2.2\,kWh = 11\,kWh$) is often regarded as a good compromise between reliability and cost. But of course it depends on the application; a holiday home is by no means a crucial case, and many "professional" systems demand far higher reliability to avoid risking serious inconvenience, economic penalties, or even danger to life. In such cases the amount of battery storage may have to be raised greatly, perhaps to 15 days or more. Alternatively, a reliable standby power source such as a diesel generator may be incorporated.

When the number of days of storage N has been decided, the capacity C of the battery bank can be calculated:

$$C = \frac{NE_D}{D\eta_{inv}} \tag{5.4}$$

where (as before) E_D is the daily electricity requirement, D is the allowable depth of discharge of the battery bank, and η_{inv} is the efficiency of the inverter, assuming an AC supply is required. Note that the usable capacity of the battery bank is less than its nominal rated capacity because complete discharge must be avoided. In our example we will assume 5 days of storage, battery discharge up to 80% of nominal capacity, and inverter efficiency of 90%. Hence,

$$C = \frac{5 \times 2.2}{0.8 \times 0.9} = 15.3 \ \text{kWh} \tag{5.5}$$

As with the PV array, it may be sensible to oversize the battery bank—or to treat it as modular, with the option of upgrading it later.

To summarize, the stand-alone system for the holiday home in southern Germany should be able to supply the desired amount of electricity between March and September using a PV array rated at $1.05 \ \text{kW}_p$ with a battery bank of capacity 15.3 kWh (assessed at the 100-hour discharge rate normal for PV systems). If the batteries are connected to give 24 V DC, which is quite common for a system of this size, then the required charge capacity is $15\,300/24 = 638 \ \text{A h}$.

This specification could be met, with a reasonable amount of oversizing, by an array of, say, eight PV modules rated at $150 \ \text{W}_p$ each ($1.2 \ \text{kW}_p$ total), together with a bank of, say, eight 12 V batteries rated at 175 A h each (16.8 kWh total). The electricity yield, and hence system reliability, could be further improved at a modest cost by specifying an MPPT charge controller. The modules could be connected in series or series–parallel; the batteries as four parallel strings of two units each to give 24 V DC. The main components of the system are illustrated in Figure 5.17.

We end this section with some further remarks on reliability. First, it must be admitted that choosing a holiday home to illustrate system sizing makes life rather easy because it allows a somewhat cavalier approach toward possible supply failures. We are making use of the relative unimportance of failures in this "leisure" application and have assumed that homeowners are flexible over their use of appliances. All this changes dramatically in the case of a "professional" system with stringent load and reliability criteria, and serious thought must given to how often a failure of supply can be tolerated.

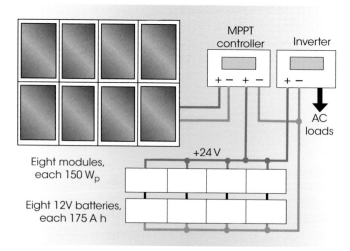

Figure 5.17 A suitable system for the holiday home.

The many uncertainties of system design mean that the problem can only be discussed sensibly in terms of probabilities. A measure known as *loss-of-load probability* (*LLP*) is widely used. Basically, LLP denotes the probability that, at any point in time, the PV system is unable to satisfy the demand for electricity. It may also be interpreted as the proportion of total time that the system is unavailable (which should include estimated maintenance and repair outages). LLP = 0 implies that the system is 100% available; LLP = 1 that it is permanently out of action. We normally hope for and expect low LLP values, say, between 0.0001 and 0.1, but it depends very much on the importance of reliability in a particular application. The smallest values of LLP, increasingly difficult and expensive to achieve, are typically found in PV systems used on space missions or in vital telecommunication links, the largest ones in leisure applications (it is hardly a disaster if solar-powered garden lights fail to work every evening!).

However it is much simpler to explain the LLP concept than to calculate its value for a particular system or design a new system to meet a customer's LLP specification. The basic difficulty is that it depends on so many factors, some fairly obvious, while others obscure or random in nature. Our previous discussion makes clear, for example, that reliability is generally increased (and LLP reduced) by specifying a more powerful PV array and/or a larger battery bank—although there is, in fact, a subtle interaction between them.[2] Sunlight statistics plays a major role. For example, occasional lengthy periods of cloudy weather, untypical of the local climate, can result in a battery bank's SOC being depleted to such an extent that supply cutoff is inevitable. Unfortunately rare and isolated weather events cannot be predicted from averaged meteorological data.

Yet in spite of the difficulties, various theoretical ways of incorporating LLP into stand-alone system design have been developed in the past 25 years,[2] and many sophisticated computer programs for system sizing and simulation are available.[6] Indeed, the complexity of the task more or less demands the use of computer software, even though it may be hard for the newcomer to understand its details. A straightforward quantitative approach to sizing, such as what we have introduced in this section, seems a good antidote to over-reliance on computer software. A few simple calculations at least allow us to check that the numbers churned out by a computer program are reasonable!

5.5   Applications

The variety of applications for stand-alone PV systems is extraordinary. Almost any need for electricity in isolated, remote, or independent locations can, in principle, be met by solar cells. We have already mentioned a number of examples in this book, from solar-powered watches and calculators to space vehicles, but our main focus has been on electricity supply for remote buildings far from an electricity grid. This has provided a chance to describe typical units that make up medium-power systems, including PV arrays, battery banks, charge controllers, and inverters, in a setting that most of us can easily imagine. It is now time to move out into the wider world—and beyond—to discuss a number of key application areas where PV has made, and continues to make, major contributions. It is hard to select just a few examples from the large number of possibilities, so we have chosen four distinctive topics, each important in its own way, that illustrate a wide range of issues and challenges in PV system design.

5.5.1 PV in Space

For more than half a century, spacecrafts have relied on solar cells for their power supplies. In the early years of the modern PV age, solar electricity was so expensive that space exploration provided its only significant market. The costs of designing, manufacturing, and launching vehicles into space are so large that the price of cells to power them is relatively unimportant, the main criteria being technical performance and reliability in the harsh space environment. Although the total amount of PV power launched beyond the Earth's atmosphere is tiny compared with today's gigawatts of terrestrial installations, solar cells remain vital to modern spacecraft including those used for satellite communications, TV broadcasting, weather forecasting, and mapping—and, of course, space exploration. You may wish to follow up this brief introduction with an approachable and authoritative account given elsewhere.[5]

We may summarize the special features of the space environment that impact on the design and deployment of solar cells and arrays with a few key points:

- Radiation in space tends to damage solar cells.
- Sunlight in space, unfiltered by the Earth's atmosphere, has a different spectrum from that received by terrestrial PV cells.
- Spacecraft, including satellites in Earth orbit, experience dramatic changes in sunlight intensity and temperature as they move in and out of shadow, causing high thermal stresses in solar cells and modules.
- PV modules and arrays need to be kept as small, neat, and light as possible to avoid adding unnecessarily to the launch payload.
- Sustained technical performance and reliability are paramount, especially on long missions.

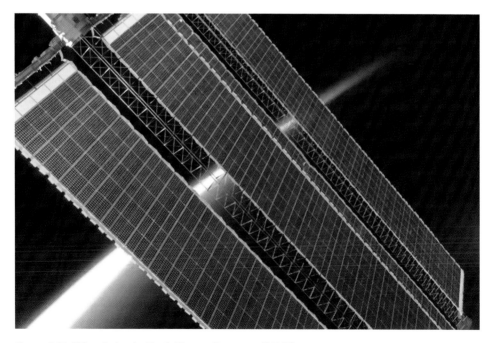

Figure 5.18 PV encircles the Earth (*Source:* Courtesy of NASA).

Each of these will now be discussed in more detail.

Radiation damage to solar cells in space is a major challenge to PV designers. The risk of damage by high-energy electrons and protons is particularly serious for satellites in *mid-Earth orbits* (*MEOs*), defined as 2000–12000 km above the Earth, which pass through the *Van Allen* radiation belts. The neighborhood of Jupiter is also a high-radiation environment. Special types of cover glass are effective at reducing the steady and cumulative degradation of cell performance over the lifetimes of long missions. The susceptibility of standard silicon solar cells to radiation damage was recognized in the early years of space exploration and much effort has been put into design improvements to mitigate the effects and raise cell conversion efficiencies, presently approaching 20%. Although high-efficiency silicon cells are in widespread use, a major advance in recent years has been the development of multi-junction III–V cells based on gallium arsenide (GaAs) and related compounds, which are much less susceptible to radiation damage and offer even better conversion efficiencies. We first described these cells in Section 2.5.1, and since they are so important to space PV, we will mention them again toward the end of this section.

In this book we have often referred to standard "strong sunlight" received by solar cells and modules at the Earth's surface. This is defined as having an intensity of 1 kW/m^2 and the AM1.5 spectrum typical of sunlight after passing through the Earth's atmosphere. Sunlight in space, unfiltered by the atmosphere, is described by the Air Mass Zero (AM0) spectrum (both spectra were illustrated in Figure 1.6). The intensity also varies according to the distance from the sun. For example, near Mercury it is almost double than near Earth, and near Jupiter only about one-thirtieth. Clearly, solar cells and modules have to operate satisfactorily and be calibrated for use under such conditions.

Figure 5.19 Wide horizons for PV: the Solar System (*Source:* Courtesy of NASA).

Closely related to changes in light intensity are changes in operating temperatures. Whereas terrestrial solar cells and modules are normally required to work between, say, −20°C and +70°C, conditions in space can be far more demanding. Spacecraft in orbit around the planets experience extremes of temperature as they pass in and out of the sun's illumination. Cell temperatures in *low Earth orbit* (*LEO*) may get down to −80°C in shadow, but in orbit around distant Jupiter they must work at −125°C, even when illuminated, around Mars, at up to +140°C. Sudden transits from shadow to sunlight can produce big power surges as well as exposing cells and modules to high thermal stresses.

The size and weight of space PV arrays is extremely important, as is their ability to deploy successfully on reaching zero-gravity conditions. The smaller and neater an array, the easier it is to integrate into the spacecraft's structure during launch; the lighter it is, the lower the payload and launch cost. For a given peak power, an array's area is proportional to cell efficiency, favoring the most efficient cells and technologies. In the early 1970s the most powerful PV system in space was that of the *Skylab 1* satellite, delivering about $16\,kW_p$. The *International Space Station*, launched in 1998 and continuously expanded and developed over the following decade, generates more than $100\,kW$ of average power from its silicon solar cells, which are mounted in eight double arrays with a total area of over $3000\,m^2$. This is large-scale PV, with exciting technical challenges for project teams in mechanical engineering and materials science as well as solar technology.

The efficiency of solar cells designed for use in space is important for several interrelated reasons. For a given peak power requirement, improvements in cell efficiency reduce the area, weight, and payload costs of a PV array. As we mentioned earlier, one

Figure 5.20 The International Space Station, photographed in 2009 (*Source:* Courtesy of NASA).

of the most important advances in recent years has been the commercial development of triple-junction cells based on GaAs and related compounds, which now attain 30% efficiency under AM0 conditions, reducing array areas by over a third compared with high-efficiency silicon. They also have better radiation resistance. Typically, a triple-junction device consists of a "sandwich" of layers of gallium indium phosphide (GaInP), GaAs, and germanium (Ge), each carefully chosen to absorb a portion of the solar spectrum—you may like to refer back to Section 2.5.1 and Figure 2.38 for a fuller explanation. Research continues apace, with even more efficient four-junction devices in prospect, and an increasing interest in concentrator systems to reduce the area and cost of these highly specialized cells.

It goes without saying that technical performance and reliability, sustained over long periods, are paramount in space systems. On manned missions there may be limited potential for carrying out maintenance and repair, but on long unmanned missions solar cells and arrays are quite literally on their own—surely the most extreme example of stand-alone systems. It is hardly surprising that PV power systems in space cost hundreds of times more per peak watt than their earthbound counterparts; but without them spacecraft would, quite literally, be lost.

5.5.2 Island Electricity

Providing a small island community with an economic, convenient, and reliable electricity supply can be a major challenge. Traditionally, islanders in the developed world have installed diesel generators and depended on fuel deliveries from a mainland depot. But diesel engine maintenance is expensive—fuel costs always seem to be rising—and there is a noise and pollution problem that people who cherish their natural environment would rather avoid. Most islands have a valuable wind resource, and many have lots of sunshine and free-running rivers or streams. Such plentiful flows of natural energy act as a strong incentive to generate renewable electricity, and when several different energy sources are available, it makes good sense to consider a hybrid system and distribute the electricity using an island mini-grid.

Such systems are still "stand-alone" in the sense of being unsupported by large conventional electricity grids. So are mini-grids serving isolated communities on the mainland. Their major advantage compared with that of the individual stand-alone systems for each user is that integration of various energy sources with different daily and seasonal peaks can provide a more consistent, reliable, and economic supply for a whole community. Although backup diesel generators are generally still needed to ensure a reliable 24-hour service throughout the year, they can be started up for short periods only when necessary—and then run hard and at high efficiency.

The Isle of Eigg, 6 × 4 km in extent, is one of the jewels of the Inner Hebrides. Lying off the west coast of Scotland to the south of Skye, it has an equable climate, thanks to the Gulf Stream, a generous wind resource, lots of sunlight in summer, a few streams, and just under a hundred inhabitants. Like many Scottish islands, Eigg has a harsh history behind it, including the 19th-century depopulation and more recent absentee landlords, but in 1997 funds were raised to purchase the island and set up the Isle of Eigg Heritage Trust to manage it for the inhabitants and their wonderful environment. Determined to update their electricity supply from reliance on aging diesel generators to a modern "green" alternative, they raised capital grants totaling £1.6 m for a hybrid

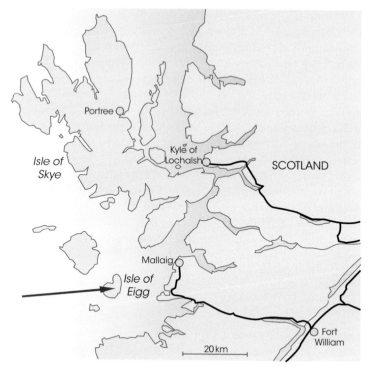

Figure 5.21 The Isle of Eigg lies off the west coast of Scotland.

system comprising PV, wind, and hydroelectric power, with diesel backup.[6] Early in 2008 all 37 households and 5 businesses on Eigg were connected to the new island grid, achieving celebrity status for a state-of-art renewable energy system that is providing inspiration to other island communities in Scotland and around the world (Figure 5.21).

Eigg's system, as installed in 2008, is illustrated in Figure 5.22, which summarizes the generation and consumption of electricity. A key feature is that all generators and loads are interconnected by an island-wide AC grid. Transmission is at 11 kV for long cable runs and at 230 V for short ones (from the PV and diesel generators), with transformers inserted where necessary. Power sources that generate DC (the wind turbines and PV) feed into the grid via inverters. An advanced load management system monitors the balance between supply and demand, bringing in the diesel generators when necessary and controlling energy flow to and from the battery banks via a set of bidirectional inverter–chargers. The batteries, PV, and wind turbines are the only DC components; homes and businesses are supplied with 230 V AC. Grid frequency is set by the inverter–chargers or by the diesel generators when they are running. We now comment further on the various items:

- *10 kW$_p$ of PV.* The Hebridean islands have a better sunshine record than the mainland, where higher mountains tend to increase cloud formation and precipitation. Eigg, at latitude 57°N, has plenty of sunlight in the summer months, with up to 18 hours between sunrise and sunset in June, so PV can make a valuable contribution when wind and hydropower tend to be at their lowest. In this system the output from 60 PV

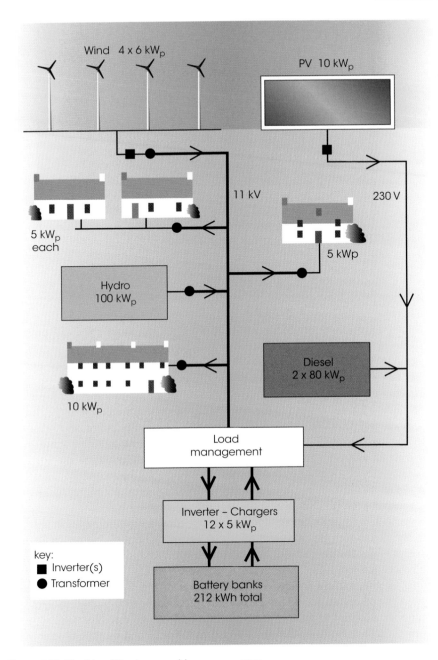

Figure 5.22 The Isle of Eigg's renewable energy system.

modules, connected in 6 strings of 10, is converted to 230 V AC by adjacent inverters. Since the first edition of this book was published, the PV array has been upgraded from 10 to 30 kW$_p$, showing confidence in Eigg's ability to generate solar electricity.

- *24 kW$_p$ of wind energy.* A group of four wind turbines, each rated at 6 kW$_p$, is sited at one of the island's windiest locations. Although wind turbines are generally rated in

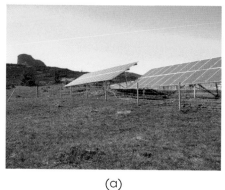 

(a)   (b)

Figure 5.23 (a) PV on Eigg (*Source:* Reproduced with permission of Wind & Sun Ltd); (b) wind power on Eigg (*Source:* Reproduced with permission of Isle of Eigg Heritage Trust).

kilowatts at a standard high wind speed, we have used kW_p units in the figure to emphasize that they rarely operate at full output—even though the months October–April are highly productive on an island subject to Atlantic storms. In fact the Eigg wind turbines make a valuable contribution throughout the year. Their DC outputs are inverted and transformed to 11 kV for transmission.

- *100 kW$_p$ of Hydropower.* The most powerful contributor to the renewable energy portfolio is a new run-of-river water turbine rated at 100 kW$_p$ supplied by a substantial stream (there are also two much smaller preexisting turbines in other locations rated at 9 and 10 kW$_p$, not shown in the figure). However on a small island the flow of streams closely follows current rainfall and tends to be intermittent and seasonal. Hydroelectric generation on Eigg is therefore variable, much stronger in winter than summer, with an average value far lower than the nominal ratings of the turbines (Figure 5.23).
- *2 × 80 kW$_p$ of diesel.* Two new diesel generators provide backup to ensure 24-hour service throughout the year. In an average year the renewable sources are expected to provide over 95% of total electricity demand, so the total diesel contribution is small. Typically, the generators are run hard for short periods to boost charge the battery bank on days when the renewables are unable to meet the full load demand. They generate power at 230 V AC.
- *Load management.* A comprehensive hybrid system of this kind, involving various energy sources and domestic and business loads, justifies a sophisticated control system. Its aims are making the most of available renewable generation, deciding between the various sources in times of surfeit, ensuring that the battery bank is neither over-charged nor over-discharged, transmitting electricity efficiently to the various loads, and bringing in the diesel generators when necessary.
- *12 × 5 kW$_p$ inverter–chargers.* Arranged as four 3-phase clusters, each with its own battery bank, the bidirectional inverter–chargers are at the heart of the system. When the renewable generation is insufficient to meet demand, they take energy from the batteries and invert it to augment the AC supply. When generation exceeds demand, they rectify the AC and charge the batteries. If the batteries are fully charged and excess energy is being generated, the inverters raise the frequency, and additional

(a)   (b)

Figure 5.24 (a) The battery banks and (b) some of the main inverter–chargers (*Source:* Reproduced with permission of Wind & Sun Ltd).

"opportunity" loads such as heaters in community buildings (not shown in the figure) detect the increase and switch on automatically. If there is still surplus energy, the frequency is increased further, and the various generators respond by backing off to prevent battery overcharging.

- *4 × 53 kWh battery bank.* The batteries are arranged in four 48 V banks and located in the power house with the inverter–chargers and diesel generators. The banks are normally kept above 50% SOC to prolong their life. The quoted total capacity is therefore half the full nominal capacity of 424 kWh and equates to approximately 1 day's electricity usage on the island. Additional days of storage are not needed in this case because of the diesel backup.
- *Households and businesses.* 37 households and 5 businesses were initially connected to the island grid and supplied at 230 V AC. Householders agreed to limit peak demand to 5 kW$_p$ each, while businesses to 10 kW$_p$. All consumers are provided with an energy meter to monitor the amount of electricity being used. The islanders have adapted well to the new system and are far better informed about electricity usage and energy conservation than most people on the mainland (Figure 5.24).

Eigg's electricity grid is an excellent example of a modern hybrid system. The PV component, being essentially modular, may be increased even further in the future to provide more summer electricity. In any case, the principles of design and implementation are of widespread relevance, even though the relative contributions from PV, wind, hydro, and diesel backup are bound to vary from one island system to another.

5.5.3   PV Water Pumping

Infectious diseases caused by tainted drinking water and primitive sewage disposal are largely unknown to those of us who live in the developed world. We tend to take the benefits of pure water for granted. But who should we thank for this blessing? It has been said that the civil engineers of the 19th century did more to improve public health than all the doctors and surgeons put together, by designing and building the infrastructure for modern water supplies.

Figure 5.25 Clean and accessible: PV-pumped water (*Source:* Reproduced with permission of EPIA/ Schott Solar).

The situation can be very different elsewhere. In rural areas of some of the poorest countries in the world, millions of people, especially women, spend hours each day fetching and carrying water, sometimes from polluted streams or pools. Yet new village wells can transform lives and health and, if equipped with automatic pumps, eliminate the daily grind of water collection (Figure 5.25).

Water pumping is one of the most successful applications of stand-alone PV in developing countries. By the year 2000 over 20 000 PV-powered systems were in use worldwide and the pace of installation continues. Of course, small water pumps can be worked by hand, larger ones by windmills or diesel engines. But the PV alternative, in addition to its cleanliness, reliability, and long life, often proves economic for medium-sized systems. Water pumping is also used for crop irrigation and stock watering.[2]

A typical scheme for village water supply is illustrated in Figure 5.26. A submersible pump/motor, protected by installation underground, raises water to a storage tank whenever sunlight falling on the PV array is sufficiently strong. From there it is fed by gravity to one or more taps. In previous sections we have often discussed the need for battery storage in stand-alone systems. But one major feature of water pumping is that the water tank replaces batteries as the energy store, using PV electricity directly to increase the potential energy of the raised water.

Although the scheme is simple in principle, a number of technical choices must be made:

- *Type of pump.* Of the many types of pump on the market, *centrifugal* designs are widely used to raise water against pumping heads up to about 25 m (the height difference between the water table and tank's input pipe). Multistage versions can cope with higher heads. A centrifugal pump has an impeller that throws water against its

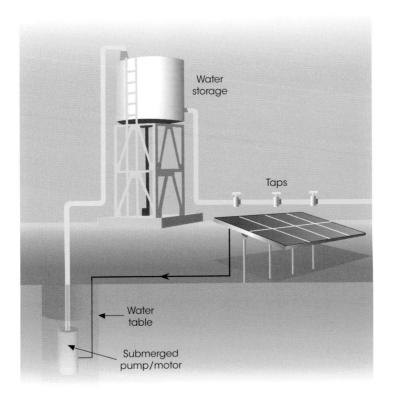

Figure 5.26 A system for village water supply.

outer casing at high speed, the kinetic energy then being converted to a pressure head by an expanding output pipe. Centrifugal pumps are compact, robust, and well suited to PV applications, but they are not normally self-priming and must therefore be kept submerged. This makes them suitable for pump/motors positioned below the water table. Alternative *displacement* or *volumetric* pumps including various self-priming types are more suitable for lower flow rates from very deep wells or boreholes.

- *Type of motor.* DC motors are generally more efficient than AC ones, but more expensive. AC motors are very rugged and need little or no maintenance so are suitable for submersion at the bottom of a well, but inverters are needed to convert PV electricity to AC, adding to the capital cost. Among DC motors the *permanent magnet* type is often preferred; but all conventional designs use carbon brushes that must be periodically adjusted or replaced, making submersion awkward. Modern *brushless* DC motors overcome this difficulty, at a cost.
- *Matching the motor and PV array.* Ideally, the PV array should be operated close to its MPP in all sunlight conditions. Unfortunately the resistive load offered by most motors does not allow this to happen, so an MPP tracking controller based on a DC-to-DC converter may be inserted to improve matching and increase efficiency.

From the PV perspective the most important task is to size the array to pump the desired amount of water. A well-known hydraulic equation is a good starting point.

The hydraulic energy E_h required to raise $1\,m^3$ of water against a head of H_w meters is given by

$$E_h = \rho g H_w\,J \tag{5.6}$$

where ρ is the density of water $(1000\,kg/m^3)$ and g is the acceleration due to gravity $(9.81\,m/s^2)$. In this case H_w is the height of the holding tank's input pipe above the water table.

In this book we have generally used the kilowatt hour (kWh) as our unit of energy. We note that 1 joule is equivalent to $1\,W\,s$, or $1/3.6 \times 10^6\,kWh$. Therefore if we wish to raise a volume V_w cubic meters of water per day, the required hydraulic energy is

$$E_h = V_w H_w \left(\frac{1000 \times 9.81}{3.6 \times 10^6}\right) = 0.0027\,V_w H_w \quad kWh/day \tag{5.7}$$

For example, suppose that a village population of 300 needs an average of $50\,l$ of water per person per day—a total of $15\,000\,l$ or $15\,m^3/day$—and that the tank's inlet pipe is $20\,m$ above the water table. The hydraulic energy required is

$$E_h = 15 \times 20 \times 0.0027 = 0.81\,kWh/day \tag{5.8}$$

We can now estimate the size of the PV array using the peak sun hours concept first mentioned in Section 3.3.2. Basically, this involves compressing the total daily radiation

Figure 5.27 PV for a village water supply in Niger (*Source:* Reproduced with permission of EPIA/Photowatt).

received by the array into an equivalent number of hours of standard "bright sunshine" $(1 \, kW/m^2)$. The peak power of the array is then approximately given by

$$P_{PV} = \frac{E_h}{S_p \eta} \tag{5.9}$$

where S_p is the number of peak sun hours for the particular location and η is the overall system efficiency. The peak sun hours are normally chosen for the "worst" month to ensure continuity of supply throughout the year. The system efficiency must take account of electrical losses in the motor and cabling, hydraulic and friction losses in the pump and pipework, and mismatch between the motor and the PV that prevents the array from working at its MPP. An average efficiency of about 40% is fairly typical for a centrifugal pump and, together with other losses, gives a typical system efficiency of around 25% (0.25). So, as an example, if the location has an insolation equivalent to 3 peak sun hours per day in the "worst" month, then the PV array needs a peak power:

$$P_{PV} = \frac{0.81}{3 \times 0.25} = 1.08 \, kW_p \tag{5.10}$$

Although peak powers up to a few kilowatts are fairly typical of systems supplying water to individual villages in sunshine countries, considerably larger PV arrays are sometimes installed to serve larger communities—for example, a cluster of villages obtaining water from a single source that is distributed by hand or pipe. A good example is the Ouarzazate scheme in Morocco, consisting of more than 20 autonomous stand-alone systems supplying a total population in excess of 10 000 people. One of the smaller PV systems in this scheme has already been shown in Figure 5.2; a much larger one, incorporating a substantial roof-mounted PV array, is shown in Figure 5.28—an impressive example of PV water pumping in action.

5.5.4 Solar-Enabled Water Desalination

In addition to pumping water from the ground, electricity from PV can be used for sustainable production of potable water from sea water. Producing fresh water via water desalination is essential for arid, water-scarce regions, but expensive and energy intensive. The cost of energy can account for half the total cost of producing water, and the use of fossil fuels to power the desalination plants causes emissions of CO_2 and other hazardous pollutants.

Availability of clean, potable water is challenging for almost 25% of the world's population, and it is projected that by 2030, 47% of the global population will face water scarcity. Solar energy could be an alternative energy source for water desalination technologies, but it is not widely used as there are not yet many concepts and system designs that can handle the deterministic and stochastic variability of renewable energy resources. However, the decline in the price of PV is catalyzing developments in membrane-based and thermal desalination systems that require a lot of energy and are typically needed in regions of high solar insolation.

The long-term sustainability of the desalination infrastructure in arid and sunny countries requires a transition to more energy-efficient desalination technologies and gradual displacement of fossil fuels with renewable energy (Figure 5.29).

Figure 5.28 A large PV water pumping station in Morocco (*Source:* Reproduced with permission of EPIA/Isofoton).

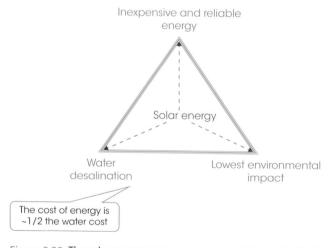

Figure 5.29 The solar energy-water-environmental Nexus (Vasilis Fthenakis).

PV-powered reverse osmosis (RO) is currently the least expensive RE desalination option in many areas.[7] The wholesale cost of electricity from PV is currently in the range of 4–8 US cents/kWh in areas of high irradiation. This is lower than the cost of subsidized grid electricity from fossil fuels in several sunny and arid regions, and solar electricity can help reduce the full cost of producing fresh water from seawater in these areas. PV electricity is cost competitive in many high insolation regions based on direct costs (levelized cost of electricity (LCOE)) and is a lot cheaper than fossil fuel electricity taking proper account of resource (fuel and water) depletion and environmental impacts.

Solar energy can be integrated into desalination plants, but for the transition to happen, the current fossil fuel subsidies should be phased out or given equally to the solar industry. Perhaps the most important enabling factor will be the empowerment of government agencies to take holistic, comparative views of energy costs and act on them.

5.5.5 Solar-Powered Boats

Boats powered by sunlight represent one of the most successful and attractive applications of PV in the field of sustainable transport. Less well-known to the public than the solar car races that have achieved international fame in Australia and the United States, solar boating has recently made headlines with a growing number of international events and a circumnavigation of the globe. Unlike road vehicles, boats do not have to climb hills or travel at high speed, and they require surprisingly little power for propulsion in calm conditions. This makes solar-powered boating on lakes, rivers, and canals relatively inexpensive and opens up a new market for PV in an important leisure industry.

The low-power levels needed to propel boats at modest speeds in calm water can be nicely illustrated with a historical example. Two hundred years ago Britain was in the middle of a canal-building frenzy. The heavy materials of the early industrial revolution, including coal and iron, needed to be transported over considerable distances for which the road network was totally inadequate. So the English narrow canals, with locks just over 2 m wide and 22 m long, were carved through the countryside by gangs of "navvies" (derived from the word navigation) using picks, shovels, wheelbarrows, and human muscle power. This extraordinary feat of civil engineering revolutionized inland transport and allowed cargos up to about 30 tonnes to be carried in individual barges, the so-called narrowboats that just squeezed into the locks. And how did the boats move in those early days? They were towed, often two at a time, by a single horse! Admittedly at slow speed, typically 2–3 km/h, but it was a vast improvement on existing methods of transport by land.

This example suggests that a single "horsepower" (HP), nowadays taken as equivalent to 746 W, is enough to shift many tonnes of boat at modest but useful speeds. And if careful attention is paid to design by making hull, motor, and propeller as efficient as possible, we now know that one or two HP can propel a modern leisure craft with several passengers at realistic speeds—say, up to 10 km/h (6.4 mph) in calm water. The quest for efficiency mirrors that of solar car design with its emphasis on streamlined bodywork and high-performance motors, transmission, and tyres. But in the case of boats, the power levels, and therefore costs, tend to be much lower.

Electric boats are a novelty to many people. For the last 100 years, most motorboats have used petrol or diesel engines for propulsion, helping to deplete the Earth's valuable fossil fuels, making a lot of noise, and polluting the waterways. But it was not always so. In the period from the 1880s up to the start of the First World War in 1914, there were plenty of battery-powered electric boats on the lakes and rivers of Europe, including some that could carry over 50 passengers. The river Thames in England boasted a scheduled passenger service, with electric charging stations along the bank. However the advent of internal combustion engines proved nearly fatal, and by 1930 electric boating was in severe decline. Half a century later it began to emerge again, largely due to increasing environmental awareness, and today represents a small, but flourishing sector of the leisure boating industry. The essential components—batteries, control circuits, electric motors, and propellers—are constantly being developed and refined, giving wonderfully silent cruising with minimal disturbance to wildlife and riverbank.

Solar electric boats are even more of a novelty. We are not talking about the many boats that use a PV panel or two to power their electronic equipment and cabin lights, but true electric boats that use PV for propulsion. These exciting craft literally "cruise on sunlight." Today there are many examples on the inland waterways of Europe, North America, and Australia, and the number rises year by year. The combination of a virtually silent, nonpolluting electric drive and solar energy is extremely attractive.

As already noted, quite a lot can be achieved with a propulsive power of 1 HP, equivalent to 746 W. In fact, the range 200 W to 3 kW covers most modern electric leisure boats at normal cruising speed, and there are a few larger craft, including passenger ferries, that require considerably more. We are referring to the mechanical power needed to propel the boat forward; more electrical power is required because of combined motor and propeller losses, typically amounting to 40%.

We now describe three recent boats with different design criteria, specifications, and passenger accommodation. The first, 6.2 m catamaran *Solar Flair III*, cruises on inland waterways in England. Designed as an experimental boat to test various combinations of PV modules, motors, and propellers, she also appears at boat shows and rallies, helping to promote PV and solar boating and convince the public of its viability, even in the British climate. She carries six 75 W_p monocrystalline silicon PV modules in front of a small cabin, plus two more behind (not visible in the photo), giving a total of 600 W_p to charge batteries that power an electric outboard motor. A smaller additional motor, mounted below the front module, acts as a bow thruster to aid sharp turning on narrow canals and rivers. The main motor takes about 450 W of input power to attain a cruising speed of 8 km/h in calm conditions. Average summer sunshine produces enough PV electricity to move *Solar Flair III* about 32 km (20 miles) per day at this speed. The design aims at technical performance and a streamlined appearance rather than passenger accommodation (Figure 5.30).

Our second example, the 6.7 m (22 ft) pontoon boat *Loon*, has been designed and developed in Ontario, Canada, as a spacious canal and river cruiser able to accommodate up to eight passengers in comfort (Figure 5.31). Raising the 1 kW_p of PV modules on a canopy greatly increases passenger space and gives protection against rain—and maybe also the sun! The input motor power to achieve 8 km/h is about 1 kW, and PV provides enough electricity, in the Canadian summer months, to travel an average of about 24 km (15 miles) per day at this speed. On long cruises the boat's batteries may be fully recharged by plugging into shore power electricity. The same company is currently also offering a twin-engined solar boat, the *Osprey*, capable of taking 30 passengers.

The third example, known under the project name *PlanetSolar*, is the largest solar-powered boat in the world (Figure 5.32). Built in Germany after extensive model testing in wind tunnels and water tanks, it is registered in Switzerland and completed the first solar circumnavigation of the globe with a crew of six in May 2012. During the voyage it broke two other records: the fastest crossing of the Atlantic Ocean by a solar boat, and

Figure 5.30 Solar-powered catamaran *Solar Flair III* (Paul A. Lynn).

Figure 5.31 The pontoon boat *Loon* (*Source:* Reproduced with permission of Tamarack Lake Electric Boat Company).

Figure 5.32 *PlanetSolar* approaches Monte Carlo after its circumnavigation of the world (*Source:* Reproduced with permission of Planer Solar).

the greatest distance ever covered by a solar electric vehicle. The 31 m craft is covered by 537 m^2 of solar panels rated at 93 kW, fitted with 8.5 tons of lithium-ion batteries, and driven by electric motors in the twin hulls. The craft is entirely solar driven, absolutely no diesel engine!

The hull's fine lines allow it to reach speeds of up to 14 knots (26 km/h). *PlanetSolar* was given a rapturous reception at many international ports during its 584-day circumnavigation, which ended, as it had begun, in Monte Carlo. Registered in Switzerland, it has recently been used as a floating marine research laboratory by Geneva University.

The catamaran or pontoon form of hull is clearly very popular for solar-powered boats, with sleek twin floats providing a good stable platform for PV, especially when raised on a canopy. However there is nothing to stop designers from using conventional monohulls; the main criterion is an efficient low-drag hull that creates minimal wash and uses the precious PV energy to best advantage.

Finally, we consider the question, "What exactly makes a boat solar powered?" Exaggerated claims are sometimes made; it is easy to stick a PV module or two on a boat and claim that it is powered by the sun. But it does PV no good to overstate its performance and capabilities, leading to disappointment and skepticism. One answer is to use a simple measure known as the *solar boat index* (*SBI*) to quantify performance and allow sensible comparison of a wide variety of leisure boats used on lakes, rivers, and canals.[8]

The SBI is based on the peak sun hours concept introduced in Section 3.3.2. We have also used it to size PV arrays for water pumping in Section 5.5.3. It involves compressing the daily radiation received by an array into an equivalent number of hours of standard "bright sunshine" ($1 \, kW/m^2$). In this case the most relevant radiation data is that for a horizontal surface (most PV modules on boats are mounted horizontally) during the summer months of the boating season. An array rated at peak power P_{PV} watts and receiving an average S_p peak sun hours per day is expected to yield about $S_p P_{PV}$ watt hours per day. If the boat needs an input motor power P_M watts to cruise at a standard speed (normally taken as $8 \, km/h$) in calm conditions, then the SBI is defined as

$$\text{SBI} = \frac{\eta S_p P_{PV}}{P_M} \tag{5.11}$$

where η is a system efficiency that accounts for the PV generally operating away from its MPP, and for battery storage losses. Using typical figures of 80% (0.8) for the PV and 75% (0.75) for the batteries, the system efficiency $\eta = 0.8 \times 0.75 = 0.6$. If we now assume $S_p = 5$ (typical daily peak sun hours for midsummer in western Europe), Equation (5.11) becomes

$$\text{SBI} = \frac{3 P_{PV}}{P_M} \tag{5.12}$$

This is easy to remember and is in fact used in the United Kingdom to quantify the performance of solar-powered boats.

SBI has a simple interpretation. It represents the approximate number of hours per day, in average summer weather, that a boat can travel at standard speed on its PV electricity. For example, if a boat's SBI is unity, this means it can travel about 1 hour a day, or 7 hours a week at $8 \, km/h$, to give a range of $56 \, km$. Most inland leisure boats are weekend boats, for which this amount of cruising is fairly typical. Therefore it seems reasonable to describe leisure boats with SBI values of 1.0 or above as "solar powered" in the west European and similar climates; otherwise they are "solar assisted." Although the SBI is only approximate, it does provide a simple quantitative measure of a boat's cruising range on sunlight and allows the solar performance of different boats to be compared. The SBIs for our first two examples are as follows:

Solar Flair III: 4.0   *Loon:* 3.0

Clearly, these values need sensible interpretation because the patterns of use of the two boats are different and so are the solar climates in which they operate. What we can say is that, if they met together on a European lake, their SBIs should give a good indication of relative solar performance.

Worldwide, there are a number of competitions for solar-powered boats that act as good catalysts for new ideas and designs, encouraging young people to get involved. A good example is the Dutch Solar Challenge,[9] held biannually on canals and lakes in the Netherlands. Such events do an excellent job of bringing to public attention the exciting future of solar-powered boats with their silence, lack of pollution, and minimal environmental impact.

5.5.6 Far and Wide

The applications described in previous sections represent a broad range of technical, economic, and social objectives. Yet the scope and geographical spread of stand-alone PV systems stretch much wider. We end this chapter with a few more photographs and captions to illustrate some of PVs past and present successes and help stir the imagination for its future potential.

On Land and Sea

Figure 5.33 Two solar-powered cars, entered by the universities of Michigan and Minnesota, speed over 100 km/h along a Canadian highway during the 2005 North American Solar Challenge (Wikipedia).

Figure 5.34 It has become commonplace for sailors to install PV modules on the decks of ocean-going yachts to power cabin lighting, services, and navigation equipment. There is now growing interest in making the sails themselves "photovoltaic" (*Source:* Reproduced with permission of EPIA/Shell Solar).

In Heat and Cold

Figure 5.35 This installation in the Libyan Desert provides *cathodic protection*, an important application of PV that helps minimize corrosion of metal structures including pipelines (*Source:* Reproduced with permission of EPIA/Shell Solar).

Figure 5.36 A PV array produces electricity for a meteorological station in Greenland. In this high northern latitude, the vertical array captures much of the available sunlight, and solar cell efficiency is enhanced by the very low temperatures (*Source:* Reproduced with permission of EPIA/Shell Solar).

For Education and Information

PV school in China
© IT Power

Figure 5.37 An increasing number of schools worldwide use PV arrays to generate valuable electricity and stir their students' imagination for the future of renewable energy (*Source:* Reproduced with permission of EPIA/IT Power).

Figure 5.38 Another example of a large PV array in a remote location: this one helps to transmit information by telecommunications link (*Source:* Reproduced with permission of EPIA/Shell Solar).

Self-Assessment Questions

Q5.1 What features are required in a PV storage battery compared with a standard vehicle battery? Why is the voltage of a lead-acid battery not a straightforward indicator of its state of charge (SOC)?

Q5.2 Figure 5.6 illustrates series charge control. Explain clearly the function of the switch S, the diode, and the switch LVD.

Q5.3 What advantages do MPPT charge controllers have over simple series and shunt controllers?

Q5.4 Why are some large stand-alone PV systems backed up by diesel generators?

Q5.5 Why is the "sizing problem" often considered the most difficult aspect of system design in a stand-alone PV system?

Q5.6 Figure 5.15 shows a range of appliances powered by a stand-alone PV system in a holiday home. To what extent might the total energy requirement of 2200 W h/day be reduced by purchasing today's most efficient low-energy lights, TV, computer, and refrigerator?

Problems

5.1 A mountain refuge in the Swiss Alps is to be supplied with a stand-alone PV system for low-energy lighting, refrigeration, and occasional use of a microwave cooker. The predicted electricity requirement is 1 kWh/day. The lowest average number of daily peak sun hours on a south-facing, tilted, PV array is 1.8 in the month of November. Assuming a system efficiency of 65% (low because of snow coverage) and allowing 50% "oversizing" of the PV array, estimate the power rating of the array in peak watts.

5.2 The system in 5.1 needs a battery bank with 7 days of storage. Assuming a maximum discharge equal to 80% of nominal capacity and an inverter efficiency of 90%, how many batteries rated at 1 kWh each are required?

5.3 On a mountain cabin we have a refrigerator, microwave, five lamps, TV, and water well. The refrigerator uses 1.2 kWh/day. The microwave requires 1000 W power and we use it 5 min a day. The TV is 70 W and we use it 4 hours a day. The lamps are 20 W and we use them 4 h/day, and the well has a 200 W pump that is used 1 hour a day. If the average solar irradiation is 5.5 kWh/m²/day and we use 17% efficient PV modules, what is the power rating and the area needed for PV with a performance ratio (derated factor) of 0.80?

5.4 A PV water pumping scheme is needed for a village in Chile. The water table is 25 m below the storage tank's inlet pipe, and the daily water requirement is 10 000 l. The location has an insolation of four peak sun hours in the "worst" month of the year, and the overall system efficiency is 20%. Calculate: (a) the daily hydraulic energy required and (b) the approximate peak power of the PV array.

5.5 Batteries would be needed for the system described in 5.3 to work 16 hours per day. Assuming 12 V 225 Ah batteries with a level of discharge of 80%, how many batteries would be needed for two days of electricity storage?

5.6 If the PV array (including inverter) costs $2/Wp and is expected to last 30 years, each battery costs $100 and is expected to last 6 years and a power control unit with a life expectancy of 10-years costs $700, what would be the levelized, over 30 years, cost of supplying water from this pumping system? Assume an operating and maintenance cost equal to 1% of the initial capital cost. Use the Levelized Cost of Electricity (LCOE) concept discussed in Chapter 7 assuming that 50% of the cost of investment would be borrowed at an interest rate of 5%; also assume a discount rate of 10%.

5.7 What would be the cost of the same system under average U.S. irradiation conditions, thus 4.9 kWh/(m² day) insolation?

5.8 What would be the cost of supplying water from the same pumping station if instead of PV and PV with batteries you use a Diesel generator costing $5000/kW lasting 10 years with fuel price of $2/gal and an inflation rate for diesel of 5%?

Assume a 40% conversion efficiency of primary energy to electricity and a life expectancy of 10 years for the diesel generator. Compare the LCOE and the LCOW of this system with those powered with PV and PV with batteries.

5.9 You want to design a water pumping system with PV and batteries that requires 1 kW of power. Assume that the PV system can be sized for such power generation and that, with batteries you want to operate the pumping system continuously for 16 hours every day. What is the necessary number of batteries when each gives 12 V and 225 Ah and the depth of discharge is 80%? You can assume storage of one day.

5.10 Solar boat *Sunseeker* is a 6 m catamaran. She has four PV modules rated at 100 W_p, each mounted at deck level, and needs an input motor power of 600 W to cruise at 8 kph. Solar boat *Sunlight* is a 5 m monohull carrying six PV modules rated at 150 W_p each, mounted on a canopy. She needs an input motor power of 1 kW to cruise at 8 kph. Which boat would you expect to have the better solar performance, according to the *solar boat index (SBI)*?

Answers to Questions

Q5.1 A vehicle's battery provides large current, typically hundreds of ampers, in a very short time to start the car engine. In contrast, a PV supporting battery needs to supply small amount of current but for long time. Also stationary batteries need to handle a large number of deep discharges without degradation or damage. The voltage of a lead acid battery changes during its usage and SOC it is not a reliable indicator of SOC as the latter depends on the history of charge and discharge.

Q5.2 Switch S: controls charging of the battery by the PV. Diode: ensures reverse current cannot flow back into the PV at night. Switch LVD: prevents battery damage if voltage falls below a critical value.

Q5.3 They allow the PV to operate at a different voltage from the battery, extracting more energy and improving system efficiency.

Q5.4 To overcome the intermittency of solar power, especially providing energy at night.

Q5.5 Because of the variability of supply and demand.

Q5.6 The answer depends on the specifications of modern equipment.

References

1 A. Luque and S. Hegedus (eds.). *Handbook of Photovoltaic Science and Engineering*, 2nd edition, John Wiley & Sons, Ltd: Chichester (2011).

2 F. Antony *et al. Photovoltaics for Professionals*, Earthscan: London (2007).
3 NASA. *Surface Meteorology and Solar Energy Tables* (2010). eosweb.larc.nasa.gov/sse (Accessed on August 26, 2017).
4 S. Silvestre. Review of System Design and Sizing Tools, in T. Markvart and L. Castaner (eds.). *Practical Handbook of Photovoltaics*, Elsevier: Oxford (2003).
5 S. Bailey and R. Raffaelle. Space Solar Cells and Arrays, in A. Luque and S. Hegedus (eds.). *Handbook of Photovoltaic Science and Engineering*, 2nd edition, John Wiley & Sons, Ltd: Chichester (2011).
6 Eigg Electric. *Isle of Eigg Electrification Project* (2016). http://www.isleofeigg.net/eigg_electric.html (Accessed on August 26, 2017).
7 V. Fthenakis *et al.* New prospects for PV powered water desalination plants: a case study in Saudi Arabia, *Progress in Photovoltaics: Research and Applications*, 4, 543–550 (2016).
8 P.A. Lynn. What is a solar boat? *Electric Boat News*, 18(4) 13 (2005). See also http://www.eboat.org.uk (Accessed on August 26, 2017).
9 Solar Sport One. *Dutch Solar Challenge* (2016). http://www.dutchsolarchallenge.nl (Accessed on August 26, 2017).

6

Photovoltaic Manufacturing

In this chapter we will discuss the production of silicon and thin-film cadmium telluride (CdTe) and copper indium diselenide (CIGS) photovoltaics (PV) that comprise most of today's market.

6.1 Production of Crystalline Si Solar Cells

The production of silicon solar cells requires silicon with high purity (typically 99.9999%, six 9s) and high crystallinity, called solar-grade silicon. In order to produce pure and crystalline Si, the production goes over three basic manufacturing stages. The first produces metallurgical-grade Si (MG-Si, typically 98.5% pure) that is also used in the production of several metal alloys. The second step purifies the MG-Si to polysilicon with five to six 9s (i.e., 99.999–99.9999%) purity, and the third stage of manufacturing is to melt poly-Si and recondense it in a controlled fashion so that a crystalline structure is formed.

6.1.1 Production of Metallurgical Silicon

MG-Si is produced from silicon, the most abundant element on Earth. Silicon in nature is bound with oxygen in the form of quartz (silica, SiO_2). Separation of oxygen from silicon is energy intensive, requiring the reduction of SiO_2 with carbon in a carbon arc furnace, producing CO_2. This takes place in gigantic electric arc furnaces (Figure 6.1).

Modern factories generally use between two and six of these furnaces. The electricity is fed through three carbon electrodes that are submerged in a mixture of quartz, carbon (in the form of coal, charcoal, and petroleum coke), and wood chips. The electricity then cuts a path to the furnace's floor, which acts as a counter electrode. Since the mixture of raw materials is a poor conductor of electricity, it heats to as high as 2500°C. The heat causes the quartz to transfer its oxygen to the carbon forming gaseous carbon monoxide, which is burned as a carbon dioxide once it escapes the reaction mixture and is combined with atmospheric oxygen:

$$SiO_2 + 2C + energy \rightarrow Si + 2CO, \quad T = 2000°C \qquad (6.1)$$

The wood chips act as a loosening material so that the gas can escape. Liquid silicon and nonvolatile impurities gather at the bottom of the furnace. Every 2 hours the furnace

Electricity from Sunlight: Photovoltaic-Systems Integration and Sustainability, Second Edition.
Vasilis Fthenakis and Paul A Lynn.
© 2018 John Wiley & Sons Ltd. Published 2018 by John Wiley & Sons Ltd.
Companion website: www.wiley.com/go/fthenakis/electricityfromsunlight

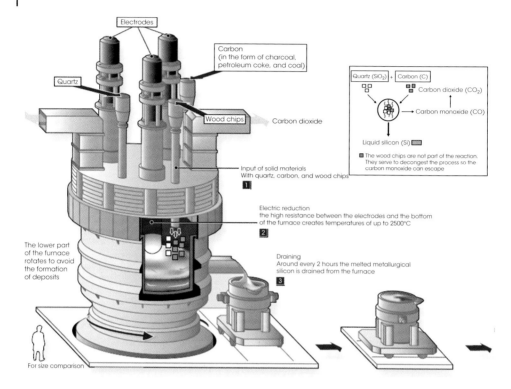

Figure 6.1 Manufacturing metallurgical silicon using an electric arc reduction furnace (*Source:* Reproduced with permission of Photon International magazine[1]).

is tapped and the liquid silicon drawn off (Figure 6.2). It then cools and hardens in a flat trough. The finished slabs are usually broken and sorted by particle size according to the customer's request.

The electricity used for producing MG-Si is about 12 000 kWh/metric tons of silicon produced. The global MG-Si production (average purity of 98.5%) is about two million tons. Most of it goes to the aluminum and chemical industries. The high purity demand by the PV and semiconductor industries absorb about 10% of the production.

Phosphorus, boron, titanium, chromium, and other elements have to be removed in additional steps to obtain the purified silicon. The impurities in MG-Si make it ineffective in solar cell production as they would lead to crystal defects and trapping levels that affect the cell power output.

The semiconductor-grade Si is typically seven to nine 9s, whereas a purity of five to six 9s is sufficient for solar-grade silicon. Typically 1.3–1.6 tons of MG-Si is needed to produce 1 ton of purified silicon, and some old, inefficient, or small-size facilities can require as much as 6 tons.

6.1.2 Production of Polysilicon (Silicon Purification)

The next step in the refining process is to react the MG-Si with hydrochloric acid, commonly in a fluidized-bed reactor, to form $SiHCl_3$ (trichlorosilane (TCS)) (Figure 6.3). This reaction, however, also produces by-products such as SiH_2Cl_2, $SiCl_4$, and chlorides

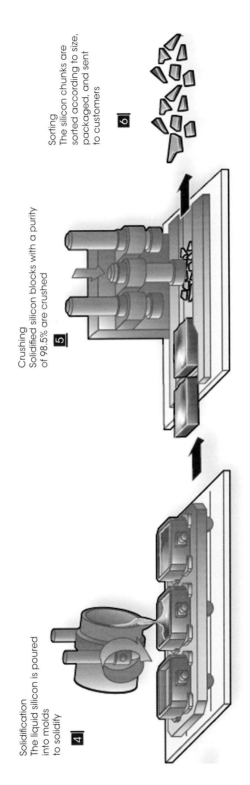

Solidification
The liquid silicon is poured into molds to solidify
4

Crushing
Solidified silicon blocks with a purity of 98.5% are crushed
5

Sorting
The silicon chunks are sorted according to size, packaged, and sent to customers
6

Figure 6.2 Subsequent steps in manufacturing metallurgical silicon using an electric arc reduction furnace (*Source*: Reproduced with permission of Photon International magazine[1]).

Figure 6.3 Production steps from MG-Si to polysilicon by the Siemens method.

of impurities. Each of these liquids has its own boiling point, so they are undertaken fractional distillation to separate the $SiHCl_3$ from the other components. The next step is to extract the silicon from the TCS. This is done by reacting $SiHCl_3$ with hydrogen in a Siemens reactor (shown in Figures 6.4 and 6.5):

$$2Si + 7HCl \rightarrow SiHCl_3 + SiCl_4 + 3H_2 + energy, \quad T = 350\,^\circ C \tag{6.2}$$

$$2SiHCl_3 + H_2 \rightarrow Si + SiCl_4 + 2HCl + H_2$$

$$SiCl_4 + H_2 + energy \rightarrow SiHCl_3 + HCl, \quad T > 1000\,^\circ C \tag{6.3}$$

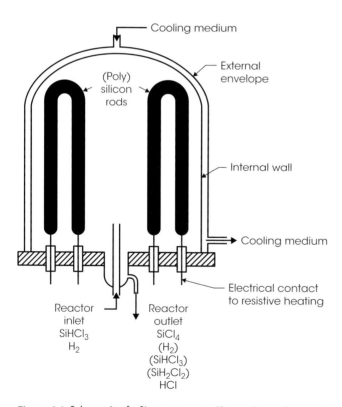

Figure 6.4 Schematic of a Siemens reactor (*Source:* Reproduced with permission of MEMC).

Figure 6.5 Interior of Siemens reactor with18 rods of polycrystalline Si (*Source:* Reproduced with permission of MEMC).

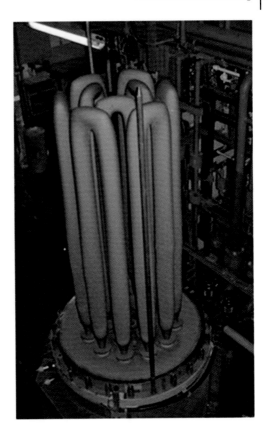

In this process, the TCS is fed into a reactor along with hydrogen, where thin rods of electronic-grade silicon, of about 8–10 mm diameter, are electrically heated to over 1100°C. Silicon from TCS or silane separates on to the hot silicon rods, reaching a diameter of 15–20 cm (Figure 6.5). For this to happen, it takes several days about 50–60 hours. The Siemens process is highly energy intensive, consuming about 40–50 kWh/kg of polysilicon produced. A lot of this heat is wasted to thermal losses through the reactor wall. To reduce losses, increase productivity, and reduce costs, polysilicon manufacturers have developed large reactors that produce up to 72 rods in each batch; Figure 6.6 shows reactors with 48 rods of polysilicon.

The gases that leave the poly (Siemens) reactor contain TCS, tetrachlorosilane (TET), dichlorosilane (DCS), hydrogen (H_2), and hydrogen chloride (HCl), and optimizing the reuse of them is an interesting systems engineering problem. TCS and TET can be separated with absorption and sent to the distillation processes before the Siemens reactor (Figure 6.3). TET is a by-product that needs to be either recycled or processed to silica for use by the silicon industry. Recycling into the poly-Si process can be achieved by converting it to TCS at high temperatures ($T > 1000\,°C$) in the presence of H_2 (Figure 6.7). However such recycling does not currently use all the TET output, and the industry is trying to reduce the TET concentration in the reactor output to eventually close the chlorosilane loop, according to reaction 6.3:

$$SiCl_4 + H_2 \rightarrow SiHCl_3 + HCl \tag{6.4}$$

Figure 6.6 Siemens reactors with 48 rods (*Source:* Reproduced with permission of Photon International).

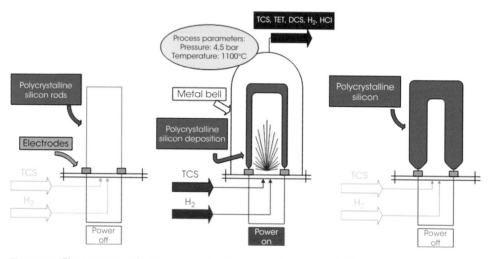

Figure 6.7 The operation of a Siemens reactor. The reactor inputs are trichlorosilane (TCS, $SiHCl_3$) and hydrogen (H_2); the outputs are trichlorosilane (TCS), tetrachlorosilane (TET - $SiCl_4$), DCS, H_2, and HCl (*Source:* MEMC).

Due to the high energy requirement, alternative poly-Si manufacturing processes are being investigated. Fluidized-bed reactor method is one such process that has been utilized in manufacturing by MEMC, REC Silicon, and others. In this process, TCS or silane is fed into the bottom of a fluid-bed reactor along with hydrogen. The gas stream

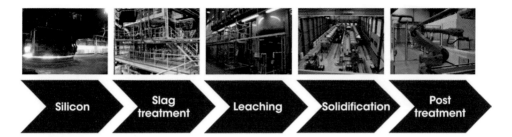

- *Silicon:* metallurgical silicon is produced from quartz in an electric arc furnace, at temperatures above 2000°C.

- *Slag treatment:* a purification process, in which the molten silicon is mixed with slag, in order to extract further impurities, especially boron.

- *Leaching:* a "wet" chemical refining process that, removes phosphorous and metallic impurities from silicon in solid form.

- *Solidification:* the silicon is melted and directionally solidified through which impurities are segregated and thereafter removed in the subsequent post-treatment process.

- *Post treatment:* surface washing and cutting.

Figure 6.8   The metallurgical refining process (*Source:* Reproduced with permission of Elkem Solar).

blends with small particles of silicon fluidizing them and creating what visually looks like a slow boil. Silicon from the TCS or silane deposits onto the particles and thus forms larger granules. Small seed polysilicon particles are fed into the top of the reactor, and product is withdrawn from the bottom of the reactor. Seed addition, gas feed rate, and withdrawal rates are used to maintain the targeted average particle size in the reactor. This requires process temperatures of 700°C for silane and 1000°C for TCS.

In both the Siemens process and fluidized-bed reactor method, the highly volatile liquid TCS is produced by combining MG-Si and hydrogen chloride and used as a feedstock. Depending on the reactor technology, it then further purified through distillation and if necessary converted to silane. The fluidized-bed reactor method consumes significantly less energy than the Siemens process, and it can be operated continuously.

Another promising alternative is directly purified Si (Elkem Solar grade). Such production starts with a carbothermic reduction of quartz to silicon metal (silicon production furnace) and then proceeds with a high-temperature slag treatment, a hydrometallurgical low-temperature cleaning, and in the end a directional solidification (DS) where the cleaning steps are fulfilled by cutting away side, bottom, and top; all cutoffs are returned back into the process. This route requires only a fourth of the energy required by the Siemens process; it is pioneered by Elkem Solar (Figure 6.8).

Figure 6.9 shows the basics of these three process paths for producing polysilicon.

6.1.3. Production of Crystalline Silicon

In addition to high purity, a semiconductor should provide clear paths to the movement of electrons and holes so that they do not recombine. So in addition to high purity this requires a certain degree of crystallinity. To produce single-crystal or multicrystalline silicon from the polycrystalline material, the silicon must be melted and recrystallized.

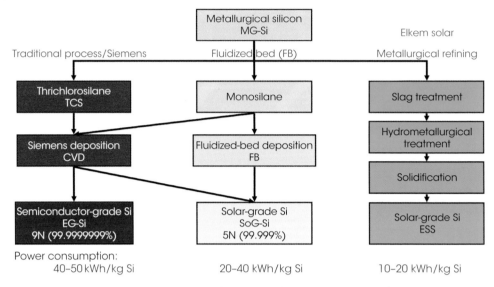

Figure 6.9 Polysilicon production processes for the PV industry (*Source:* Elkem Solar).

Monocrystalline material is produced by dipping a silicon seed crystal into the melt and slowly pulling it out (Czochralski method), and multicrystalline silicon is typically produced by DS.

6.1.3.1 Single-Crystal Silicon

The Czochralski (Cz) process starts with chunks of polysilicon stacked inside a graphite crucible lined with high purity quartz, which is inside a furnace (Figure 6.10). The machine is tightly sealed, purged to eliminate gas impurities, and fired. The silicon chunks are melted at about 1450°C and a controlled amount of dopant is added to the melt to form either p- or n-type silicon; a boron precursor is used in the production of p-type crystals and arsenic or phosphorus in the production of n-type crystals. Thus, one of the two dopants required for the formation of a p–n junction is introduced here, and the second will be introduced after the ingot is cut into wafers (Section 6.1.3.6, fabrication of the junction).

As the furnace heater ramps to temperature, the crucible begins clockwise rotation. Once the melt has reached the desired temperature, a counterclockwise rotating silicon

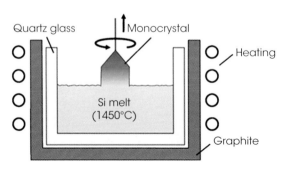

Figure 6.10 The Czochralski method for producing pure monocrystalline Si (*Source:* Mertens[2]. Reproduced with permission of Wiley).

"seed" crystal is lowered into the molten polysilicon. The melt is slowly cooled to the recipe's temperature as crystal growth begins around the seed. The seed is then slowly raised (pulled) from the melt, allowing controlled cooling and solidification. The temperature is controlled tightly so that silicon solidifies at the interface between the seed and the melt and the atoms arrange themselves according to the crystallographic structure of the seed. The crystal thus grows both vertically and laterally, aided by a rotation movement, yielding a cylindrical ingot of single-crystal silicon. The level of purity increases during the growth process since most impurities tend to segregate toward the liquid phase. The growth rate in the Czochralski method (Cz) method is about 5 cm/h, and the cylindrical ingots are typically 1 m long, 15–30 cm in diameter. The crystal is allowed to cool before it is extracted from the crystal puller for further processing. Figure 6.11 shows the top piece of a Cz ingot.

Figure 6.11 Top piece of Czochralski ingot (*Source:* Reproduced with permission of MEMC).

Float-zone (FZ) process is another method for producing monocrystalline silicon. It also starts from when polysilicon is slowly melted in a cylindrical bath by a coil using radio-frequency induction. By moving the coil along the bath, the silicon adopts the crystalline structure. A very high purity and also high structural quality can be achieved by performing several passes of the coil, thus maximizing impurity segregation. The typical growth rate is 15–30 cm/h, and the typical ingot is 15 cm in diameter and 1 m in length. FZ silicon is the preferred material for the fabrication of high-efficiency laboratory record cells, but it is not typically used in commercial production.

6.1.3.2 Multicrystalline Silicon

A simple method of producing multicrystalline silicon entails melting poly-Si into a crucible and carefully controlling of the cooling rate. During cooling a DS takes place and relatively large crystals grow in columns (Figure 6.12). The nucleation of the silicon

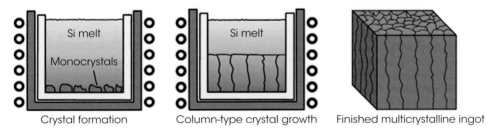

| Crystal formation | Column-type crystal growth | Finished multicrystalline ingot |

Figure 6.12 The directional solidification method for producing pure multicrystalline Si (*Source:* Mertens[2]. Reproduced with permission of Wiley).

Figure 6.13 Products in the basic production steps starting from quartz, metallurgical silicon chunks (or in the case of Elkem Solar their ESS® billet), followed by the mono- or multicrystalline ingot, wafers (ingot slices), solar cells, and modules (*Source:* Reproduced with permission of Elkem Solar).

atoms commences in many places simultaneously, leading to a myriad of crystals (or grains) of arbitrary shape and crystallographic orientation. Because of its multiple-grained structure, the material is called multicrystalline.

Each grain is several millimeters to centimeters across, and internally it has the same structure as single-crystalline silicon. The boundaries between the different grains are the most obvious imperfection in the material, but they are not the only ones. Microdefects are also common and contamination from the crucible can happen as well, not to mention the possible impurities present in the starting silicon. These crystallographic defects and impurities mean that mc-Si typically has lower electronic quality than the material produced by the CZ method, leading to a typical loss of efficiency of 1% absolute and a wider spread in the production statistics; this difference is, nevertheless, narrowing rapidly. The typical crystallization rate is 3.5 kg/h, and the growth cycle of a complete 160 kg ingot takes 46 h. This is, nevertheless, faster than the CZ method.

Figure 6.13 shows the major steps in the solar-grade mc-Si production followed by wafering, cell, and module production. The starting point is quartz, then polysilicon, crystalline solar-grade silicon, wafers, solar cell, and modules. Shown in this figure are metallurgical silicon chunks produced by the Siemens or fluidized-bed Si production, and the alternative Elkem Solar Silicon(R) blocks, both of which are precursors to the multicrystalline solar Si production.

6.1.4 Ingot Wafering

Wafering refers to slicing the cylindrical or rectangular brick of mono- or multicrystalline material produced from the crystallization phase to the thickness needed for solar cells. As we discussed in Chapter 2, effective absorption of photons by an indirect semiconductor requires a least a 100 μm thickness, although tricks exist for implementing smaller thickness.

The silicon block becomes what is called an "ingot" after the seed end (the top, shown in Figure 6.11) and the tapered end (the bottom) are cut off. These ends are sometimes discarded; however, to avoid a complete material loss, some ends are remelted and used in future crystal specifications. Then the ingot is cut into shorter sections and after quality testing is transferred to the slicing equipment to make thin (e.g., 130–180 μm) wafers. Before slicing, ingots have to be shaped to meet dimensional specifications. The cylindrical CZ ingots are usually reduced to a quasi-square shape. This implies a loss of about 25% of the material but is necessary if a high packing factor of the cells in the module is required. The large cast silicon parallelepipeds (Figure 6.13) are sawn into smaller bricks. In the case of mc-Si ingots, the shaping is also used to discard the peripheral regions that are usually heavily contaminated by the crucible, which represents approximately 15% of the ingot. Typical wafer sizes are $11.4 \times 11.4 = 130 \, cm^2$ and $15 \times 15 = 225 \, cm^2$. Multi-wire slurry saw (MWSS) machines can cut simultaneously whole blocks (Figure 6.14). An abrasive slurry (typically with silicon nitride) helps the steel wires cut through silicon, a very hard material indeed, and the cutting is very slow (e.g., ~8 hours). Slicing remains as one of the most costly and wasteful steps of the whole silicon solar cell fabrication. Even if very thin wires are used, approximately 30% of the silicon is wasted as saw dust, called "kerf loss." However, the industry starts implementing diamond cutting, and this reduces kerf losses.

Since circular wafers mounted in a module leave a large amount of empty space between the wafers, often times the edges of the wafers are trimmed to make the wafers closer to a square. Also the sawing causes significant surface damage, so the wafers are chemically etched to remove any residual slurry and restore the surface. Then, the wafers proceed into a series of refining steps to make them stronger and flatter. First, the sharp, fragile edges are rounded or "profiled" to provide strength and stability to the wafer. This will ultimately prevent chipping or breakage in subsequent processing. Next, each wafer is laser-marked with very small alphanumeric or bar code characters. This laser-mark ID gives full traceability to the specific date, machine, and facility where the wafers were manufactured. The wafers are then loaded into a precision "lapping" machine that uses pressure from rotating plates and an abrasive slurry to ensure a more

Figure 6.14 Multi-wire saw for cutting Si bricks or ingots into wafers.

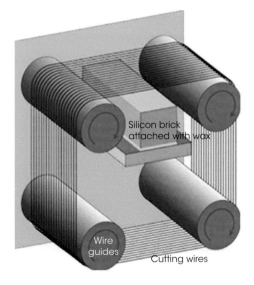

uniform simultaneous removal of saw damage present on both front and backside surfaces. This step also provides stock removal and promotes flatness uniformity—a critical foundation for the polishing manufacturing process.

The following process would be chemical etching or the removal of residual surface damage caused by lapping; it also provides some stock removal. During the etching cycle, wafers progress down another series of chemical baths and rinse tanks with precise fluid dynamics. These chemical solutions produce a flatter, stronger wafer with a glossy finish. All wafers are then sampled for mechanical parameters and for process feedback.

6.1.5   Doping/Forming the p–n Junction

The next step in producing single-crystal silicon cells is to create the p–n junction. This is accomplished with diffusing into the wafer the dopant that was not introduced during the crystal formation. Most commonly a p-type dopant (e.g., boron) was introduced in the melt of the Cz crystallization process, and an n-type dopant (e.g., phosphorus) is diffused into the wafer. The conventional way to diffuse P is to use a diffusion furnace where a P source (typically liquid $POCl_3$) is transferred by bubbling nitrogen. The diffusion can be performed using a tube furnace in a batch process where approximately 300 wafers are loaded in a quartz boat as shown in Figure 6.15. A phosphorus-containing gas, phosphoryl chloride ($POCl_3$), is introduced into the quartz tube furnace, where it reacts with O_2 to form phosphorus pentoxide (P_2O_5) that deposits on the oxidized wafer surface, providing the source of phosphorus atoms that diffuse into the wafer to form an n-type layer on the wafer's surface. A typical tube furnace diffusion process takes about 50 minutes and is performed at temperatures in the range of 800–900°C.[3]

Figure 6.15 Tube furnace showing a quartz boat loaded with silicon wafers in preparation for phosphorus diffusion. (*Source:* Lennon and Rhett[3]. Reproduced with permission of Wiley).

6.1.6 Cleaning Etch

Phosphorus diffuses into all the wafer surfaces and is necessary to clean the front surface from phosphorus atoms to prevent the creation of shunting pathway between the junction and the rear surface. The shunting resistance has to be large to prevent electron losses as shown in Figure 2.23. Usually such cleaning is done by etching the rear surface by HF and HNO_3.

6.1.7 Surface Texturing to Reduce Reflection

Wafers cut from a single crystal of silicon (monocrystalline material) need to be textured to reduce reflection (as discussed in Chapter 2). This can be done by etching pyramids on the wafer surface with a chemical solution. While such etching is ideal for monocrystalline CZ wafers, it relies on the correct crystal orientation and so is only marginally effective on the randomly orientated grains of multicrystalline material. For multicrystalline Si texturing could be done with either lasers or isotropic chemical etching.

6.1.8 Antireflection Coatings and Fire-Through Contacts

Antireflection coatings are particularly beneficial for multicrystalline material that cannot be easily textured. Two common antireflection coatings are titanium dioxide (TiO_2) and silicon nitride (SiN_x). The metal contacts can be applied at the top of the antireflective coating using screen printing with a paste containing cutting agents.

6.1.9 Edge Isolation

There are various techniques for edge isolation such as plasma etching, laser cutting, or masking the border to prevent diffusion from occurring around the edge in the first place.

6.1.10 Rear Contact

An aluminum layer is printed on the rear on the cell and with subsequent alloying through firing, it produces a back surface field (BSF) and improves the cell bulk through trapping impurities; this thermal treatment is called "gettering" and can be done with either P or Al at temperatures between 700 and 950°C under a controlled O_2 atmosphere, which allows P or Al to diffuse throughout the Si layers and to "getter" eventual metal impurities toward the phosphorus- or aluminum-doped layer. A second print of Al/Ag may be required for solderable contact. In most production, the rear contact is simply made using an aluminum/silver grid printed in a single step.

6.1.11 Encapsulation

The solar cells in numbers of 36 or higher are typically interconnected in series and encapsulated to form a module. The encapsulated materials and processes should protect the module for outdoors operation of 30 years. Typically tempered glass is used for the front cover, ethylene vinyl acetate (EVA) as an adhesive, and Tedlar as a back cover.

These layers are laminated by applying heat at about $150\,^{\circ}$C and pressure under vacuum. The edges are sealed with a neoprene gasket and most often protected with an Al frame. An alternative encapsulation scheme, employed in First Solar's thin-film module production, is using a second glass sheet as back cover and a sealing compound without a metal frame.

6.2   Opportunities and Challenges in Si PV Manufacturing

Let us reflect on the status of Si PV manufacturing and discuss challenges and opportunities for improvements in individual manufacturing stages.

Feedstock and Ingot Growth

Until about 10 years ago, purification techniques for silicon have been dictated the industry that employs electronic-grade silicon with purity levels of less than 1 ppb. A significant shift occurred on or about 2008 when the PV industry surpassed the integrated circuit (IC) industry as the largest consumer of refined silicon. Despite decades of steady growth, this transition caused short-term shortages and price spikes. However, silicon foundries responded by increasing capacity and examining strategies to develop "solar-grade" silicon. Electronic-grade silicon is produced primarily using the energy-intensive Siemens process. As we discussed earlier, new production strategies such as fluidized-bed technology are being investigated to upgrade MG-Si, potentially reducing these energy requirements by up to a factor of four. Though less expensive, these techniques often retain higher levels of metals such as Fe and Al than electronic-grade silicon. While these impurities would be catastrophic in IC manufacturing, such low levels may be tolerable in solar cells. An active area of research is focused on establishing the impurity levels that can be tolerated by solar cells, as well as devising processing strategies to mitigate and/or passivate these defects. Such improvements should reduce both the energy payback time and ecotoxicity associated with silicon production (discussed in Chapter 7).

Wafering

The cost of the silicon feedstock remains a large contributor in manufacturing costs, and kerf loss during cutting to form wafers significantly impacts this cost. To reduce the required amount of polysilicon, the wafer thickness was gradually reduced from 500 μm in 1979 to an average of 180 μm in 2016. The kerf loss, the silicon lost due to sawing, has decreased as well, to about 150 μm/wafer today. With further advances in the MWSS process, it is expected that 120 μm wafers with 120 μm kerf lost would be achieved by 2020. Aside from kerf loss, the MWSS process has the drawbacks of water contamination, high breakage, and high material consumption. To address these drawbacks, alternative wafering methods started being implemented, most notably a wire saw using a fixed diamond abrasive on the wire, instead of slurry. The kerf from diamond-wire cutting is recyclable in contrast to kerf from the slurry-based process that is not. However, the cost of the wire is much higher, and this can stand in the way of increasing

PV market penetration in markets dominated by artificially cheap fossil fuels. To avoid kerf loss altogether, kerfless cutting and the development of ultrathin silicon (ut-Si) are pursued.[4] These approaches are briefly reviewed in the following section.

Kerfless Wafers

Improvements in wire-saw technology have enabled the reduction of wafer thicknesses to 180 µm at present. However, over 50% of the silicon is lost as silicon sawdust or "kerf." While this material can be recycled if diamond-wire cutting is implemented, it would be desirable if the wire-sawing step could be eliminated altogether. Techniques for the direct production of wafers from the melt include ribbon technologies and ultrathin Si technologies.

The ribbon techniques are edge-defined film-fed growth (EFG) and string ribbon silicon technologies. In the EFG process the Si wafers are pulled out from the melt through a graphite die using capillary action. Ribbon silicon is produced by pulling a pair of high-temperature strings through a crucible of molten Si. These two growth techniques produce vertical sheets of mc-Si approximately 300 µm thick and up to a 100 mm wide. These technologies were early commercialized, but they became obsolete with the advent of high-volume, high-throughput thin wafer production from Asia.

ut-Si is a promising way to accomplish a kerfless production of Si. This refers to solar cell technology where the photon-absorbing silicon layer is on the order of 5–50 µm thick. Because silicon has an indirect bandgap, it is often assumed that silicon must be thicker than 100 µm to effectively absorb light. However, theoretical studies showed that with new emitter designs and improved surface passivation strategies, 19.8% efficient devices could be obtained with only 1 m of single-crystal silicon. If one could produce such materials using a kerfless process, it would result in an order of magnitude reduction in material cost with respect to today's state-of- the-art wafers: thus, ut-Si would merge the benefits of crystalline silicon with those of thin-film solar cells. Thin silicon is also amenable to use of bifacial architectures, which harvest light from both directions.

There are a few general approaches to the fabrication of ut-Si. The first strategy employs hetero-epitaxial growth followed by lift-off or removal of a sacrificial substrate. Material grown at high rates is typically polycrystalline, necessitating the use of post-processing techniques such as laser annealing or rapid thermal processing (RTP) to produce the desired material quality. One complication is identifying low-cost substrates with appropriate properties for these types of processes. A related approach involves deposition a-Si directly on glass followed by thermal recrystallization. A third approach involves "peeling" ut-Si layers off from silicon ingot. For example, Silicon Genesis introduced a process where this is achieved through a combination of ion implantation and thermal treatment, producing kerf-free wafers as thin as 25 µm. Ion implant technology accelerates protons and shoots them into crystalline silicon, where they settle at a finite depth. The hydrogen ions line up at that finite depth, where they are then heated, resulting in a wafer that cleaves right off the substrate along the crystalline plane. The proton energy determines how deep the protons go and therefore sets the thickness of the wafer (Figure 6.16).

Substantial challenges remain once ut-Si is produced. Achieving high efficiency will require the use of the most advanced technologies for both surface passivation and light

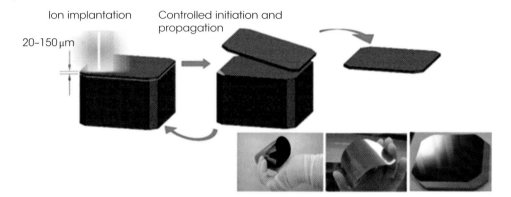

Figure 6.16 Silicon Genesis' kerfless wafering approach (*Source:* DOE SunShot web site).

trapping. Also commercialization of ut-Si technologies will need to address the nontrivial challenge of mechanically handling these ultrathin wafers while maintaining high throughput and low cost.[5]

Surface recombination (discussed in Section 2.2.3.4) becomes more important as the wafer thickness decreases, so very-high-quality passivation schemes are needed. Excellent results have recently been achieved on thermal SiO_2/plasma-enhanced chemical vapor deposition SiN_x stacks, but thermal growth of SiO_2 might be less favorable for very thin wafers. An effective low-temperature approach is to use amorphous silicon (a-Si), as in a-Si:H/crystalline silicon heterojunction (HIT) cells. The HIT and effective passivation increase the open-circuit voltage (V_{oc}) significantly: an efficiency of 23.7% has already been achieved on a 98 mm thick wafer.

Because front-to-rear interconnection and soldering of interconnects will induce too much stress on thin wafers, the International Technology Roadmap for Photovoltaics (ITRPV) roadmap expects 35% of all cells to be rear contact by 2020.[6] There are three main approaches to rear-contact cells: metal wrap-through (MWT), emitter wrap-through (EWT), and back-junction (BJ). In the first two approaches, the emitter is still at the front of the device, but holes are laser drilled through the wafer that transports carriers to the rear, through either the metal contacts (MWT) or the emitter (EWT). The main difference between MWT and EWT is that the MWT still has grid lines (but no bus bars) on the front surface. In a BJ cell, the emitter is located at the rear surface, typically in an interdigitated fashion with the BSF. A BJ has the benefit that the contacts can cover almost the whole rear side of the cell, greatly reducing series resistance. All three approaches reduce contact shading, although this is especially true for the EWT and BJ types. So far, large-area efficiencies of 24.2% have been reached on BJ solar cells and over 20% on MWT cells. An interdigitated back-contact (IBC) silicon HIT cell (IBC-HIT) has been reported at an efficiency of 20.2%, but simulations show that 26% conversion efficiency is achievable.

Traditional screen printing is reaching its limits too. It is a "hard contact" technique, which creates yield issues on thin wafers. Furthermore, the aspect ratio of fingers is rather low, resulting in either large contact area (which are high surface recombination areas) or small contacts with high resistance. Finally, the cost of silver in the metallization paste is becoming quite high, at 6–14 ¢/W_p today. All in all, this implies a move away from silver screen printing in the coming years. A first step may be the introduction of stencil printing, providing higher aspect ratios than screen printing, but the preferred option on the somewhat longer term seems to be plating of copper contacts. Reducing the wafer thickness relaxes the demands on the bulk diffusion length for efficient carrier collection but at the same time reduces the amount of light absorbed if no additional light trapping measures are taken. Random pyramidal textures are shown to perform quite well even at 50 μm wafer thickness, with demonstrated efficiencies of over 19%.

Encapsulation

The encapsulation step is responsible for most material and labor costs: as of 2010 it accounted for 30–40% of the total costs per W_p. Hence, there is much to gain by cutting down on material consumption and through automation and increasing throughput.

Front-to-rear interconnection through stringing and soldering is responsible for a large fraction of the yield losses today, and with decreasing wafer thickness, this may get worse. There are reports that interconnection yield losses are prevented in a fully auto-mated high-throughput module assembly line. Because for BJ cells most processing is on the back of the cell, it is possible to attach the wafers to a glass superstrate after processing of the front. This then allows for monolithic processing of the cells, which increases the yield of very thin wafers during cell processing.

6.3 Thin-Film PV Manufacturing

Thin-film PV are formed by depositing the semiconductor and metal compounds on the whole area of a substrate, typically glass, but it could also be metal or plastic, and then vertically scribing the films so that solar cells are formed. The two configurations shown in Figure 6.17 refer to the position of the substrate where the films are grown.

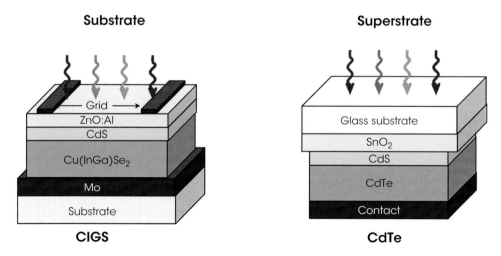

Figure 6.17 Configurations of CIGS and CdTe photovoltaics (*Source:* Courtesy of William Shafarman, U. Delaware).

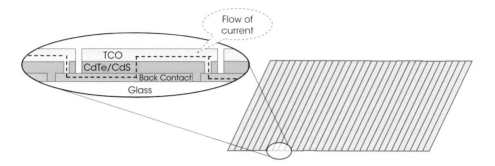

Figure 6.18 Laser scribing of thin-film layers; CdTe example (Vasilis Fthenakis).

The films in CIGS PV are deposited from the bottom up, whereas in CdTe are deposited on the top (sun facing) glass that called superstrate. Glass substrate device module needs transparent, UV-resistant moisture barrier on top, typically a second glass sheet, whereas superstrate device module needs a rear moisture barrier (either second glass or polymer).

The monolithic interconnection of the cells in thin-film modules offers cost advantages in comparison with Si PV module manufacturing. These include reduced number of processing steps, easier automation, less handling of materials, and less breakage. On the other hand, thin-film deposition requires large area uniformity for high yield. Forming solar cells from the thin films deposited on the substrates takes typically three scribing steps are taken place; those for CdTe solar cells are shown in Figure 6.18. TCO is the top electricity contactor; the bottom contactor is deposited on the glass superstrate. The first scribing separates the bottom conductor. The CdTe and CdS are then deposited and a second scribing is taken place very close to the first one. Subsequently CdS and TCO is deposited and a third scribing all the way down to the back contact is performed. A polymeric interlayer, for example, EVA, will fill in the last scribing shown in Figure 6.16 so that one cell is isolated from the other. Thus the top conductor from one cell contacts the bottom of the next cell and that will contact with the top of the next and so on. Connection in series-negative–positive–negative—is the same as batteries in series.

In summary, thin-film technologies where films are applied over the whole module area and then divided by scribing into individual cells have the inherent advantage over silicon cell technologies based on interconnecting small cells to form a module. The difference in the number of manufacturing process steps is shown in Figure 6.19.

6.3.1 CIGS Thin-Film Manufacturing

The first copper chalcopyrite PV devices were introduced in 1976 in the form of copper indium diselenide ($CuInSe_2$) by Larry Kazmerski and his coworkers at NREL. CIS has a suboptimal bandgap of 1 eV. However, the bandgap may be continuously engineered over a very broad range (1–2.5 eV) by substituting either Ga for In or S for Se. The abbreviation CIGS(S) is now used to describe this material as current manifestations often involve all five elements. The development of CIGS followed a combination of

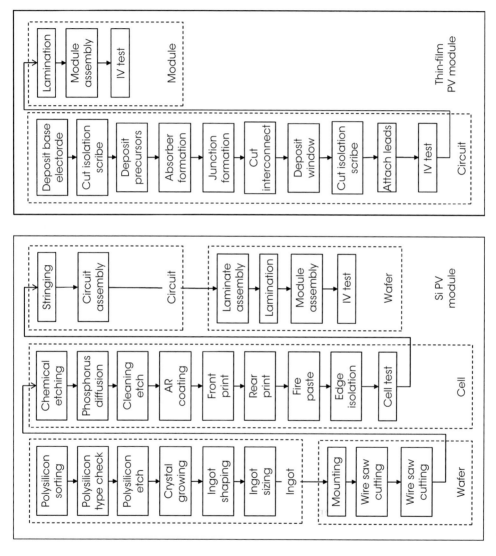

Figure 6.19 Process sequence for manufacturing Si and thin-film PV modules.

Substrate/Mo/Cu(InGa)Se$_2$/CdS/ZnO/TCO/grid

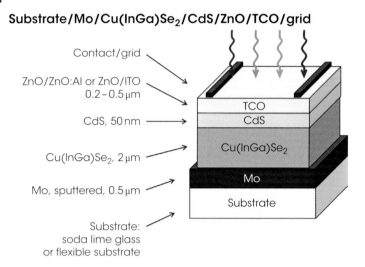

Contact/grid

ZnO/ZnO:Al or ZnO/ITO
0.2–0.5 μm

CdS, 50 nm

Cu(InGa)Se$_2$, 2 μm

Mo, sputtered, 0.5 μm

Substrate:
soda lime glass
or flexible substrate

TCO
CdS
Cu(InGa)Se$_2$
Mo
Substrate

Figure 6.20 Configuration of CIGS PV (*Source:* Courtesy of William Shafarman, U. Delaware)

starts and stops. The early 1990s brought a rapid succession of improvements that elevated device efficiencies to over 16%. Since then cell efficiencies in the laboratory reached a record of 21%. Several companies started production in 2007–2009 with 10–30 MW/year capacities, but as of 2016 only one company, Solar Frontier, Japan, was able to grow and maintain high-capacity manufacturing.

Substrates used in CIGS manufacturing include soda lime glass, metal foils, or polyimide (PI). The latter has garnered substantial interest for applications where flexible substrates have an advantage like building-integrated PV (BIPV) and portable power. However, in the case of deposition on flexible substrates, it is critical to match the coefficient of thermal expansion, with highest efficiencies obtained on titanium and stainless steel foils. The insulating nature of PI is advantageous for monolithic integration, but process temperatures are limited to less than 450°C, which limits efficiency.

The basic structure of the CIGS device is quite similar across manufacturers. Fabrication begins with the deposition of a Mo back contact followed by the p-type CIGS absorber (1–3 μm), a thin buffer layer (50–100 nm), with doped ZnO serving as the transparent front contact (Figure 6.20). Here the similarities end. Scores of firms worked to commercialize this technology, and each appeared to employ a somewhat unique strategy, particularly with respect to the formation of the CIGS absorber. An issue of progress in PV was completely dedicated to the topic of chalcopyrite thin films,[5] and the reader is directed to the papers in that issue for a more comprehensive overview of these topics. At present, the performance of commercial modules is 60–70% of the efficiency of champion cells, with much of this difference attributed to the quality of the absorber layer.

The approaches to CIGS fabrication may be classified into three basic categories: co-evaporation, selenization/sulfurization of metal films, and non-vacuum solution-based techniques. Here we assess the major advantages and issues associated with each and conclude this section by addressing the other major issues that impact CIGS manufacturing.

6.3.1.1 Co-evaporation

Co-evaporation is the process that has produced world records. This process alter-nates between copper-rich and copper-poor conditions to produce the large grains and graded Ga/In profile characteristic of high-efficiency material. Companies such as Q-Cells and Global Solar are pursuing in-line production using co-evaporation. There are a number of important practical challenges involved in the manufacturing of CIGS solar cells. Evaporation sources typically have a cosine flux distribution, and it is difficult to introduce sharp changes in composition or maintain uniformity over large areas under the diffuse conditions of high vacuum. In addition, sources must be mounted in a top-down configuration in order for large glass substrates to be supported and heated to 600 °C (Figure 6.21).

Another challenge with co-evaporation is that the relatively unreactive Se must always be supplied in great excess, leading to practical concerns related to condensation and material management.

6.3.1.2 Metal Selenization/Sulfurization

Another method for synthesizing CIGS films is selenization and sulfidization of a stack or alloy of the constituent metal films predeposited on a substrate in a predetermined stoichiometry.

Essentially this is a two-step process where in the metals (Cu, In, Ga) are sputtered onto the substrate and then converted to CIGS through annealing in a selenium- and sulfur-containing environment; either H_2Se and H_2S or vapors of elemental Se and S are used in these processes. The first are more reactive and speed up the process, but they create concerns due to their high toxicity. Practitioners include Showa Shell Solar/SolarWorld that as of late 2016 is the only large-scale manufacturer (stated module production capacity of ~1 GW/year). The best module efficiency obtained to date with this process is 15.7%.

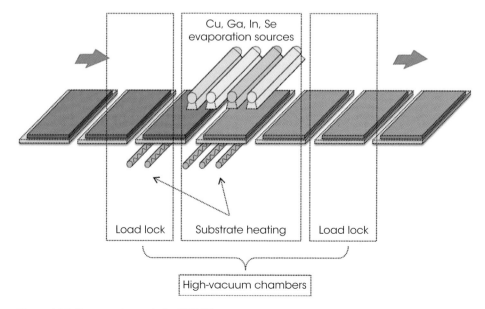

Figure 6.21 Process schematic for CIGS PV.

6.3.1.3 Non-Vacuum Particle or Solution Processing

The third general approach to CIGS manufacturing is to eliminate vacuum processing. In general, these are also two-step processes, application of a coating followed by a high-temperature step for annealing or sintering. Ostensible advantages include reduced capital requirements, improved materials utilization, potentially lower energy requirements, and compatibility with roll-to-roll (shown in Figure 2.29) processing. A general challenge with the non-vacuum approaches is the potential of contamination introduced by either the compounds themselves or the solvents employed. As such, it has been much harder to produce dense, homogeneous absorber layers. It is also more challenging to produce chemically graded structures with this technique. Record cell efficiencies trail co-evaporation and metal selenization, but values up to 14% have been obtained by a number of techniques.

The non-vacuum strategies typically include particulate deposition and solution processes. The particulate route is currently the most actively pursued, with variations employing particles composed of CIGS, metal, metal oxides, and/or metal selenides. In all of these methods, a coating of particles is first formed on the substrate surface and reacted and/or sintered at high temperature to form the final film. It was found that CIGS particles required excessive temperature for sintering and problems with handling and premature oxidation have limited this approach. The best results were obtained using slurries (also called inks) containing mixtures of metal oxide or selenide powders. Nanosolar in a pilot commercial operation based on this approach printed CIGS inks on aluminum foil in a roll-to-roll process. During a ten-year development effort, they only produced about 50 MW of modules at efficiencies of 10–11%. The MWT architecture chosen by Nanosolar was found to be unsuitable for low-cost production; also production yield was not satisfactory, and the company ceased to operate in 2013. It is noted that during 2010–2013, there were several CIGS companies that failed as they were not successful in quickly ramping up a low-cost production and could not catch up with the influx of inexpensive Si modules from China, as well as the fast-improving CdTe production from First Solar. The least of failed companies include Solyndra who launched an expensive commercial operation with a questionable[7] module design comprised of cylindrical cells, whereas companies with sustainable technologies like Global Solar, MiaSolé, HelioVolt, and Ascent were bought by either Chinese or Korean companies and are continuing their development and commercialization efforts. It is noted that Solyndra was always a dubious technology proposition; it combined the challenges of depositing CIGS films uniformly at high speeds with a module shape that it was both hard to manufacture and expensive to ship.

Commercial production of CIGS remained a challenge, as with five elements and several binary and ternary phases, the CIGS system presents much greater complexity than the CdTe that is discussed in the following text.

6.3.2 CdTe PV Manufacturing

CdTe has a number of intrinsic advantages as a light absorber. First, its bandgap of 1.45 eV is well positioned to harness solar radiation. Its high optical absorption coefficient allows light to be fully captured using only 2 μm of material. Like many II–VI compounds, CdTe sublimes congruently: it vaporizes homogeneously, and the compound's thermodynamic stability makes it nearly impossible to produce anything other than stoichiometric CdTe.

Figure 6.22 Conceptual schematic of lab-scale sublimation/vapor transport deposition of CdTe (First Solar).

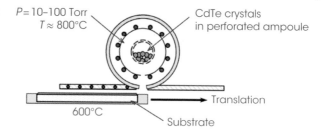

$P = 10\text{–}100$ Torr
$T \approx 800°C$

CdTe crystals in perforated ampoule

Translation

600°C

Substrate

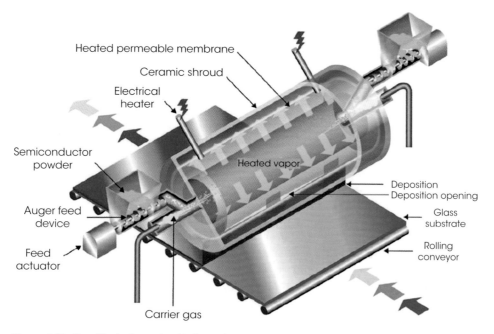

Heated permeable membrane

Ceramic shroud

Electrical heater

Semiconductor powder

Heated vapor

Auger feed device

Deposition
Deposition opening

Glass substrate

Feed actuator

Rolling conveyor

Carrier gas

Figure 6.23 Simplified schematic of industrial vapor transport deposition in CdTe PV manufacturing (First Solar).

Close-space sublimation employs diffusion as the transport mechanism; this is typically used in laboratory scales (Figure 6.22), while very high rates (>20 µm/m) are obtained in industrial scales using convective vapor transport deposition (VTD) (Figure 6.23).

As shown in Figure 6.21, there is a feeder providing a stream of powder into a gas flow, which then carries the powder into the hot zone. The powder vaporizes very fast and then through the use of appropriate geometries is directed to a substrate in the form of a vapor flow. The Auger feed device controls powder transport. The inert gas transports powder into the hot zone. The hot zone is created by the heated permeable membrane, and the ceramic shroud is used to control vapor flow toward the substrate as illustrated. Temperatures tend to be higher than in the lab equipment to faster vaporize the CdTe and CdS powders.

Standard CdTe-based devices employ a superstrate configuration: production begins with a glass substrate followed by the successive deposition of the transparent

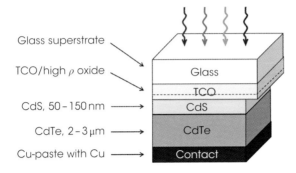

Figure 6.24 Configuration of CdTe PV (Courtesy William Shafarman, U. Delaware).

Glass superstrate

TCO/high ρ oxide

CdS, 50–150 nm

CdTe, 2–3 μm

Cu-paste with Cu

conducting oxide (TCO), the n-type window layer (CdS), the p-type CdTe absorber, and finally the back contact (ZnTe/Cu/C) (Figure 6.24). Part of First Solar's success has been due to their ability to integrate these various process steps into an in-line manufacturing process.

Another general element of the CdTe PV success is that the crystallinity of the deposited films can be improved by post-deposition thermal processing a $CdCl_2$ vapor; such post-processing options are not available for CIGS.

The thermal $CdCl_2$ processing promotes grain growth in the CdS and CdTe films and improves the uniformity and, therefore, the production yield of the modules.

With low manufacturing costs established, the biggest opportunities for CdTe lie in the improvement of device efficiency. Champion cells convert just over 70% of their SQ potential, while commercial modules are at 16.5% power conversion efficiency. Improving efficiency will require enhancements in both current and voltage. The former is perhaps the most straightforward route, as much of the blue region of the solar spectrum is absorbed in the TCO and CdS layers that make up the front contact. In record laboratory cells the FTO may have been replaced with advanced TCOs such as cadmium stannate and ITO. Likewise, the CdS window layer (2.6 eV) absorbs a significant fraction of the blue light. Integration of advanced front contacts into manufacturing appears to be the near-term strategy. This will not be trivial because ITO is expensive and cadmium stannate is a complex material. Furthermore, it is not clear what might be used to substitute for CdS though sulfides of zinc and indium have attracted significant interest. A new generation of CdTe devices is emerging that instead of CdS incorporate the alloy of CdSeTe in a graded structure with CdTe and ZnTe at the top of the metal contact layer.

The more daunting challenge is improving the voltage. The Voc of champion CdTe cells is well below that of similar bandgap PV materials. For example, the best Voc obtained in CdTe is 230 mV short of GaAs that has a similar bandgap. Short carrier lifetimes are the root of this limitation. The combined effect of defects and grain boundaries limits minority carrier lifetimes in polycrystalline CdTe to few nanoseconds.

A number of fundamental questions must be solved for CdTe PV to move beyond current records and go beyond the current 22.1% record efficiency. At present, the issue of extending carrier lifetime is partially addressed by chemical passivation. Examples include the introduction of O_2 during CdTe growth, post-deposition $CdCl_2$ treatments, and controlled diffusion of Cu from the back contact. Fundamental research in understanding these defects and how to passivate them would be transformative, leading to improvements in one of the most promising solar cell technologies.

A final area that deserves attention is the back contact. It is difficult to contact CdTe because it has low conductivity. Moreover, the back contact has been implicated as a potential contributor to degradation. The issues discussed earlier are nontrivial and will require substantial investment and fundamental research to resolve.

Self-Assessment Questions

Q6.1 What is called kerf loss in the slicing of silicon ingots onto wafers, and how can be reduced or eliminated?

Q6.2 Why silicon kerf loss from multi-wire saws cannot be recycled, whereas the same from diamond saws can be recycled?

Q6.3 What compounds can currently be recycled from the slicing of silicon ingots onto wafers?

Q6.4 What purity of silicon is required for solar cells?

Q6.5 What is the purity of metallurgical-grade Si and how is it further purified to solar-grade Si?

Q6.6 Why crystalline Si PV modules require more processing steps that thin-film PV modules (e.g., CIGS and CdTe)?

Q6.7 What processing steps in the manufacture of Si PV modules are intended to improve the optical properties of the module?

Q6.8 What processes are used for the deposition of CdTe and CdS layers in current commercial CdTe PV manufacturing?

Q6.9 What processes are used for the deposition of CIGS and CdS layers in current commercial CIGS PV manufacturing?

Q6.10 What are the thicknesses of the CdTe, CdS layers in CdTe PV modules, and those of CIGS and CdS in CIGS PV modules?

Q6.11 What is the function of copper chloride treatment in CdTe PV manufacture?

Q6.12 What is the approximate processing time of CdTe PV modules from glass entering the deposition champers to the final product ready for shipping?

Q6.13 What quality control steps are required for PV modules before being shipped out from a manufacturing site?

Q6.14   What type of testing is required for PV modules to ensure their reliability and longevity in the environmental?

Q6.15   What type of testing is required for PV modules to ensure their breakage resistance in the field?

Problems

6.1   Why is silicon opaque? For what range of wavelengths will Si appear transparent? What can be said about the bandgap of (transparent) glass? (You would need to find some Si physical properties.)

6.2   The density of Si is $2.3\,\mathrm{g/cm^3}$. Assume you are making square mc-Si wafers, $15 \times 15\,\mathrm{cm^2}$, $180\,\mu\mathrm{m}$ thick, leading to solar cells of 16% efficiency (assume wafers have 16% efficiency also). Assume $150\,\mu\mathrm{m}$ kerf losses (due to slicing) and sunlight of $1\,\mathrm{kW/m^2}$. How much power can you generate per kilogram of Si (i.e., what is the $\mathrm{kW_p}$ for wafers per kilogram of Si ingot)?

6.3   What volume and weight of Te and Cd is needed to produce a $100\,\mathrm{MW}$ CdTe PV plant if the CdTe thickness is $2.5\,\mu\mathrm{m}$. Assume equal mole ratio for Cd and Te in CdTe and PV module efficiency of 16%. (You would need to find Te and CdTe densities.)

6.4   Repeat the calculations above for material utilization rates of 90% and DC to AC and thermal conversion losses of 10%.

6.5   What are the weights of indium and gallium in a CIGS module of $1.2\,\mathrm{m^2}$ area? Assume a CIGS layer of $1.5\,\mu\mathrm{m}$ and that Indium occupies 13% of the layer's volume and Ga 5% of the same.

6.6   How many modules of the above specifications and 16% efficiency will be produced in a $500\,\mathrm{MW/yr}$ manufacturing facility? What weighs of In and Ga are needed in such facility if the material utilization is 90%?

6.7   What quantity of metallurgical silicon is needed to produce 1 metric ton of polysilicon? Assume the stoichiometric reactions depicted in 6.2 and 6.3.

6.8   What is the quantity of silicon tetrachloride generated in the polysilicon production per tonne of polysilicon production and what can be done with it?

6.9   What is the weight of multi-crystalline bricks needed to produce 10 000, 4-inch square-wafers with thickness of $180\,\mu\mathrm{m}$ each, using multi-wire slurry saws, when the kerf loss per cut is $60\,\mu\mathrm{m}$? Assume brick dimensions of 3 ft by 3 ft by 1.2 ft.

6.10   What is the number of wafers with the same dimensions produced from the same mass of multi-crystalline bricks when diamond saws are used instead of multi-wire slurry saws?

Answers to Questions

Q6.1 The silicon mass loss when slicing ingots to wafers.

Q6.2 Because the saw lubricants contaminate it.

Q6.3 Silicon carbide used for coating the wires of the saws.

Q6.4 five to six 9s; thus 99.999–99.9999%

Q6.5 ~98.5%. By processing to polysilicon

Q6.6 Si module production requires several earlier steps in wafer and silicon manufacturing before assembling cells onto modules.

Q6.7 Texturing and SiN_3 antireflective layering

Q6.8 Vapor transport deposition (VTD) for both CdTe and CdS.

Q6.9 Co-evaporation or reactive sputtering for CIGS and Chemical Batch Deposition for CdS.

Q6.10 2–3 μm for CdTe and CIGS; 0.2–0.5 μm for CdS

Q6.11 Listed in figure 6.19.

Q6.12 less than 2.5 hours

Q6.13 Mechanical strength, encapsulation, electrical.

Q6.14 Thermal cycling and moisture cycling.

Q6.15 Hail-impact simulations.

References

1 D. Sollmann. A highly coveted raw material, Photon International Magazine, pp. 136–141 (January 2009).
2 K. Mertens. *Photovoltaics: Fundamentals, Technology and Practice*, John Wiley & Sons, Ltd: Chichester (2013).
3 A. Lennon and E.R. Rhett. Manufacturing of Various PV Technologies, in A. Reinders *et al.* (eds.). *Photovoltaic Solar Energy from Fundamentals to Applications*, John Wiley & Sons, Ltd: Chichester (2017).
4 C. Wolden *et al.* Photovoltaic manufacturing: present status and future prospects. *Journal of Vacuum Science and Technology A: Vacuum, Surfaces and Films*, 29(3), 030801-1–030801-16 (2011).

5 A.N. Tiwari *et al. Prog. Photovoltaics*, 18, 389(2010). Special Issue: Chalcopyrite Thin Film Solar Cells, Volume 18, Issue 6 (edited by: A.N. Tiwary, D. Lincot, and M. Contreras).

6 S.A. Mann *et al.* The energy payback time of advanced crystalline silicon PV modules in 2020: a prospective study. *Progress in Photovoltaics Research and Applications*, 22, 1180–1194 (2014).

7 Scientific American. How Solyndra's Failure Promises a Brighter Future for Solar Power. https://www.scientificamerican.com/article/how-solyndras-failure-helps-future-of-solar-power/ (Accessed on August 26, 2017).

7

PV Growth and Sustainability

During 2000–2015, photovoltaics (PVs) enjoyed an average growth of approximately 45% per year. Is such a rate of growth **sustainable** in the long term, and what is the maximum level it could reach in the foreseeable future? To answer this question, we need to reflect on what sustainable development is all about. We may think of it as "development that meets the needs of the present without compromising the ability of future generations to meet their own needs." PVs, as fuel-free energy sources, are inherently sustainable unless they are too expensive to produce, are manufactured using materials that are depletable, or are environmentally unsafe. Measurable aspects of sustainability include **cost**, **resource availability**, **and environmental impact**. The question of cost concerns the affordability of solar energy compared with other energy sources throughout the world. Environmental impacts include local, regional, and global effects, as well as the usage of land and water, which must be considered in a **comparable context** over a long, multigenerational horizon. Finally, the availability of material resources matters to current and future generations under the constraint of affordability. More concisely, PV must meet the need for generating abundant electricity at competitive costs while conserving resources for future generations and having environmental impacts much lower than those of current modes of power generation, preferably lower than those of alternative future energy options. The challenges vary among different PV technologies.

For example, first-generation crystalline silicon PVs make use of abundant silicon, but its costs are relatively high. By comparison, second-generation thin-film technologies are cheaper to manufacture, but they use materials that are not so abundant. Early on, thin-film modules were not as efficient as the crystalline silicon, but the efficiency of cadmium telluride PV has now reached that of multicrystalline silicon. As discussed in Chapter 2, cadmium telluride thin-film modules can be produced at lower cost than crystalline silicon modules as they require fewer processing steps and their production throughput of about 2.5 hours is faster than other PV types. However, there are some concerns about the availability of tellurium and the toxicity of cadmium used as a precursor to CdS and CdTe. Copper indium gallium (di)selenide (CIGS) technologies share these concerns about material availability[1] (i.e., gallium, indium), and some high-performance silicon technologies have been using potent greenhouse gases (GHGs) (e.g., SF_6, NF_3) for reactor cleaning.[2]

Electricity from Sunlight: Photovoltaic-Systems Integration and Sustainability, Second Edition.
Vasilis Fthenakis and Paul A Lynn.
© 2018 John Wiley & Sons Ltd. Published 2018 by John Wiley & Sons Ltd.
Companion website: www.wiley.com/go/fthenakis/electricityfromsunlight

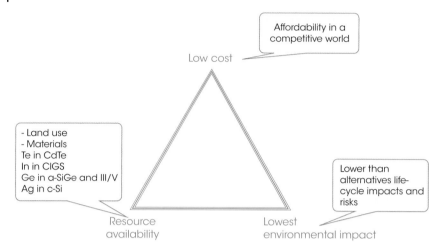

Figure 7.1 The three major pillars of PV large growth sustainability (Concept by Vasilis Fthenakis).

Assessing the sustainability of the rapid growth of PVs necessitates undertaking a careful analysis because PV markets largely are enabled by its promise to produce affordable and reliable electricity with minimum environmental burdens. Let us discuss in some detail the three pillars of sustainability shown in Figure 7.1.

7.1 Affordability

7.1.1 Costs and Markets

One of the most encouraging aspects of the current PV scene is the steady reduction in costs. Continuing improvements in cell and module efficiencies are making a substantial contribution, but above all it is the sheer volume of production in state-of-the-art factories using highly automated facilities that is driving down costs. Right back in Section 1.5, we introduced the "learning curve" concept to illustrate how, for a wide range of manufactured products, costs tend to fall consistently as cumulative production rises. Figure 1.14 showed that PV costs have fallen for more than two decades by around 20% for every doubling of cumulative production—and the trend continues. The long-held, cherished ambition of the PV community to produce modules at "one US dollar per watt" was finally achieved in 2009 in the case of high-volume thin-film CdTe manufacturing, and by 2017, the module prices for most commercial technologies had fallen to $0.40–$0.70/W.

Of course the cost of a PV system also depends heavily on balance-of-system (BOS) components, and there are design, installation, and maintenance charges to consider. Fortunately, most of these are also falling broadly in line with cumulative PV production and today typically represent—as they have in the past—about half of total system costs.

The speed of market penetration by a new technology normally depends greatly on economics. Potential purchasers of grid-connected PV systems, which have come to dominate the global market, wish to know how much generating solar electricity would cost them. For example, if you are considering installing a rooftop PV system, how does

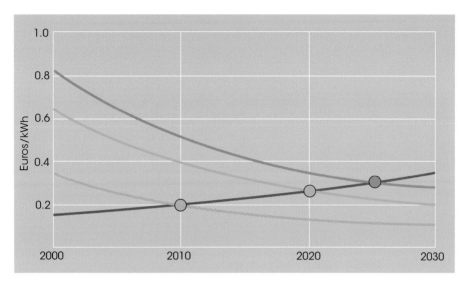

Figure 7.2 Toward grid parity in Europe.

the cost of a unit of electricity (1 kWh) compare with the price charged by the local utility, and does it look like an attractive investment? In the case of stand-alone PV systems, there are different criteria since grid electricity is not generally available as an alternative; comparisons are more likely to be made with diesel generators, and decisions affected by environmental concerns, including noise and pollution.

It is important to bear in mind that, in many cases, the installation of a PV system is not only about money. Companies may be concerned to demonstrate their green credentials, schools to educate and inspire their pupils, and individuals to "do their bit" to reduce carbon emissions. You may know someone who, instead of buying an expensive new vehicle, settled for a cheaper model that burns less fuel and spent the rest of the money on a rooftop PV system. For citizens in developed economies, it can be as much a lifestyle choice as a purely economic one.

As far as the economic case is concerned, Figure 7.2, although necessarily speculative, illustrates some important trends. Predicted costs of PV electricity are plotted up to year 2030 for electricity supplied by utilities to domestic customers in Europe (red curve) and for electricity generated by rooftop grid-connected PV systems in various countries (orange, green, and blue curves). Most experts expect that the increasing global demand for energy, together with falling fossil fuel reserves, will result in real price rises for conventional electricity in the coming years. This is shown by the red curve, assuming an annual increase of 2.5% compound. By contrast, the price of solar electricity is expected to fall as cumulative PV production soars. In sun-drenched European locations such as southern Spain and Italy (orange curve), the current cost is roughly competitive with conventional electricity because PV arrays are highly productive. In less sunny northern Germany and England (green curve), PV is expected to achieve "grid parity" by about 2020, and in Norway and Sweden (blue curve), perhaps 5 years later. But whatever the detailed timescales, the trends seem clear and inevitable—even if the citizens of northern Europe will need a bit more patience!

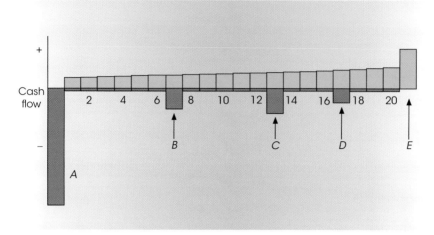

Figure 7.3 Positive and negative cash flows for a PV system.

In many ways this picture is oversimplified. First, the costs of PV systems and the prices paid by consumers for grid electricity are not uniform between different countries. Second, price increases for grid electricity over the coming years cannot be predicted with any certainty. And additional factors will surely influence the cost of PV electricity—a cost that is by no means dictated solely by the choice of modules and the amount of sunlight. To understand this, we need to consider the capital and income components of a PV project; we will discuss the residential and commercial or utility cases separately.

Let us again imagine investing in a residential rooftop PV system. It is helpful to start by estimating expected *cash flows* over the life of the system, say, 20 years, as in Figure 7.3. This is the key ingredient of what is known as *life-cycle cost analysis*.[1, 2] Negative cash flows (expenditure) are shown red; positive ones (income) are shown blue. A major feature of PV systems is that the initial capital cost (A) produces by far the largest negative cash flow. This is followed by many years of positive cash flows representing the value of electricity generated (or savings due to electricity not purchased) and small negative ones to pay for routine system maintenance. Generally, it is also prudent to allow for additional capital expenditure to replace worn-out or damaged BOS components such as charge regulators or inverters, or batteries in a stand-alone system (B, C, and D). And finally we may hope to obtain an end-of-life scrap value for the system (E).

We are now in a position to assess the financial viability of the project. Of various measures, the easiest to understand are the simple *payback period*, the number of years it takes for the total costs to be paid for by the income derived from the system, and the *rate of return*, the percentage annual return on the initial investment. But it is hard to know how long the system will last or to allow for additional capital injections that may be needed as time goes by (items B, C, and D above).

An even more important limitation is that the simple payback period and rate of return take no account of the "time value" of money—a major consideration for a long-term project. In a nutshell, a cash flow expected in the future should not be given the same monetary value today. For example, would you rather have $100 today, or the

Figure 7.4 Investing in the future: PV for a school in South Africa (*Source:* Reproduced with permission of EPIA/IT Power).

expectation of $150 in 10 years' time? Your answer will probably depend on predicting future interest rates (you could put the money in the bank), or the confidence you have about future payments, or you may prefer to purchase something for $100 today. A proper life-cycle cost analysis takes this into account by referring all future cash flows to their equivalent value in today's money using a *discount rate*. This is the rate above general inflation at which money could be invested elsewhere, say, between 1 and 5%. In this way the *present worth* of a complete long-term project can be estimated and compared with alternatives, allowing a more realistic investment decision to be made. As you may imagine, a positive value of present worth is generally taken as a good indication of financial viability.

So far so good, provided we recognize that the decision, even when based on careful life-cycle cost analysis, contains uncertainties about technical performance, system and component lifetimes, interest rates, and the future price of electricity. And, as we have previously noted, it may also be based on environmental and social factors.

Investment in a commercial or industrial system carries less uncertainty as it is often linked with a power purchase agreement (PPA) with a utility that guarantees revenue that, over the course of the investment period, carries all initial and recurrent expenses. Solar as well as other renewable energy systems may have a high up-front cost, but they use free solar or wind "fuel" during the many years they operate. On the other hand, the operating costs of coal and natural gas-burning power plants are high and uncertain as they depend on the price and availability of fuel. Thus to compare the cost of energy technologies with different operating conditions, we need to sum the capital and operating costs and assign them to the amount of electricity produced over a period of time. For this purpose, the levelized cost of electricity (LCOE) is defined as a constant unit cost, per kilowatt-hour or megawatt hour, of a payment stream over a period of time, which has the same present value as the total cost of constructing and operating a power-generating plant over its life. It represents the constant level of

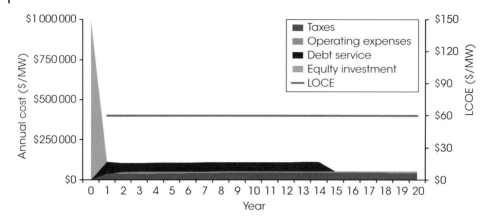

Figure 7.5 Initial investment, annual costs, and levelized cost of electricity (LCOE).

revenue necessary each year to recover all expenses over the life of a power plant. In simple words, LCOE represents the total life-cycle cost divided by the revenue from the total lifetime electricity production. The costs include the initial capital investment and the lifetime operating expenses, less the end-of-life value discounted into the present value. The revenue from the energy production depends on the location (solar irradiation), the performance ratio (PR), and the life of the system. A detailed discussion of these factors in the economics of PV systems is given by Campbell.[3] A simplified LCOE equation is shown in the following text and a calculator can be found at the book resource website: *www.wiley.com/go/fthenakis/electricityfromsunlight*.

An important constituent in the LCOE is the cost of financing (debt service) of the relatively high equity investment needed up front. The schematic in Figure 7.5 illustrates this concept. The cost of financing is linked to the real and/or perceived risk of the asset. Long-term PPA with utilities and government feed-in-tariff (FIT) programs reduce the risk to investors. An equally important consideration is that modern PV power plants are reliable, resilient, and utility friendly.[4] Deployment of PV power plants removes the uncertainty of fuel supply and prices, and, as more field experience is gained, these plants are expected to last longer than thermoelectric power plants, because they lack the high-temperature operating moving parts of steam and gas turbines. As PV grows and becomes more familiar, these attributes are more and more realized by the markets.

We have summarized the ideas behind conventional life-cycle cost analysis, with its positive and negative cash flows and levelized cost of electricity that averages all costs over a certain period. But what if the economics are affected by a government decision to offer capital grants to offset the initial purchase price or suddenly to change or terminate grants that are presently available? And what if the price paid for renewable electricity is bolstered by special tariffs that may be altered or removed by a change of government? Over the years there have been many such stop–go incidents in countries as wide apart as Australia, Spain, and the United States. One of the biggest threats to rational decision making and steady growth in the PV market is uncertainty about government policy, and one of the biggest benefits is consistent long-term support. We shall discuss support schemes in the next section.

Levelized Cost of Electricity (LCOE)

$$LCOE = CR + PTI + O\&M + FC$$

with

$$CR = \frac{OCC}{CF \times 8760} \times \frac{WACC(1+WACC)^{t}}{(1+WACC)^{t} - 1}$$

where

CR: capital recovery
OCC: overnight capital cost
O&M: annualized operating and maintenance expenses (variable and fixed)
PTI: annualized property taxes and insurance
FC: annualized fuel costs
CF: capacity factor of the plant
WACC: weighted average cost of capital
t: economic life of the plant

You may be wondering why governments offer financial support to PV in the first place. There are two principal reasons. First, the products of a new high-tech industry tend to be very expensive at the start, before cumulative production gathers pace. If governments wish to pursue urgent policy objectives such as the reduction of carbon emissions, they may decide to stimulate market development with financial incentives. Second, Figure 7.5 makes clear that PV, like other renewable energy technologies including wind and wave, has its major costs "up front," with no fuel charges. This is quite different from conventional electricity generation based on fossil fuels. Projects with high initial costs that must be set against future income are commonplace for large corporations but tend to be far more problematic for small businesses, organizations, and individuals who find it hard to raise the initial capital.

Government support, although generally welcome and necessary for PV, tends to distort the market and prevents it from behaving according to the assumptions of

(a)   (b)

Figure 7.6 Diverse markets for rooftop PV systems: (a) an elegant home in the developed world and (b) a "mobile" home in Mongolia (*Source:* Reproduced with permission of EPIA/Shell Solar, EPIA/IT Power).

classical economics. Realistic life-cycle cost analysis becomes more problematic. In effect the global PV market becomes split into a number of sub-markets with different characteristics. As an extreme example, the decision of an organization to install a large grid-connected system on its office building is likely to be influenced by very different financial criteria and incentives from that of a family in a developing country struggling to find initial funds for a solar home system (SHS). This is not to say that economic analysis is worthless, just that it should be approached and interpreted with caution. If you refer back to some of the photographs in earlier chapters, you will see plenty of examples of PV systems based on a wide range of investment criteria—political, economic, environmental, and social.

7.1.2 Financial Incentives

We have already noted that PV, an exciting new technology with major environmental benefits, both justifies and deserves the support of governments wishing to accelerate market growth and counter the effects of air pollution and global warming. The solar PV market grew very rapidly in recent years, mostly driven by government policies supportive of renewable energy; these early policies created a significant market that catalyzed technological improvements that reduced costs. In turn, cost reductions enabled worldwide deployment and further cost reductions.[5]

Japan showed the way in 1994 with a 70 000 solar roofs program. Germany, after succeeding with its own 100 000 roofs program, went from strength to strength after 2004, thanks to improvements in its groundbreaking renewable energy legislation. Spanish government legislation led to an extraordinary burst of activity in 2008 when $2.7\,GW_p$ of PV capacity was installed in a single year (you may like to refer back to Section 4.7 on large PV power plants). The United States, held back during the years of the Bush administration, surged ahead during the Obama years. In spite of a certain amount of stop–go in all these programs and difficulties due to the global economic recession, many other governments around the world have now joined the pioneers by offering substantial financial incentives to install PV systems.

Of the various ways in which governments have sought to provide financial incentives for the installation of grid-connected PV systems, three key ones are particularly relevant to our discussion here:

- Capital grants, in the form of rebates and investment tax credits (ITCs), to offset the initial cost of the system
- Special tariffs for the electricity generated, which is either used on-site or fed into the grid
- Financing options and loan warrantees

7.1.2.1 Capital Grants
Referring back to Figure 7.3, the capital grant route is designed to reduce a project's initial negative cash flow, denoted by the letter *A* in the figure. Such grants, often covering 30% or more of the purchase price, are funded out of general taxation and are therefore paid for by all taxpayers. One disadvantage is that the money is paid up front, generally with no redress if the system is poorly maintained and fails to produce the expected amount of electricity. Another is that governments normally "cap" the total

Figure 7.7 Rooftop arrays on the Reichstag building in Berlin exemplify the German government's support for PV (*Source:* Reproduced with permission of EPIA/Engotec).

amount of money available, which can lead to an initial rush of grant applications that rapidly exhausts the fund—a perfect recipe for stop–go market development, unless the scheme is constantly reviewed and reactivated. Such rush to the market, incentivized by generous rebates, exhausted the markets in Spain and Greece. More sustainable rebate programs are those paid on expected performance. For instance, the rebate program for solar PV in California under the California Solar Initiative (CSI) aims to support 3 GW of solar PV in 2017 using "buy-down" or "performance-based incentives." For systems 50 kW or smaller, the buy-down level is calculated from the system's expected performance, taking into account tilt, location, and orientation. The subsidy is referred to as "expected-performance-based buy-down."

Tax credits are also a powerful financing incentive; different variations of tax credits have been implemented in several countries. These credits are in the form of percentages of capital investments made, such as ITCs, or in the form of dollars per each unit of electricity generation as are the production tax credits (PTCs). Whatever the form of the credit, the idea is to reduce the levelized cost of solar electricity to be on par with that of other forms of electricity generation. The 30% ITCs significantly leveraged the development of solar energy in the United States.

Despite their instrumental role in promoting solar energy, the proponents of conventional energy resources often criticize the ITCs for placing additional burdens on the budgets of the countries. However, ITC programs can generate positive returns for the governments in the form of direct payroll taxes and other revenues. An analysis by the US Partnership for Renewable Energy Finance demonstrates that, over the lifetime of the solar assets, leases and PPA-financing structures can deliver a nominal 10% internal rate of return (IRR) to the federal government on the federal ITC for residential and commercial solar projects. Accordingly, a $10 500 residential solar credit can deliver a $22 882 nominal benefit to the government, while a $300 000 commercial solar credit can create a $677 627 nominal benefit in lease and PPA scenarios over 30 years. These government returns are generated by the direct participants in a solar transaction, that is, the developer (or an investment fund established by the developer), the system installer, and the energy user.

7.1.2.2   Special Tariffs

The second approach, tariffs for the electricity generated, increases the amount of income received over the life of the system. It therefore encourages the purchase of high-quality systems that are carefully installed and maintained. Often taking the form of Feed-in Tariffs (FITs), the subsidies are financed by requiring utilities to buy renewable electricity at above normal market prices. The tariff is based on the cost of the electricity plus some reasonable return for the investor. From a government viewpoint, FITs are generally "revenue neutral." The cost is spread over all customers who must pay a small annual percentage increase in their electricity bills. Their major advantage is the guaranteed income payments offered over timescales of 20 or 25 years, reducing uncertainty and increasing investor confidence.

FITs are implemented in many jurisdictions including EU countries, Australia, Brazil, Canada, and China and in California. The FIT has played a major role in boosting solar energy in Germany and Italy. In Germany, a renewable energy law passed in 2000 introduced a FIT that proved extremely effective at stimulating a range of renewable energies. The PV tariffs were tweaked in 2004 to compensate for the termination of the German 100 000 roofs program, providing payback times of around 8–10 years. This resulted in a veritable boom in PV installations. Huge numbers of PV arrays were put on domestic and commercial buildings, farmers placed PV on barns and in fields, and many large PV power plants were commissioned. In 2005 total installed capacity in Germany exceeded $1\,GW_p$, and in 2008 it reached $6\,GW_p$. Of course, a generous FIT can become unsustainable if continued too long, so in many cases tariffs for new installations are lowered by a certain percentage each year to take account of PV's expected "learning curve." Guaranteed payments over a period of years offer investors sufficient confidence to fund the initial capital in solar projects. However, the challenge with FIT designs is that it can over- or under-compensate solar projects as there is no market pricing mechanism in many programs. As the costs of FITs usually are passed on to consumers, it is essential to design incentives that attract sufficient investment while yet permitting adjustment of subsidies for new additions to capacity as technology costs fall, so avoiding unnecessary increases in electricity prices and maintaining public acceptance.

An extra generous Spanish FIT started in 2004 propelled Spain on an exciting journey into the gigawatt era. However, a few years later, the Spanish government introduced annual caps, and this dampened a market that had surged beyond expectation.

More than 60 other countries have now entered the FIT arena, and many are no doubt learning from the operational experiences of the pioneers. And in spite of the negative effects of the global economic recession that started in 2008, most commentators believe that PV and other renewable energy technologies will ride the storm relatively unscathed and continue to attract the support of governments increasingly focused on the dangers of global warming.

Figure 7.8 This PV factory is in Malaga, Spain (*Source:* Reproduced with permission of EPIA/Isofoton).

7.1.2.3 Financing Options

The third approach for directly assisting the growth of PV deployment is to provide financing options that reflect the high societal value of solar installations. The cost of financing ("debt service" in Figure 7.5) could amount to one-third of the LCOE, and early-on financing of solar projects has been especially challenging given the relatively high up-front costs and the lack of business models for financing in the private sector. There were also early reliability and life expectancy concerns, but those have been addressed by the PV industry; in our days PV power plants have earned the approval of many utilities worldwide and the appreciation of investors like Warren Buffett who made multi-billion-dollar investments in large 500 MW PV power plants in south California. Such big investments were made possible by loan financing and loan warrantee programs. Several governments have loan financing programs available for solar energy projects. In India, for example, Shell Foundation worked with two investment banks to develop renewable energy financing portfolios. This project helped the banks put in place an interest rate subsidy, marketing support, and a vendor qualification process. Within 2.5 years, these programs had financed nearly 16 000 SHSs.

In the United States, the Energy Policy Act of 2005 authorized the Department of Energy (DOE) to issue loan guarantees for projects that "...avoid, reduce or sequester air pollutants or anthropogenic emissions of greenhouse gases; and employ new or significantly improved technologies as compared to commercial technologies in service in the United States at the time the guarantee is issued."

In addition to these direct financial incentives, important policy issues that can create markets for renewable energy are renewable portfolio standards (RPS) and carbon programs.

7.1.2.4 Renewable Portfolio Standards

In the RPS programs, renewable energy production or consumption targets are set, and the electricity suppliers (utilities or the load-serving entities) are obliged to meet those targets or pay fines. Most of the time, the suppliers are required to meet a certain percentage of their retail electricity sales through renewable energy. In some cases, there are installed capacity targets for renewable generation (e.g., the Texas market). The RPS programs create a trading regime wherein utilities with low achievements in renewable energy can buy from those with levels above their requirements.

RPS especially has emerged as a popular form of policy in the United States; 31 out of the 50 states have some form of RPS programs. The standards range from 10 to 40%. New Jersey became the first state to create a carve-out in its RPS program for solar energy and elevated NJ to the number 2 or 3 solar market in the United States. This is a fair policy; although PV is improving fast, having the best learning curve among renewable energy technologies, it is still less mature than wind and small hydro, and it deserves the early push that other technologies enjoyed in the past.

Quota obligation schemes based on tradable green certificates (TGCs) have become a popular policy instrument in the Nordic countries and Poland. Also, India has launched a new renewable energy certificate (REC) scheme that is linked to its existing quota policies.

7.1.2.5 Carbon Fees/Programs

Greenhouse Gas (GHG) emission programs with carbon fees per ton of CO_2 emissions help in levelizing the cost of solar power to that of fossil-fired power generation. Solar PV generation benefits indirectly from incremental carbon costs as the fossil fuel power generation that PV displaces emits high volumes of CO_2. The primary goal of charging for carbon emissions is to reduce GHG emissions, causing global warming.

Assuming the low end in future estimates of incremental CO_2 costs, \$20/Mt would increase the levelized cost of electricity from coal-fired generation by 2 cents/kWh, and it would level the cost of electricity from natural gas-fired generation by 1 cent/kWh. Ideally, these taxes would be counterbalanced with reduction on personal income taxes, and revenues from those would be used in electricity grid and renewable energy developments.

While the deliberations following the COP21 conference in Paris continue regarding agreed binding targets, it is important that individual jurisdictions move forward with their GHG programs. Let us now look at policies in individual countries and what we learned from them.

Japan

Japan was the worldwide market leader in installed solar generation capacity until the end of 2004 despite its scarcity of wide open fields suitable for installing large-scale PV systems and relatively low solar irradiance throughout the year. That success was driven by long-term Japanese PV research and development (R&D) programs, as well as market implementation that started in 1994.

In 2008, the Japanese government announced "Action Plan for Achieving a Low-carbon Society" that targets increasing by ten-fold the installations of solar power generation systems by 2020 and 40-fold by 2030. That same year, "Action Plan for Promoting the Introduction of Solar Power Generation" announced measures to support the development of solar technology and promote installation of solar in selected sectors. As directed by these action plans, the Ministry of Economy, Trade and Industry (METI) announced its FIT policy in July 2010, which took effect in 2012. Under this FIT scheme, if a renewable energy producer requests an electric utility to sign a contract to purchase electricity at a fixed price and for a long-term period guaranteed by the government, the electric utility is obligated to accept this request. In the 2012 FIT scheme, solar PV generation was given a 42 yen/kWh fixed price ($\approx$50 cents/kWh) of 20 years for projects greater than 10 kW and of 10 years for projects smaller than 10 kW. The high tariff rates of solar energy mostly are necessitated by the low solar irradiance and are justified by the high costs of imported natural gas and oil in Japan.

Since the surplus electricity purchase system that allows customers to sell their excess solar electricity back to the power grid was established in 2009, the introduction of residential PV power generation increased greatly in Japan.

Germany

In 2000, the German government introduced a large-scale FIT system under the "German Renewable Energy Sources Act" (EEG). It resulted in explosive growth of solar PV deployment. In 2011, Germany produced 14% of its energy from renewable sources that has been attributed to the success of its comprehensive FIT system. German FIT payments are technology specific, such that each renewable energy technology type receives a payment based on its generation cost, plus a reasonable profit. Each tariff is eligible for a 20-year fixed price payment for every kWh of electricity produced. Germany's FIT assessment technique is based on a so-called corridor mechanism, which sets a corridor for the growth of PV capacity installation that is dependent on the PV capacity installed the year before; this results in a decrease or an increase of the FIT rates according, respectively, to the percentage that the corridor path is exceeded or unmet. As PV capacity installations were above those planned by the government in 2010, the FIT rates were decreased by 13% in January 2011.

Germany's generous FIT system has been criticized for not producing the desired results in accord with its total costs of nearly $30 billion euros between 2000 and 2010. In its report on German energy policy, the IEA suggested that "policies funding R&D activities can be more effective in promoting PV than the very high feed-in tariffs," on the ground that "the government should always keep cost-effectiveness as a critical component when deciding between policies and measures." The absence of German PV manufacturers from the list of top solar PV companies shows that Germany worked mostly as a "pull" market during the last decade. Nevertheless, Germany created a large market that enabled the drastic cost reductions the whole world is enjoying.

United States

The United States has a combination of "pull" and "push" policies toward developing solar energy, some of which were discussed previously. Among various state- and federal-level "pull" incentives, perhaps the most effective one is the federal ITC for solar PV projects that is equal to 30% of expenditures on any equipment that employs solar energy to generate electricity. The ITC program has been a key driver in the increase of solar PV deployment in both commercial and residential applications. It started in 2006, and then in 2008 it was extended to 2016 with bipartisan support. The multi-year extension of the ITC helped annual solar installation grow by over 1600% from 2006 to 2015. The ITC was lately extended to 2018 at the 30% level; after this it would gradually be reduced to 10% by 2022. Such predicable policy would greatly assist the continuation of PV growth in the United States.

An interesting development in the United States was the Clean Power Plan (CPP) ruling promulgated by the Environmental Protection Agency (EPA), which categorized CO_2 as a hazardous pollutant, a designation that gave it the authority to regulate it. The CPP targeted a 32% reduction of CO_2 emissions in the power industry by 2030. It has come up with a number of options for curbing the carbon emissions, such as energy efficiency initiatives at power plants, shifting away from coal-fired power to natural gas, investing in renewable energy, and implementing carbon capture and sequestration (CCS) technologies. According to the plan, states must develop their own regulations to meet these carbon emission reduction goals.

However, the coal industry and some utilities created a legal argument against the authority of the EPA to regulate CO_2, and the initiative was blocked and then withdrawn by executive order of president Trump. Once America has its national CO_2 emission targets in place, solar PV projects will get a major boost to their growth as they reach parity with the grid cost faster.

The DOE's loan guarantee program is a good example of "push" policy incentives, which supports the manufacturing side of the industry. Increased R&D support also is designed to sustain a technological advantage that is a prerequisite for the continuation of PV evolution.

China

The rapid development of the PV industry and market in China primarily reflects governmental support. Programs for rural electrification were the driving force for expansion of the solar PV market in China in the last two decades. Most PV projects were government sponsored with international aid, or within the framework of government programs at the national or local levels. China's energy policy is developed through a two-step approach. The central government first sets up broad policy goals in its five-year plans. Ministries, agencies, and the National People's Congress then use those plans to design specific and targeted programs and policies.

The major supporting programs are the Brightness Program Pilot Project, the Township Electrification Programs, and the China Renewable Energy Development Project (REDP). The plans in the Brightness Program Pilot Project provided electricity to 23 million people in remote areas using 2300 MW of energy from wind, solar PV, wind/PV hybrid, and wind/PV/diesel hybrid systems. The Township Electrification Programs, launched in 2002, installed more than one thousand small hydro and PV/wind hybrid systems in 2005. The China REDP, also established in 2002 and supported by the World Bank's Group Global Environmental Facility (GEF) grant, afforded a direct subsidy of $1.5/W to PV companies to help them market, sell, and maintain 10 MW of PV systems in Qinghai, Gansu, Inner Mongolia, Xinjiang, Tibet, and Sichuan.

India

In India, the primary policy driver is all-in FITs of around 15 cents/kWh for solar PV and thermal projects commissioned after March 2011 for up to 25 years. Solar PV projects in remote locations even receive higher subsidies. One such program that aims to establish a single light solar PV system in all non-electrified villages covers 90% of costs of projects. For below-poverty-level families, state governments underwrite 100% of the system costs. India plan to develop 100 GW of solar electricity by 2022.

The countries we discussed previously are a small sample of the approximately 110 countries that in 2015 had some type of renewable energy policy. More than half of them are developing countries or emerging economies. Of all the policy instruments that were detailed in the earlier section, FITs and RPS are the most common.

One of the main drivers of solar energy development especially in developing countries is public investment. Emerging economies (e.g., China, India) host several government- and donor-funded projects to support solar energy under their rural electrification programs.

7.1.3   Rural Electrification

So far we have been concentrating on the economic aspects of grid-connected systems and the ways in which governments in developed nations encourage the development of PV markets. Passing reference has been made to stand-alone PV systems, noting that the chief competitor for supplying electricity in remote areas is generally the diesel generator. But all this relates to relatively wealthy nations including those that have driven PV's spectacular growth over the last decade.

There is another important dimension to the terrestrial PV story, and it concerns the provision of relatively small amounts of solar electricity to families and communities in the developing world, who have little prospect of buying and maintaining diesel generators and no prospect of connection to a conventional electricity grid in the foreseeable future. This challenging yet worthwhile activity is referred to as *rural electrification*.

A major aspect of rural electrification is the supply of *SHSs* to individual families, and we shall concentrate on it here. Other applications include irrigation and water pumping (see Section 5.5.3), refrigeration of vaccines and medicines in remote hospitals, and the supply of PV systems to small businesses and institutions. It is sobering for those of us who live in developed countries to realize just how little electricity is needed to provide valuable services to people who otherwise have none. For example, the average electricity consumption of a household in Western Europe is around 10 kWh/day. The stand-alone system for a holiday home that we designed in Section 5.4.1 (see also Figure 5.15) assumed a consumption of 2.2 kWh/day, sufficient to run a good range of modern electrical appliances if used with care. But when we consider an SHS based on a single PV module, typically rated between 30 and 60 W_p, the figure is more likely to be 0.2 kWh/day—one-fiftieth of the electricity taken for granted by most European families. This modest amount can power a few low-energy lights and a small TV, offering genuine improvements to rural living standards and a contact with the wider world (Figures 7.9 and 7.10).

Like other PV systems, SHSs have most of their costs up front. A system comprising a small PV module, charge controller, 12 V battery, cabling, switches, and some low-energy lights may retail for a few hundred US dollars or euros. This may not seem much

Figure 7.9 Pride of ownership: A family in China (*Source:* Reproduced with permission of EPIA/Shell Solar).

in America, Australia, or Europe, but to many families engaged in subsistence farming in the developing world, it looks like an unattainable fortune. The SHS market—or perhaps we should say "markets," because conditions vary widely from one country to another—therefore needs its own financing arrangements.

Efficient and convenient lighting is arguably the most important service offered by an SHS. Families without electricity often spend a substantial proportion of their disposable monthly income on kerosene lamps, candles, or dry cell batteries, so this money is in principle available to pay for a PV system (Figure 7.10).

Typical financing schemes for SHSs include:

- A short-term loan to cover all or most of the initial cost, paid back with interest over a period between 1 and 3 years
- A leasing arrangement whereby an SHS is installed and maintained by an organization or company in exchange for monthly *fee-for-service* payments

Figure 7.10 PV modules and low-energy lights replaced kerosene lighting. Sharedsolar user Uganda, 2012 (*Source:* Courtesy of V. Modi, Columbia University).

A wide variety of official, commercial, and aid organizations are involved in the financing of SHS programs around the world. In addition to national governments, local banks, and leasing companies, the United Nations and the World Bank are actively involved and so are various nongovernment organizations (NGOs) and aid agencies.

Among the many countries with impressive rural electrification and SHS programs we might mention the following:

- *In Asia:* China, India, Sri Lanka, Bangladesh, Thailand, and Nepal
- *In the Americas:* Mexico, Brazil, Argentina, Bolivia, and Peru
- *In Africa:* Morocco, South Africa, Kenya, and Uganda

We end this section with a few comments on cultural and social issues surrounding the introduction of high-tech products into developing countries. In many cases a small stand-alone PV system represents the only contact of a rural family with 21st-century technology (Figure 7.11). Proper system maintenance can be a problem, and education

Figure 7.11 Enthusiasm for PV (*Source:* Reproduced with permission of EPIA/NAPS).

Figure 7.12 Sharedsolar mini-grid installation during construction phase. Mali, 2011 (*Source:* Courtesy of V. Modi, Columbia University).

is a very important part of the package. Although PV modules are normally very reliable, the lead-acid batteries used in SHSs need regular topping-up and occasional replacement, modules must be kept reasonably free of dust and bird droppings, and electrical connections must remain tight and corrosion-free. Such tasks are far removed from the experience of many rural communities (Figure 7.12). When SHSs are financed as part of a community electrification project, there may be problems of management and control. A great deal has been learned over the past 30 years about the cultural pitfalls of rural electrification, where failures tend to occur for reasons other than the purely technical. But any such difficulties should not detract from the great social benefits of rural electrification, which is surely one of PV's most admirable achievements. The challenge remains for providing affordable and reliable electricity to 1.5 billion people in developing countries (Figure 7.13).

Figure 7.13 Education, a very important part of the package (*Source:* Reproduced with permission of EPIA/Shell Solar).

7.1.4 External Costs and Benefits

PV technologies carry one very important existential attribute that often is neglected in discussions of energy policy; namely, by displacing conventional fossil fuel-based power generation, they prevent a spectrum of acute and chronic health and environmental impacts that carry a great economic cost to society as a whole. Monetization of the external costs of energy life cycles, including those of PV, is well documented in reports by the National Research Council (NRC), the Harvard School of Public Health, and others. These studies show that the greatest health and environmental effects in the life cycle of coal power generation are those from toxic air pollutants during combustion, followed by the impacts of mining, and GHG emissions. More specifically, the Harvard study estimated that the total external costs of coal-fired electricity during extraction, transport, processing, and combustion are $345 billion or 18 cents/kWh of electricity produced. These estimates are based on health costs, health insurance, and damage costs that are not included in the electricity costs but are paid by the society at large. If these were fully accounted for, then the price of electricity from coal in the United States would have been much higher than the LCOE of PV. Carbon dioxide capture with carbon sequestration (CCS) is advocated widely, enabling the continuation of coal burning for power generation, but CCS, while reducing GHG emissions, will lead to increases of toxic emissions and environmental health and safety (EHS) impacts in mining regions, as coal consumption per unit of electricity output would increase. The same applies to the proliferation of natural gas from gas-shale resources, as hydraulic fracturing increases the impacts of extracting gas on both the air and water pathways, and conversion to LNG for exportation further increases the upstream safety risks.

7.1.5 Policy Recommendations for Further Growing Solar Energy

The growth of solar PV industry will be driven by a multi-year sustained policy mix that guarantees attractive returns on solar PV investments and addresses the technical and regulatory requirements for solar energy. This mix comprises FIT, federal ITCs, loan guarantees, RPS, and REC. This mix can be deemed successful by looking at the recent increase in solar PV deployment in developed countries. However, when trends in the solar PV industry are analyzed further, it becomes evident that there are issues with market stability and some gaps in the existing policy mix, hindering the further deployment of solar PV. The needed improvements in the federal policy mix include R&D funding, solar financing flexibility, reduction of permitting burdens/costs, markets for clean energy credits, and imposing fees for carbon emissions.[3]

7.1.5.1 R&D Funding

R&D is the backbone of efforts to reduce the costs and improve the reliability of solar energy. Establishing long-term roadmaps and maintaining core competence in R&D centers are essential to the PV industry's competitive advantage. The stability and continuity of R&D funding is as important as is market stability to the success of a renewable energy policy. The private sector is likely to underinvest in R&D given the risk of knowledge spillover when intellectual property rights are not protected, and the returns on investments are unpredictable. Therefore, governments should subsidize R&D activities in the solar industry in the form of government-sponsored laboratories or direct funding to the private sector. Furthermore, governments have the responsibility of safeguarding and improving public health and the environment, and, therefore, continuing funding is needed for proactive research on the EHS impacts of new technologies and large scales of deployment. An example of a successful paradigm of government support to EHS is funding for Sematech (consortium of semiconductor companies in the United States), of which 10% was allocated by statute to R&D on EHS impacts and pollution control.

7.1.5.2 Solar Financing Flexibility

To achieve a significant market share for solar energy comparable with other generation sources, retail financing must be flexible. Innovations in solar financing and business development are critical for the growth of solar energy.

At the utility scale, merchant solar projects are considered as financially risky projects as the generation output of solar panels over the long term is uncertain, and also the dispatch profile of solar generation varies, both of which increase the expense of financing solar projects. Long-term power purchase contracts with all-in prices lower the risk of utility-scale solar projects by reducing the revenue uncertainty over the project's lifetime. However, at the residential level, there is no option of signing a contract with the utilities or states.

In the United States, solar projects historically were financed by energy sector players, banks, and the federal government; however, this pattern is rapidly changing. Recently, new business models are emerging for residential solar systems that emphasize third-party financing. Companies like SunCap Financial and Clean Power Finance provide 20-year financing plans for installing rooftop solar panels. Therefore, customers need to pay monthly loan payments for their rooftop solar rather than monthly electricity bills. That business model is convenient and its economics work in some states.

New business models are introduced in the energy sector where solar panels are seen as investment vehicles with reasonable returns over time. Energy sector players develop business models between utility-scale systems and rooftop solar ones (e.g., neighborhood small-scale solar fields), so maximizing total returns by increasing the system's capacity factor (by receiving higher irradiance compared with rooftop solar ones) while minimizing transmission, distribution, and storage needs. Cooperation with power generator utilities enables the transmission utilities to allocate fields for developing neighborhood solar farms, taking into consideration the transmission infrastructure. In the utility scale, flexibility in solar financing is achieved through the support of long-term contracts given by utilities and enforced by renewable or solar targets. In the residential and commercial retail solar segment, cooperation of power generators with transmission utilities and the introduction of new business models reducing the acquisition and financing costs of customers will determine the pervasiveness of solar PV.

7.2 Resource Availability

The main environmental credentials of PV are established beyond doubt: its important contribution to reducing carbon emissions, cleanliness and silence in operation, lack of spent fuel or waste, and general public acceptability in terms of visual impact. We have already referred to such advantages at various points in this book. But there are further environmental considerations as PV accelerates into multi-gigawatt annual production—can planet Earth provide the necessary quantities of raw materials, and is there enough land available for hundreds of millions of PV modules?

7.2.1 Raw Materials

We start with the issue of raw materials. Most of the PV module is glass and most of the mounting structure is galvanized steel and aluminum. One point is clear: insofar as PV's future is majorly based on silicon solar cells encapsulated in glass sheets, there is no problem. Silicon is one of the most common elements in the Earth's crust and, almost literally, as plentiful as sand on the beach. There is no future scenario in which it could become exhausted, and the only concern is that it requires a lot of energy to process it to the high purity and crystallinity required for solar cells. This energy was initially supplied by hydroelectric power, but there is not enough hydro to support the growth we are currently experiencing, so unfortunately dirty fossil fuel energy is increasingly used in silicon purification. However, the energy return (from PV electricity when the system operates) is 10–30 times higher than the energy invested in the manufacturing of Si PV, so the processing is energetically sustainable. A residual concern is logistical and financial; although the raw materials are abundant, advance scheduling is needed for

Figure 7.14 Effectively inexhaustible: Silicon for solar cells (*Source:* Reproduced with permission of EPIA/Solar World).

both silicon and glass manufacturing plants as it takes at least three years for such large plants to be completed. Our studies indicate that at a 30% annual growth, the glass needs of the PV industry by 2025 will be higher than the current total flat glass manufacturing capacity (about 8 billion square meters in 2015).[6]

However, there have been concerns that physical constraints in the availability of some materials may limit the growth of thin-film PVs, which lead the industry in cost reductions. In 2012 European Commission and US DOE reports listed gallium, indium, and tellurium (Ga, In, Te), which are used in CIGS and CdTe PV, as critical in terms of supply risk and economic importance. Their use is expected to increase because the entire PV industry is experiencing high growth. Furthermore, beyond PVs, the usage of gallium in integrated circuits and optoelectronics and indium in flat-panel displays is expected to rise. These materials are limited in supply because they are minor by-products of aluminum, zinc, copper, and lead production; accordingly, their production is inherently linked to that of the base metals, and thus, the rate of production of these base metals must be examined. The energy to extract the elements may pose an additional limitation.

Copper is the parent metal for tellurium; zinc for indium, germanium, and gallium; lead for tellurium, cadmium, and indium; and aluminum for gallium. The demand for Cu is expected to reach a peak within 50 years, and for Zn and Pb in about approximately 20 years, whereas the demand for Al is forecasted to increase through the end of the century. The US Geological Survey predicted a rate of growth in global demand for copper of 3% per year between 2000 and 2020; so far, this prediction has been correct. Most models predict a peak in copper production in 2050–2055; thereafter, demand is expected to decrease gradually or remain about constant during the rest of the 21st century, as the role of recycling becomes more significant. Zinc extraction grew by 3.2% annually between 1910 and 2002, and this trend has continued over the last 10 years. The demand growth rate for Zn till 2030 is assumed to be the same as for Cu.

Let us now discuss the production of the so-called critical metals (Te, In, and Ga), which are by-products of the base metals.[1] The main sources of tellurium are the anode slimes from copper electrorefining operations, and their average recovery rate is about 40%. In contrast, the recovery rate of copper from the same ores is 80% or better, and that of gold is over 95%. Evidently, the market drives the rate of recovery, with a higher demand and price justifying additional processing. Nothing inherently prevents recovery rates for tellurium from being as high as those for copper, or perhaps even gold, provided that the price is sufficiently high. Indeed, the concentration of gold in anode slimes typically is lower than that of tellurium.

However, there is a limit to the price of tellurium that will sustain affordable CdTe PVs. At US\$120/kg, the tellurium currently used in CdTe modules is approximately US\$0.01/W_p. Several scenarios are suitable for assessing the future availability of tellurium. All are related to projected copper production because, with very few exceptions, the quantities and prices of the minor metals do not warrant the extraction and processing of ores without the simultaneous recovery of copper. Starting with the tellurium content in copper anodes of 1250 metric tons (tonnes) per year and applying 3.1% annual growth and a gradual increase to 80% recovery from anode slimes, by 2020, the annual primary production of metallurgical-grade tellurium would be 1450 tonnes/year. Figure 7.15 shows the mass of Te available for PV, assuming that the current demand for other than PV applications remains constant. In addition to tellurium from copper mines, there are other types of smaller reserves, including tellurium-rich mineral deposits in China and Mexico from which the near-term direct mining of tellurium is economically feasible. Over the longer term, tellurium recovery from mining tailings and from refining of lead–zinc ores is also possible. In addition, massive resources of tellurium exist in ocean-floor ferromanganese nodules, reportedly as much as 9 million tonnes at mean concentrations of 50 ppm. However, because quantitative information is not available for the former and because the recovery of metals from deep ocean is not currently cost-efficient, these resources are not included in current assessments.

A secondary yet rich source of Te is found in end-of-life CdTe modules that contain up to 500 ppm of Te. The industry is already practicing Te recovery from defect and field returns of CdTe modules. The technical and economic feasibility of recycling most materials from CdTe PV is proven. Laboratory- and pilot-scale studies by Fthenakis and Wang at Brookhaven National Laboratory (BNL) achieved 99.99% separation of tellurium and cadmium from end-of-life modules at an estimated cost of US\$0.02/W_p. On an industrial scale, a 90% overall recovery rate is reported from First Solar. The collection of spent modules can be expected to be 100% from large utility installations and 80% from residential installations, provided that there are economic incentives or laws regulating disposal. Accordingly, recycling can become a significant source of secondary tellurium before the middle of the century (Figure 7.15).

CdTe PV production: The total annual production of CdTe PVs that tellurium availability in copper smelting can support will be constrained to 16–24 GW_p by 2020, 44–106 GW_p by 2050, and 60–161 GW_p by 2075 (Figure 7.16(a)). The tellurium-based limit of cumulative global production of CdTe PVs (Figure 7.16(b)) is 120 GW_p by 2020, rising to 1–2 TW_p by 2050 and 4–10 TW_p by 2100. These limits are based strictly on tellurium coproduction during copper production from known resources and do not include the potential for direct mining or the discovery of additional resources.

For the production of CdTe PVs to continue growing by 40% per year, tellurium recovery from anode slimes must increase to 80%, for which there is already a technological basis. We also assumed increase in the efficiency of the modules to 16% and decrease of the thickness of the CdTe film thickness to 2 μm.

Indium: Indium is a by-product of zinc extraction. Interestingly, it is not an especially rare element in the Earth's crust; it is actually about three times more plentiful than silver but only extracted at 1/60th the rate, emphasizing the dependence of indium

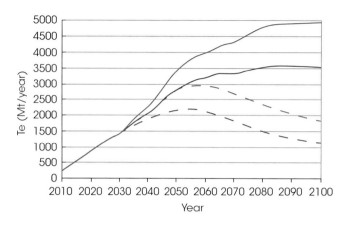

Figure 7.15 Projections of tellurium availability for photovoltaics from copper smelters (dashed lines; peaking in ~2055) and total from copper smelters and recycling of end-of-life photovoltaic modules (solid lines; continuing upward trend until 2095). The red and blue curves in each pair correspond to high and low projections, respectively.

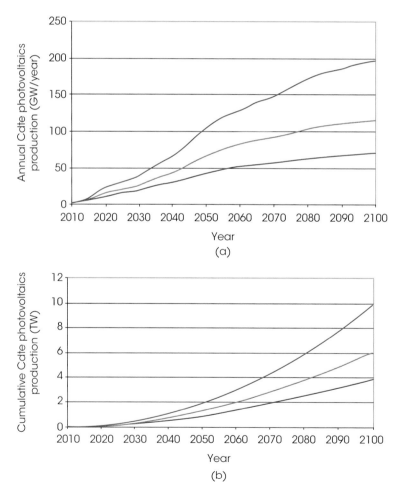

Figure 7.16 Projections of CdTe photovoltaics: (a) Annual and (b) cumulative production limits under tellurium production constraints shown in Figure 7.15. The red, pink, and blue curves correspond to the optimistic, most-likely, and conservative scenarios, respectively.

volumes on the scale of zinc mining. The supply of indium is tied to the production of zinc and is likely to remain so in the future. The price of indium reached a high of US$1000/kg in 2005 but is currently about US$600/kg. In 2010, the estimated production of indium was 1345 tons, of which 480 came from mining and another 865 tons came from recycling of used indium sputtering targets; by 2015 the annual production has increased to 1612 tons. The main use of indium is in liquid-crystal displays (LCDs), accounting for 65% of current consumption; PVs use less than estimated 5% of the primary production of indium. Competing applications of indium present extra challenges to PV growth, and further, its recovery from the zinc circuit is already high (~70–80%), leaving little room for enhancement.

Gallium: Most gallium is produced as a by-product of treating bauxite ore to extract aluminum; about 10% originates in sphalerite (ZnS), and it is produced during the purification stages of the zinc production circuit. The world resources of gallium in bauxite ore are estimated to be about 1 Mt, but evaluations of the reserves (deposits that currently can be mined economically) are lacking. Most gallium is extracted electrolytically from a solution of crude aluminum hydroxide in the Bayer process for producing alumina and aluminum. In 2010, the production of gallium was estimated to be 207 tons, of which 100 tons was derived from mining and the rest from recycling scrap. Only approximately 10% of alumina producers extracted gallium, with the others not finding it economical. The price of gallium reached a peak of US$2500/kg in early 2015 and is currently US$500–600/kg. Presently, almost all of the gallium produced is used in integrated circuits and optoelectronics, with both usages exhibiting upward trends. The estimated 2015 supply is 325 tons.

CIGS production: Under the listed assumptions on indium and gallium availability, the material-constrained growth potential of CIGS PVs has been calculated as 13–22 GW_p/year by 2020, 17–106 GW_p/year by 2050, and 17–152 GW_p/year by 2075. These estimates assume 80% extraction recoveries and use of only 50% of the growth in the supply of indium for CIGS PVs, as well as foreseeable improvements in module efficiency and material requirements. Note that the estimates for midcentury and beyond are based on the presumption that the growth of zinc extraction will follow that of copper; this is questionable because the depletion time of zinc may be shorter than that of copper. Furthermore, recovering indium and gallium from CIGS is more challenging than recouping tellurium from CdTe, as their respective concentrations are lower and their separations are more difficult.

In summary, the availability of Te, In, and Ga indeed constrains the production of CdTe and CIGS PVs. However, several comprehensive assessments have shown that there are sufficient resources to bring each of these technologies to terawatt production by midcentury; if recycling is widely adopted, such production can reach up to 10 TW by the end of the century. Recycling the end-of-life PVs is becoming increasingly important as PV deployment grows. It helps in keeping the material costs low as it provides significant secondary resources at a price lower than the primary ones, it displaces energy from material production, and it resolves concerns about potential environmental contamination from the uncontrollable disposal of PV.

It is probable that other types of cells currently in the research phase, or entirely new ones not yet discovered, will be in volume production by 2050; so any aspirational projections of PV growth, for example, the Grand Solar Plan discussed in the first chapter, could be met with a mix of PV technologies including thin films, silicon, and probably new ones. The important issue is to prepare the infrastructure for hosting these technologies, for example, to invest in grid upgrades and long-distance transmission instead of gas pipelines!

7.2.2 Land Use

PVs offer advantages for distributed power generation, and rooftop installations, which do not use any extra land, representing one-half of today's world market. However, ground-mount PV power plants occupy significant amounts of land. Data from many PVPS in the United States show land use in the range of 5–8 acres/MW_{dc} (Figure 7.17). This is the land that modules occupy to receive the solar irradiation, thus the free "solar fuel," plus the open areas needed to prevent shading, and the access roads and facilities. Now is this a lot of land? To answer this question we would need to make comparisons with the land used by the coal, natural gas, and nuclear life cycles.

Fossil fuel-based generation such as coal has a seemingly low land footprint, as power plants take up a relatively small surface area for its large power output. However, the picture changes when life-cycle land use is assessed, accounting for the direct (mining and fuel processing, plant footprint) and indirect (land usage for materials and building infrastructure needed to operate the mines) land transformed. The life-cycle land usage for different sources of electricity can be compared on a "surface area per energy unit"

Figure 7.17 Environmentally friendly use of land in Sinzheim, Germany, 1.4 MW plant
(*Source:* Reproduced with permission of First Solar).

basis, although in that case it is important to define a finite time scope as the land used for solar and wind could virtually generate electricity forever (in the case of this study: 30 years, the PV system lifetime). Land usage for PV and wind would perhaps better be described with a "surface area per power unit" as their power source for that surface area will exist for virtually infinite time. In contrast, the energy source from surface coal mines is exhausted after the fuel is extracted.

Historical data show that ground-mount solar farms in the United States often use less land during their life cycle than coal during its life cycle (Figure 7.18).[7]

For the coal fuel cycle, the direct land transformation is primarily related to the coal extraction, electricity generation, and waste disposal stages, while the indirect land use refers to the upstream land use associated with energy and materials inputs during the fuel cycle. Land-use statistics during coal mining vary with factors including heating value, seam thickness, and mining methods. Surface mining in the Western United States tends to disturb less area (per unit coal mined) than in other areas due to the thick seams, 2–9 m. Central states where seam thickness is only 0.5–0.7 m transform the largest area for the same amount of coal mined (Figure 7.19). On the other hand, underground mining transforms land for the most part indirectly. Wood usage for supporting underground coal mines accounts for the majority of indirect land transformation. Currently in the United States, about 70% of coal is mined from surface.

For operating a coal power plant, land is required for facilities including powerhouse, switchyard, stacks, precipitators, walkways, coal storage, and cooling towers. The size of a coal power plant varies greatly; a typical 1000 MW capacity plant requires between 330 and 1000 acres, which translates into 6–18 m^2/GWh of transformed land based on a capacity factor of 85%. Another study based on a 500 MW power plant located in the Eastern United States estimated 32 m^2/GWh of land transformation. Also, coal-fired power plant generates a significant amount of ash and sludge during operation. Disposing the solid wastes in the US account for 2–11 m^2/GWh, 50% each for ash and sludge.

The natural gas life cycle would use less land, but when it is extracted from the dilute formations on shale rocks, then the land occupation is expected to be as large as that of coal.

In addition to the use of the land, we should consider its transformation and possible damage. PV installations do not damage the land (Figure 7.17). Once the installation is completed, PV operation will not disturb the land, in contrast to coal mining, which often pollutes the land. Also fossil or nuclear fuel cycles need to transform certain amount of land continuously in proportion to the amount of fuel extracted. Restoring land to the original form and productivity takes a long time and often is infeasible. Accounting for secondary effects including water contamination, change of forest ecosystem, and accident-related land contamination would make the PV cycle even better than other fuel cycles. For example, water contamination from coal and uranium mining and from piles of uranium mill tailings would disturb adjacent lands. Additionally, land transformed by accidental conditions especially for the nuclear fuel cycle could change the figures dramatically. The Chernobyl accident contaminated 80 million acres of land with radioactive materials, irreversibly disturbing 1.1 million acres of farmland and forest in Belarus alone.

Land requirement for PV in the SW: 310 m^2/GWh

Land requirement for US surface coal mining: 320 m^2/GWh

Figure 7.18 The land occupied by PV plants in the southwest of the United States is smaller than the land occupied by coal mines.[7] The picture shows a PV plant in Arizona and a mountain top coal mine in West Virginia.

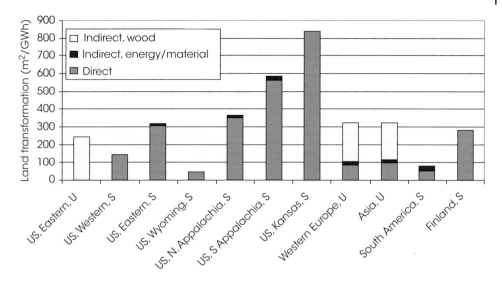

Figure 7.19 Land transformation during coal mining. S, surface mining; U, underground mining.

We now turn to the question of how much land is needed for a transition to a solar-based power infrastructure. This has already been mentioned in Section 1.5, where we suggested that an area of land 140×140 km, or $20\,000$ km^2, roughly 3 times the size of London or Paris, would be sufficient to accommodate 1000 GW$_p$ of PV modules. It seems that by 2020, or soon after, we may be approaching this huge total, some 3 times greater than global installed capacity of 350 GW in 2017, assuming that PV continues its present remarkable progress. But where would the land actually come from, and would we resent it?

If $20\,000$ km^2 sounds like a large parcel of land, consider some even larger ones: the Sahara Desert is about 850 times bigger, the Australian Outback about 200 times, and the state of Arizona about 15 times. In the United States, cities and towns cover some $700\,000$ km^2, and in many countries wide tracts of land are set aside for military uses, airports, highways, fuel pipelines, and so on. In short, if the world's PV is sensibly spread around among the world's nations, the landscapes seen by the vast majority of people will be virtually unchanged from those they enjoy today.

Of course this is far from the whole story, because PV can be installed on buildings (Figure 7.20). There are vast numbers of existing homes, offices, public buildings, factories, warehouses, airports, parking lots, and railway stations with suitable roofs and façades, and we may be sure the that tomorrow's architects will be even more aware of the possibilities. BIPV will undoubtedly provide a major part of PV's future space requirements, leaving deserts and other unproductive land to supply most of the balance. Sunshine is everywhere, high and low, city and country, and at fairly predictable levels.

7.2.3 Water Use

PVs have a major advantage against any thermoelectric power generation (e.g., coal, natural gas, nuclear, biomass, concentrated solar power) as it does not need any water for power generation. Electricity generation via conventional pathways accounts for a

Figure 7.20 No need for extra land: A rooftop PV array at Munich Airport (*Source:* Reproduced with permission of EPIA/BP Solar).

major part of water demand. According to US Geological Survey, thermoelectric power plants in the country accounted in 2009 for 49% of the freshwater withdrawal, much higher than withdrawals for agricultural irrigation that accounted for 31% of the total water demand. Most of this water is used for cooling during plant operation, and a large fraction is returned, at higher temperature, into the sources from where it was extracted. In contrast, renewable energy sources, such as PV and wind power, do not use water during their operation with the exception of water for cleaning dust from high-concentration PVs, an amount negligible in comparison with the water demand in thermoelectric power plant operation (Figure 7.21). However, every energy-generation technology does use water sometime throughout its entire life cycle. For example, during the PV life cycle, water is used for cleaning silicon wafers and glass substrates and preparing chemical solutions. In addition, a significant amount of the electricity used to purify silicon and other semiconductor materials is generated by thermoelectric power plants that rely on a water-cooling system. Conversely, as well as using water during their operation, such plants need water both directly and indirectly during fuel acquisition, plant construction, and disposal stages. The total water withdrawal during the life cycle of PV was estimated to be about $350 \, m^3/GWh$ for PV operation in SW United States, compared with $2500–120000 \, m^3/GWh$ for various coal, nuclear, and national gas plants in the region. Overall, it was estimated that PV deployment in SW United States would displace $1700–5600 \, m^3/GWh$ of water demand, when it displaces grid electricity.[8]

The water issue is region specific and is becoming increasingly important as climate change materializes with increased drought episodes.

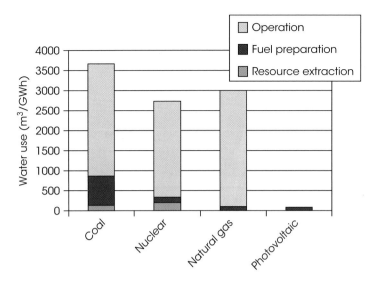

Figure 7.21 Water use in energy life cycles.

7.3   Life-Cycle Environmental Impacts

7.3.1   Life-Cycle Analysis

In the previous section we considered PV's requirements for raw materials and land—two environmental issues that surface before PV production even begins. Further important environmental questions arise during a PV system's lifetime, which starts with extraction and purification of raw materials; proceeds through manufacture, installation, and many years of operation; and ends with recycling or disposal of waste products. The whole sequence is referred to as a *life cycle*, and it is important to appreciate its environmental consequences. Note that this form of life-cycle analysis (LCA) is not the same as the classical economic life-cycle costing version previously introduced, which deals with cash flows and financial decisions. We are now moving on to something much broader, with important implications for global energy policy and society as a whole.

In this brief introduction we will consider LCA under two main headings:

- *Environmental and societal costs.* What costs, in addition to classic economic costs, are incurred or avoided?
- *Energy balance.* How does the amount of electrical energy generated over a system's lifetime compare with the energy expended in making, installing, and using it?

We start with environmental and societal costs. It is clear that all methods of energy production—whether based on oil, gas, coal, nuclear, or renewable sources—have impacts on the environment and society at large that are ignored by the traditional notion of "cost." A narrow economic view of industrial processes assesses everything in terms of money while ignoring other factors that common sense tells us should be taken into account in any sensible appraisal of value. For example, the "cost" of generating electricity in nuclear power plants has traditionally been computed without taking any account of accident or health risks; and in the case of coal-fired plants, without acknowledging their unwelcome contribution to global warming; and in the case of wind power, without placing any value on landscape.

There are two main reasons for this apparent shortsightedness. First, aspects such as health, safety, environmental protection and the beauty of a landscape cannot easily be quantified and assessed within a traditional accounting framework. We all know they are precious, and in many cases at least as important to us as money, but appropriate tools and methodologies for including them are only now being developed and accepted. It is surely vital to do this, because so many of our current problems are bound up with the tendency of conventional accounting "to know the price of everything and the value of nothing."

The second reason relates to the important notion of the *external costs* of energy generation. These costs, most of which are environmental or societal in nature, have generally been treated as outside the energy economy and to be borne by society as a whole, either in monetary terms by taxation or in environmental terms by a reduction in the quality of life. They contrast with the *internal costs* of running a business—for buildings and machinery, fuel, staff wages, and so on—that are paid directly by a company and affect its profits. If planet Earth is treated as an infinite "source" of raw materials and an infinite "sink" for all pollution and waste products, it is rather easy to ignore external costs. For example, it seems doubtful whether the 19th-century pioneers of steam locomotion ever worried much about burning huge quantities of coal or the 20th-century designers of supersonic civil airliners about fuel efficiency and supersonic bangs. One of the remarkable changes currently taking place is a growing world view that external costs should be worked into the equation—not just the local or national equation, but increasingly the global one. In other words, external costs should be *internalized* and laid at the door of the responsible industry or company. In modern phraseology, "the polluter should pay."

Many of the external and internal costs associated with industrial production are illustrated in Figure 7.22. The external ones, representing charges or burdens on society as a whole, are split into environmental and societal categories, although there is quite a lot of overlap between them. You can probably think of some extra ones. Internal costs, borne directly by the organization or company itself, cover a very wide range of goods and services, from buildings to staff wages. The distinction between internal and external costs is somewhat clouded by the fact that many items bought in by a company, for example, fuel and materials, have themselves involved substantial "external" costs during production and transport. In the case of electricity generation, a proper analysis of the environmental burdens should take proper account of all contributing processes and services "from cradle to grave," whether conducted on- or off-site. Needless to say, this is a challenging task.

One of the special difficulties facing renewable electricity generation, including PV, is that so many of its advantages stem from the *avoidance* of external costs and are therefore hidden by conventional accounting methods. Renewables tend to produce very low carbon dioxide emissions, cause little pollution, make little noise, create few hazards to life or property, and have wide public support. PV can claim all these benefits. Yet when economists and politicians talk about PV, reduction or avoidance of external costs is seldom mentioned. Fortunately, energy experts and advisers to governments are taking increasing notice of environmental LCA in their decisions and assessing the risks and

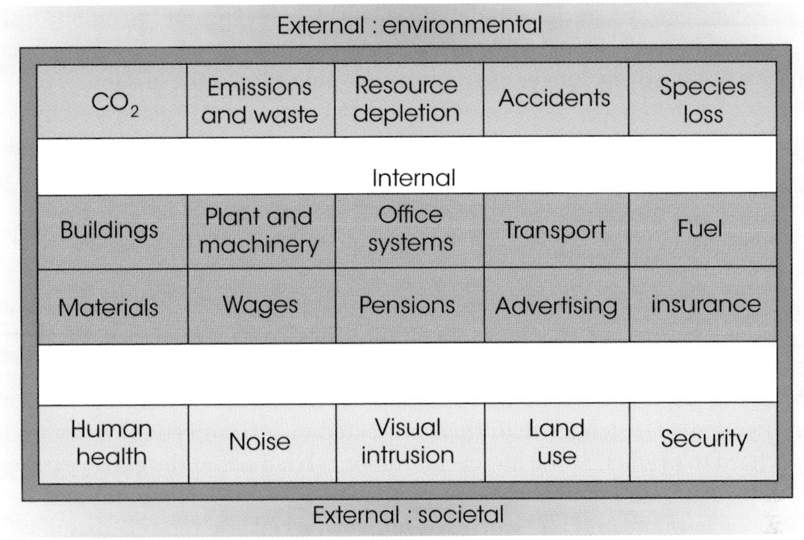

External : environmental

CO_2	Emissions and waste	Resource depletion	Accidents	Species loss

Internal

Buildings	Plant and machinery	Office systems	Transport	Fuel
Materials	Wages	Pensions	Advertising	insurance

Human health	Noise	Visual intrusion	Land use	Security

External : societal

Figure 7.22 External and internal costs.

benefits of competing technologies on a more even footing. Certainly, the PV community must be involved in countering outdated thinking about the wider benefits of its technology.

We now move on to the much-discussed topic of *energy balance*. Clearly, it takes energy to produce energy. But how does the total amount of electrical energy generated by a PV module or system over its lifetime actually compare with the input energy used to manufacture, install, and use it? Closely related to the energy balance is the *energy payback time (EPBT)*, the number of years it takes for the input energy to be paid back by the system. We naturally expect PV to have favorable energy balances and payback times, especially in view of its claims to be clean and green.

Two initial points are worth making. Firstly, energy payback is not the same as economic payback. The latter is concerned with repaying a system's capital and maintenance costs (including cost of energy consumed) by a long-term flow of income and is essentially a financial matter; energy payback is much more about the environment. Secondly, the environmental benefits of a short payback time depend on the present energy mix of the country, or countries, concerned. If the required input energy is largely derived from coal-burning power plants, it is more damaging than if it comes from, say, hydroelectricity.

Major energy inputs to a PV system occur during the following activities:

- Extraction, refining, and purification of materials
- Manufacture of cells, modules, and BOS components
- Transport and installation

Interestingly, some of the most significant energy inputs are for components such as aluminum frames and glass for modules and concrete foundations for support structures in large PV plants. Although the energy required to refine pure silicon and make crystalline silicon solar cells is considerable, the continual trend toward thinner wafers using less semiconductor material is reducing this problem. The energy input for thin-film cells is generally very small. Also, in most ground-mount installations, concrete foundations are not used any longer as supports are pinned into the ground.

The other side of the energy balance—the total electrical energy generated by a system over its lifetime—depends on a number of factors discussed in previous chapters:

- Efficiency of PV modules and other system components
- The amount of annual insolation
- Alignment of the PV array and shading (if any)
- The life of the system

Now, let us discuss the LCA in a more structured way.

The energy balance is most favorable for systems that are efficiently produced in state-of-the-art factories and installed at optimal sites in sunshine countries. Things get even better if systems last longer than their projected or guaranteed lifetimes—but of course this is hard to predict. Some life-cycle studies carried out in the early years of the new millennium painted a rather gloomy picture of PV's environmental and health impacts, due largely to the fossil fuel energy used during cell and module manufacture. However, up-to-date peer-reviewed studies that take proper account of advances in PV engineering have corrected this picture. The tremendous advances in improving the energy consumption and production of modern PV systems are illustrated at the steep decline of EPBT shown in Figure 7.23. EPBTs of current PV power plants are between 0.5 and 2 years, depending on the radiation in the place of installation and the type of technology used; this is a two-order-of-magnitude improvement from the EPBT of the early (1970) systems, and the picture is getting even better.[9]

To better understand this result, we need to discuss the elements of the LCA, which is a comprehensive framework for quantifying the environmental impacts caused by material and energy flows in each and all the stages of the "life cycle" of a product or an activity. It describes all the life stages, from "cradle to grave," thus from raw material extraction to end of life. The cycle typically starts from the mining of materials from

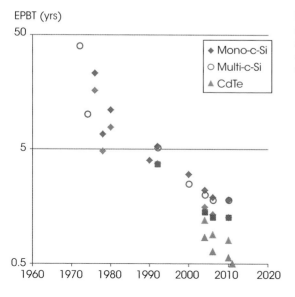

Figure 7.23 Historical evolution of energy payback times (EPBT) from 50 years down to half a year; published estimates corresponding to insolation of 1700 and 2300 kWh/m²/year.

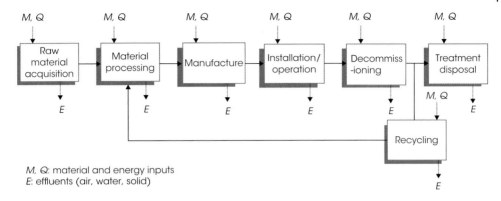

Figure 7.24 The life-cycle stages of photovoltaics.

the ground and continues with the processing and purification of the materials to manufacturing of the compounds and chemicals used in processing and manufacturing of the product, transport, installation if applicable, use, maintenance, and eventual decommissioning, and disposal and/or recycling. To the extent that materials are reused or recycled at the end of their first life into new products, then the framework is extended from "cradle to cradle." This life cycle for PV is shown in Figure 7.24.

The life-cycle **cumulative energy demand (CED)** of a PV system is the total of the (renewable and nonrenewable) primary energy harvested from the geo-biosphere in order to supply the direct energy (e.g., fuels, electricity) and material (e.g., Si, metals, glass) inputs used in all its life-cycle stages (excluding the solar energy directly harvested by the system during its operation). Thus,

$$\mathrm{CED}\left[\mathrm{MJ}_{\mathrm{PE-eq}}\right] = E_{\mathrm{mat}} + E_{\mathrm{manuf}} + E_{\mathrm{trans}} + E_{\mathrm{inst}} + E_{\mathrm{EOL}} \tag{7.1}$$

where

$E_{\mathrm{mat}}\left[\mathrm{MJ}_{\mathrm{PE-eq}}\right]$: primary energy (PE)
   demand to produce materials comprising PV system

$E_{\mathrm{manuf}}\left[\mathrm{MJ}_{\mathrm{PE-eq}}\right]$: primary energy demand to manufacture PV system

$E_{\mathrm{trans}}\left[\mathrm{MJ}_{\mathrm{PE-eq}}\right]$: primary energy demand to transport
   materials used during the life cycle

$E_{\mathrm{inst}}\left[\mathrm{MJ}_{\mathrm{PE-eq}}\right]$: primary energy demand to install the system

$E_{\mathrm{EOL}}\left[\mathrm{MJ}_{\mathrm{PE-eq}}\right]$: primary energy demand for end-of-life management

The CED of a PV system may be regarded as the energy investment that is required in order to be able to obtain an energy return in the form of PV electricity.

The life-cycle nonrenewable cumulative energy demand (NR-CED) is a similar metric in which only the nonrenewable primary energy harvested is accounted for; details are given in the IEA PVPS Task 12 LCA Guidelines report.[10]

EPBT is defined as the period required for a renewable energy system to generate the same amount of energy (*in terms of equivalent primary energy*) that was used to produce (and manage at end of life) the system itself:

$$\text{EPBT [years]} = \frac{\text{CED}}{\left(E_{\text{agen}}/\eta_G\right) - E_{\text{O\&M}}} \tag{7.2}$$

where E_{mat}, E_{manuf}, E_{trans}, E_{inst}, and E_{EOL} are defined as previously mentioned; additionally,

$E_{\text{agen}}\left[\text{MJ}_{\text{el}}/\text{year}\right]$: annual electricity generation

$E_{\text{O\&M}}\left[\text{MJ}_{\text{PE-eq}}\right]$: annual primary energy demand for operation and maintenance

$\eta_G[\text{MJ}_{\text{el}}/\text{MJ}_{\text{PE-eq}}]$: grid efficiency, that is, the average life-cycle primary energy to electricity conversion efficiency at the demand side

For systems in operation for which records exist, the annual electricity generation (E_{agen}) is taken from the actual records. Otherwise it would be estimated with the following simple equation (note the units into parentheses):

$$E_{\text{agen}}\left[\text{kWh/year}\right] = \text{irradiation}\left[\text{kWh}/(\text{m}^2\,\text{yr})\right] \times \text{Module efficiency}\left[\text{kW/kWp/m}^2\right]$$
$$\times \text{performance ratio}\left[\text{dimensionless}\right]$$

where

- Irradiation is the global irradiation on the plane of the PV.
- Module efficiency is the manufacturer rated efficiency measured under $1\,\text{kW}_p/\text{m}^2$ irradiance.
- The PR (also called derate factor) describes the difference between the modules' (DC) rated performance (the product of irradiation and module efficiency) and the actual (AC) electricity generation. It mainly depends upon the kind of installation. Mean annual PR data collected from many residential and utility systems show PR or 0.75 and 0.80, respectively.

In general, the PR increases with (i) decline in temperature and (ii) monitoring the PV systems to detect and rectify defects early. Shading, if any, would have an adverse effect on PR. This means that well-designed, well-ventilated, and large-scale systems have a higher PR.

Using either site-specific PR values or a default value of 0.75 is recommended for rooftop and 0.80 for ground-mounted utility installations; these default values include degradation caused by age.

When site-specific PR values based on early years' performance are used, degradation-related losses should be added to longer-term projections of the performance.

E_{agen} is then converted into its equivalent primary energy, based on the efficiency of electricity conversion at the demand side, using the grid mix where the PV plant is being installed. Note that E_{agen} is measured (and calculated) in units of kWh and we first have to convert it to MJ ($1\,\text{kWh} = 3.6\,\text{MJ}$) and then use η_G to convert megajoules of electricity to megajoules of primary energy ($\text{MJ}_{\text{PE-eq}}$). Thus, calculating the primary energy equivalent of the annual electricity generation (E_{agen}/η_G) requires knowing the life-cycle energy conversion efficiency (η_G) of the country-specific energy mixture used to generate electricity and produce materials. The average η_G for the United States and Europe are, respectively, approximately 0.30 and 0.31.

Energy return on (energy) investment (EROI) is defined as the dimensionless ratio of the energy generated over the course of its operating life over the energy it consumed (i.e., the CED of the system). The electricity generated by PV needs to be converted to primary energy so that it can be directly compared with CED. Thus EROI is calculated as

$$\text{EROI}\left[\text{MJ}_{\text{PE-eq}}/\text{MJ}_{\text{PE-eq}}\right] = T \times \frac{\left(E_{\text{agen}}/\eta_G\right) - E_{\text{O\&M}}}{\text{CED}} = \frac{T}{\text{EPBT}}$$

where T is the period of the system operation; both T and EPBT are expressed in years.

EROI and EPBT provide complementary information. EROI looks at the overall energy performance of the PV system over its entire lifetime, whereas EPBT measures the point in time (t) after which the system is able to provide a net energy return. Further discussion of the EPBT and EROI methodology can be found in the IEA PVPS Task 12 LCA Guidelines, and a discussion of its misrepresentation in a few publications is discussed by Raugei and colleagues.[11, 12]

As shown in Figure 7.22, the EPBT of PVs has been reduced by almost two orders of magnitude over the last three decades, as material use, energy use, and efficiencies have been constantly improving. For example, the CED energy used in the life cycle of complete rooftop Si PV systems were 2700 and 2900 MJ/m², respectively, for multi- and monocrystalline Si modules, down from 5000 and 2700 in 2006. The EPBT of these systems is 1.8 years for module efficiencies of 13.2 and 14% correspondingly in rooftop installations under Southern European insolation of 1700 kWh/m²/year, with a grid efficiency (η_G) of 0.31 and a PR of 0.75 (Figure 7.23). The corresponding EROI ratio, assuming a 30-year lifetime, is 17. In these estimates, the BOS for rooftop application accounts for 0.3 years of EPBT, but there are different types of rooftop mounting systems with different energy burdens.[13]

For CdTe PV, the primary energy consumption is 850 and 970 MJ/m², correspondingly based on actual production from First Solar's plant in Frankfurt-Oder, Germany, and Perrysburg, Ohio, United States. For insolation levels of 1700 kWh/m²/year and a grid efficiency (η_G) of 0.31 (values that correspond to average U.S. and South European conditions), the EPBT and EROI values for installed PV + BOS systems are EPBT = 0.65 years and EROI = 46 (Figure 7.25).

EPBT decreases (and EROI increases) as the solar irradiation levels increases; for example, in the southwest of the United States (latitude optimal irradiation of 2300 kWh/m²/year), the EPBTs of crystalline silicon and cadmium telluride PV in fixed-tilt ground-mount utility installations, respectively, are 1.2 and 0.5 years (Figure 7.26). In the highest solar irradiation regions (e.g., northern Chile; irradiation of 4000 kWh/m²/year) on a 1-axis sun-tracking plane, the EPBT for CdTe PV systems is only 3–4 months. Overall, PV systems are generating all the CED used in their life cycle in 0.4–2 years depending on the PV technology and the location where the system operates. With an assumed life expectancy of 30 years, their EROI is in the range of 15–90. So, depending mainly on the irradiation in the location where they are installed, PV systems generate 15–90 times more energy than the energy consumed in their life cycle, saving a tremendous amount of primary energy that can be used for building a new power infrastructure comprising PV modules and structures, power electronics, grid extension, and storage to enable the transition to a carbon-free society.[14]

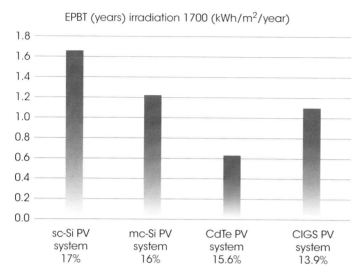

Figure 7.25 Energy payback times of PV systems installed under South European and average US irradiation conditions.

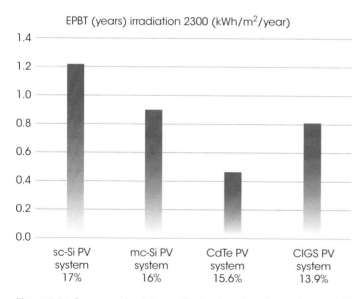

Figure 7.26 Energy payback times of PV systems installed in the South United States.

GHG emissions and global warming potential (GWP): The overall global warming potential (GWP) due to the emission of a number of GHGs along the various stages of the PV life cycle is typically estimated using an integrated time horizon of 100 years (GWP_{100}), whereby the following CO_2-equivalent factors are used: $1\,kg\ CH_4 = 23\,kg\ CO_2$-eq, $1\,kg\ N_2O = 296\,kg\ CO_2$-eq, and $1\,kg$ chlorofluorocarbons $= 4\,600–10\,600\,kg\ CO_2$-eq. It is noted that the CO_2-equivalent factor of CH_4 has recently been updated to 32; the effect of this update on the reported GHG has not been assessed yet. Also if we

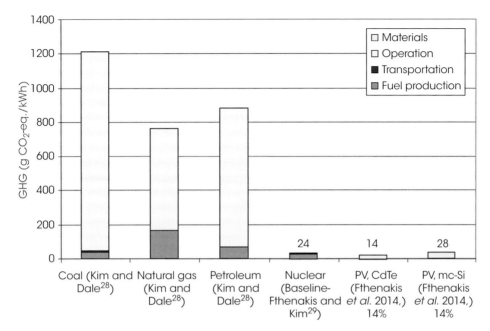

Figure 7.27 Life-cycle greenhouse gas emissions in electricity generation (as of 2014); PV continues to get better in terms of increasing efficiencies and reducing emissions (see Leccisi *et al.*[9]).

consider a 20-year time horizon, the GWP of CH_4 is a much higher 72. Electricity and fuel use during the production of the PV materials and modules are the main sources of the GHG emissions for PV cycles, and specifically the technologies and processes for generating the upstream electricity play an important role in determining the total GWP of PVs, since the higher the mixture of fossil fuels is in the grid, the higher are the GHG (and toxic) emissions.

In this chapter we present the most up-to-date estimates of EPBT, GHG emissions, and heavy metal emissions from the life cycles of the currently commercial PV technologies (e.g., monocrystalline Si, multicrystalline Si, and CdTe). The production of solar-grade silicon and of silicon and thin-film PV modules was discussed in Chapter 6. Life-cycle inventory (LCI) data on materials, energy, and emissions of these technologies and of BOS can be found in IEA PVPS Task12 reports.

In the next paragraphs, we will briefly discuss the results of comparative assessments of various technologies.

The GHG emissions of Si modules are about 28 g CO_2-eq./kWh range for a rooftop application under Southern European insolation of 1700 kWh/m²/year and a PR[1] of 0.75 (Figure 7.27). Notably, under these estimates, the BOS for rooftop applications account for approximately 5 g CO_2-eq./kWh of GHG emissions.

The GHG emissions of CdTe PV systems total 14 g CO_2-eq./kWh for irradiation levels of 1700 kWh/m²/year. The manufacturing of the module and upstream mining/smelting/purification operations comprises most of the energy and greenhouse burdens. The BOS

---

1 Ratio between the DC rated and actual AC electricity output.

share was about 5 g CO_2-eq./kWh under 1700 kWh of insolation, with 14% module efficiency and a PR of 0.75. It is noted that the PV efficiencies shown in Figure 7.26 correspond to 2014 production and the current module efficiencies are significantly higher. As of mid-2017, the average efficiencies of both CdTe and polycrystalline Si PV are approximately 16.5%; accordingly the GHG emissions during their lives would be 12 and 24 g CO_2-eq./kWh of generated electricity.

All the recently published LCA studies found CdTe PV having the lowest EPBT and GHG emissions among the currently commercial PV technologies, as it uses less energy in its material processing and module manufacturing.[9] This is explained by the lower thickness of the high purity semiconductor layer (i.e., 2–3 μm for CdTe vs. 15–200 μm for c-Si) and by the fact that CdTe module processing has fewer steps and is much faster than c-Si cell and module processing.

Figure 7.27 compares GHG emissions from the life cycle of PV with those of conventional fuel-burning power plants, revealing the environmental advantage of using PV technologies. The majority of GHG emissions come from the operational stage for the coal, natural gas, and oil fuel cycles, while the material and device production accounts for nearly all the emissions for the PV cycles. With over 50% contributions, the GHG emissions from the electricity demand in the life cycle of PV are the most impactful input. Therefore, the LCA results strongly depend on the available electricity mix. In the United States replacing grid electricity with PV systems would result in an 89–98% reduction in the emissions of GHGs, criteria pollutants, heavy metals, and radioactive species.

However, this discussion is about large penetration of PVs, and so far we accounted for the GHG emission reductions enabled by PV displacing fossil fuel generation. For a balanced comparison, we also have to account for negative impacts resulting from the intermittency of the solar resource. Fluctuations in PV generation would necessitate fuel generators' up- and down-ramping and part-load operation, causing additional emissions. The net emission savings from RE integration are due to the displaced conventional generation minus the incremental emissions from suboptimal and part-load generation. Fthenakis and Nikolakakis modeled this impact in a scenario of adding PV and wind capacities in the west zone of the NYS grid totaling 1.2 GW in the winter and 2.7 MW in the summer and determined that the subsequent non-optimum dispatch of natural gas peaker generators caused 5–8% higher emissions per kWh than when their output does not fluctuate to accommodate PV and wind penetration into the grid. More specifically the emissions of conventional generators are being raised due to RE integration from 376 to 406 g/kWh in the spring highest penetration case and from 380 to 397 g/kWh for the winter lowest penetration case. These increases are more than counterbalanced by the zero operational emissions of PV and wind.

Toxic gas emissions: The emissions of toxic gases (e.g., SO_2, NO_x) and heavy metals (e.g., As, Cd, Hg, Cr, Ni, Pb) during the life cycle of a PV system are largely proportional to the amount of fossil fuel consumed during its various phases, in particular processing and manufacturing PV materials. Figure 7.28 shows estimates of SO_2 and NO_x emissions. Heavy metals may be emitted directly from material processing and PV manufacturing and indirectly from generating the energy used at both stages. For the most part, they originate as trace metals in the coal used.

Heavy metal emissions: When considering heavy metal emissions in PV life cycles, it is important to distinguish between direct and indirect emissions. Indirect emissions are from electricity inputs during the life cycle of the PV system, while direct emissions are those released "on-site" during mining, smelting, manufacturing, and the end-of-life disposal.

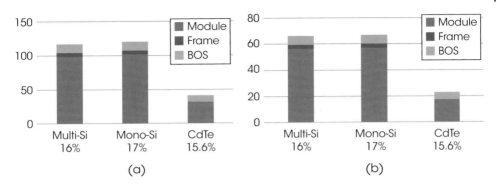

Figure 7.28 Life-cycle emissions of (a) SO_2 and (b) NO_x emissions from silicon and CdTe PV modules. BOS includes module supports, cabling, and power conditioning. The estimates are based on rooftop-mount installation, insolation of 1700 kWh/m^2/year, performance ratio of 0.75, lifetime of 30 years, and European production with electricity supply from the UCTE grid.

Electricity generated from CdTe PV has direct Cd emissions from raw material extraction and processing, module processing, and possible accidental releases of Cd if CdTe PV modules are exposed to fire. Both CdTe and crystalline silicon PV have indirect emissions from their fossil fuel-based electricity inputs. For CdTe these were found to be 10 times bigger than the direct emissions showing the importance of the source of electricity in manufacturing solar PV systems. Other heavy metal emissions such as arsenic, chromium, lead, mercury, and nickel are emitted in trace amounts, also as a result of electricity and fuel inputs into the system. The heavy metal emissions from a coal-fired power plant with state-of-the-art air pollution control systems are 90–300 times bigger than that of solar electricity on a per unit of energy basis. The issue with cadmium bears special discussion as the element is used in CdTe/CdS and to a smaller extent in CIGS/CdS PVs. Direct emissions of cadmium in the life cycle of CdTe PV have been assessed in detail. They total 0.02 g/GWh of PV-produced energy under an irradiation of 1700 kWh/m^2/year; this includes emissions during fires on rooftop residential systems, quantified in experiments at BNL that simulated actual fires.[15] These experiments were designed to replicate average conditions, and the estimated emissions were calculated by accounting for US fire statistics pointing to 1/10 000 houses catching fire over the course of a year in the United States where most houses have wood frames by assuming that all fires involve the roof. However, the indirect Cd emissions from electricity usage during the life cycle of CdTe PV modules (i.e., 0.23 g/GWh) are an order of magnitude greater than the direct ones (routine and accidental) (i.e., 0.02/GWh).[2] Cadmium emissions from the electricity demand for each module were assigned, assuming that the life-cycle electricity for the silicon and CdTe PV modules was supplied by the Union for the Coordination of the Transmission of Energy's (UCTE)

---

2 Indirect emissions of heavy metals result mainly from the trace elements in coal and oil. According to the US Electric Power Research Institute's (EPRI's) data, under the best/optimized operational and maintenance conditions, burning coal for electricity releases into the air between 2 and 7 g of Cd/GWh. In addition, 140 g/GWh of Cd inevitably collects as fine dust in boilers, baghouses, and electrostatic precipitators (ESPs). Furthermore, a typical US coal-powered plant emits per GWh about 1000 tons of CO_2, 8 tons of SO_2, 3 tons of NO_x, and 0.4 tons of particulates. The emissions of Cd from heavy-oil-burning power plants are 12–14 times higher than those from coal plants, even though heavy oil contains much less Cd than coal (~0.1 ppm); this is because these plants do not have particulate-control equipment.

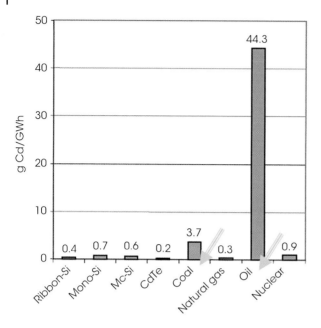

Figure 7.29 Emissions of cadmium in the life cycles of PV compared with those from fossil and nuclear power generation (as of 2014; all PV technologies continue to improve).

(European) grid. The complete life-cycle atmospheric Cd emissions, estimated by adding those from the usage of electricity and fuel in manufacturing and producing materials for various PV modules and BOS, were compared with the emissions from other electricity-generating technologies (Figure 7.29).[16] Undoubtedly, displacing the others with Cd PV markedly lowers the amount of Cd released into the air. Thus, every GWh of electricity generated by CdTe PV modules can prevent around 5 g of Cd air emissions if they are used instead of, or as a supplement to, the UCTE electricity grid. In addition, the direct emissions of Cd during the life cycle of CdTe PV are 10 times lower than the indirect ones due to use of electricity and fuel in the same life cycle and about 30 times less than those indirect emissions from crystalline PVs. The same applies to total (direct and indirect) emissions of other heavy metals (e.g., As, Cr, Pb, Hg, Ni); CdTe PV has the lowest CED and, consequently, the fewest heavy metal emission. Regardless of the particular PV technology, these emissions are extremely small compared with the emissions from the fossil fuel-based plants that PV will replace. Furthermore, the external environmental costs of PVs are negligible in comparison with the external costs of fossil fuel life cycles.

Recycling would further improve the EPBT and the EROI of PV because recycling glass, aluminum, tellurium, and other semiconductor materials requires only a fraction of the energy consumed in their primary production, and, consequently, it produces only a fraction of GHG emissions.

7.3.2 Environmental Health and Safety (EHS) in PV Manufacturing

PV technologies have distinct environmental advantages for generating electricity over conventional technologies. The operation of PV systems does not produce any noise, toxic gas emissions, or GHGs. PV electricity generation, regardless of which technology is used, is a zero-emission process. However, as with any energy source or product, there are EHS hazards associated with the manufacture of solar cells. The PV industry uses toxic and flammable substances, although in smaller amounts than many other industries, and use of hazardous chemicals can involve occupational and environmental hazards. Addressing EHS concerns was the focus of numerous studies of the National Photovoltaic EHS Assistance Center at the BNL, which operated till 2013 under the auspices of the US DOE. More than 150 articles highlighting these studies are posted in the Center's website (www.bnl.gov/pv) and at the website of Columbia University's Center for Life Cycle Analysis (www.clca.columbia.edu). This work has been done in cooperation with the US PV industry, which takes EHS issues very seriously and reacts proactively to concerns. The following text is a summary of EHS issues pertaining to the manufacture of crystalline-Si (x-Si), amorphous silicon (α-Si), copper indium diselenide (CIS), CIGS, gallium arsenide (GaAs), and cadmium telluride (CdTe), which are currently commercially available.[17]

Crystalline Silicon (c-Si) Solar Cells
Occupational Health Issues
In the manufacture of wafer-based crystalline silicon solar cells, occupational health issues are related to potential chemical burns and the inhalation of fumes from hydrofluoric acid (HF), nitric acid (e.g., HNO_3), and alkalis (e.g., NaOH) used for wafer cleaning, removing dopant oxides, and reactor cleaning. Dopant gases and vapors (e.g., $POCl_3$, B_2H_3) also are hazardous if inhaled. $POCl_3$ is a liquid, but in a deposition chamber it can generate toxic P_2O_5 and Cl_2 gaseous effluents. Inhalation hazards are controlled with properly designed ventilation systems in the process stations. Other occupational hazards are related to the flammability of silane (SiH_4) and its by-products used in silicon nitride deposition; these hazards are discussed in the a-Si section, because SiH_4 is a major feedstock in a-Si PV manufacturing.

Public Health and Environmental Issues
No public health issues were identified with the c-Si PV technology. The environmental issues are related to the generation of liquid and solid wastes during wafer slicing, cleaning, and etching and during processing and assembling of solar cells.

The c-Si PV industry has embarked upon programs of waste minimization and examines environmentally friendlier alternatives for solders, slurries, and solvents. Successful efforts were reported in laboratory and manufacturing scales in reducing the caustic waste generated by etching. Other efforts for waste minimization include recycling stainless steel cutting wires, recovering the SiC in the slurry, and in-house neutralization of acid and alkali solutions.

Finally, the content of lead (Pb) in solder in many of today's modules creates concerns about the disposal of modules at the end of their useful life. As of 1999, ASE Americas in Massachusetts had adopted a Pb-free soldering technology, developed by Dr. Ron Gonsiorawski; the company's CEO, Dr. Charlie Gay, offered the technology know-how to the whole PV industry "to preserve the good, environmental image of PV globally."[18] As of 2016, it appears that nearly 2/3 of PV module manufacturers use Pb-free solder.

Amorphous Silicon (α-Si) Solar Cells
Amorphous silicon, cadmium telluride, copper indium selenide, and gallium arsenide are thin-film technologies that use about 1/100 of the PV material used on x-Si.

Occupational Safety Issues

The main safety hazard of this technology is the use of SiH_4 gas, which is extremely pyrophoric. The lower limit for its spontaneous ignition in air ranges from 2 to 3%, depending on the carrier gas. If mixing is incomplete, a pyrophoric concentration may exist locally, even if the concentration of SiH_4 in the carrier gas is less than 2%. At silane concentrations equal to or greater than 4.5%, the mixtures were found to be metastable and ignited after a certain delay. In an accident, this event could be extremely destructive as protection provided by venting would be ineffective. Silane safety is discussed in detail by Ngai and Fthenakis.[19] In addition to SiH_4, hydrogen used in a-Si manufacturing also is flammable and explosive. Most PV manufacturers use sophisticated gas handling systems with sufficient safety features to minimize the risks of fire and explosion. Some facilities store silane and hydrogen in bulk from tube trailers to avoid frequently changing gas cylinders. A bulk ISO module typically contains eight cylindrical tubes that are mounted onto a trailer suitable for over the road and ocean transport. These modules carry up to 3000 kg of silane. Another option is a single, 450 l cylinder, mounted on a skid, which contains up to 150 kg of silane (mini-bulk). These storage systems are equipped with isolation and flow-restricting valves.

Bulk storage decreases the probability of an accident, since trailer changes are infrequent, well-scheduled special events that are treated in a precise well-controlled manner, under the attention of the plant's management, safety officials, the gas supplier, and local fire department officials. On the other hand, if an accident occurs, the consequences can be much greater than one involving gas cylinders. Currently, silane is used mainly in glow discharge deposition at very low utilization rates (e.g., 10%). To the extent that the material utilization rate increases in the future, the potential worst consequences of an accident will be reduced.

Toxic doping gases (e.g., AsH_3, PH_3, GeH_4) are used in quantities too small to pose any significant hazards to public health or the environment. However, leakage of these gases can cause significant occupational risks, and management must show continuous vigilance to safeguard personnel. Applicable prevention options are discussed elsewhere[20]; most of these are already implemented by the US industry.

Public Health and Environmental Issues

Silane used in bulk quantities in a-Si facilities may pose hazards to the surrounding community if adequate separation zones do not exist. In the United States, the Compressed Gas Association (CGA) guidelines specify minimum distances to places of public assembly that range from 80 to 450 ft depending on the quantity and pressure of silane in containers in use (CGA P-32, 2000). The corresponding minimum distances to the plant property lines are 50–300 ft. Prescribed separation distances are considered sufficient to protect the public under worst-condition accidents.

Cadmium Telluride (CdTe) Solar Cells

Occupational Health Issues

In CdTe manufacturing, the main concerns are associated with the toxicity of the feedstock materials (e.g., CdTe, CdS, $CdCl_2$). The occupational health hazards presented by Cd and Te compounds in various processing steps vary as a function of the compound-specific toxicity, its physical state, and the mode of exposure. No clinical data are available on human health effects associated with exposure to CdTe. Limited animal data comparing the acute toxicity of CdTe, CIS, and CGS showed that from the three compounds, CdTe has the highest toxicity and CGS the lowest.[21] No comparisons with the parent Cd and Te compounds have been made. Cadmium, one of CdTe precursors, is a highly hazardous material. The acute health effects from inhalation of Cd include pneumonitis, pulmonary edema, and death. However, CdTe is insoluble to water and, as such, may be less toxic than CdTe. This issue needs further investigation.

In production facilities, workers may be exposed to Cd compounds through the air they breathe, as well as by ingestion from hand-to-mouth contact. Inhalation is probably the most important pathway, because of the larger potential for exposure and higher absorption efficiency of Cd compounds through the lung than through the gastrointestinal tract. The physical state in which the Cd compound is used and/or released to the environment is another determinant of risk. Processes in which Cd compounds are used or produced in the form of fine fumes or particles present larger hazards to health. Similarly, those involving volatile or soluble Cd compounds (e.g., $CdCl_2$) also must be more closely scrutinized. Hazards to workers may arise from feedstock preparation, fume/vapor leaks, etching of excess materials from panels, and maintenance operations (e.g., scraping and cleaning) and during waste handling. Caution must be exercised when working with this material, and several layers of control must be implemented to prevent exposure of the employees. In general, the hierarchy of controls includes engineering controls, personal protective equipment, and work practices. Area and personal monitoring would provide information on the type and extent of employees' exposure, assist in identifying potential sources of exposure, and gather data on the effectiveness of the controls. The US industry is vigilant in preventing health risks and has established proactive programs in industrial hygiene and environmental control. Workers' exposure to cadmium in PV manufacturing facilities is controlled by rigorous industrial hygiene practices and is continuously monitored by medical tests, thus preventing health risks.

Public Health and Environmental Issues
No public health issues have been identified with this technology. Environmental issues are related to the disposal of manufacturing waste and end-of-life modules; these are discussed later in the section about recycling.

Copper Indium Selenide (CIS) Solar Cells
Occupational Health and Safety
The main processes for forming CIS solar cells are co-evaporation of Cu, In, and Se and selenization of Cu and In layers in H_2Se atmosphere. The toxicity of Cu, In, and Se is considered mild. Little information exists on the toxicity of CIS. Animal studies have shown that CIS has mild to moderate respiratory track toxicity.

Although elemental selenium has only a mild toxicity associated with it, hydrogen selenide is highly toxic. It has an immediately dangerous to life and health (IDLH) concentration of only 1 ppm. Hydrogen selenide resembles arsine physiologically; however, its vapor pressure is lower than that of arsine, and it is oxidized to the less toxic selenium on the mucous membranes of the respiratory system. Hydrogen selenide has a TLV-TWA of 0.05 ppm to prevent irritation and prevent the onset of chronic hydrogen selenide-related disease. To prevent hazards from H_2Se, the deposition system should be enclosed under negative pressure and be exhausted through an emergency control scrubber. The same applies to the gas cabinets containing H_2Se cylinders in use.[22]

The options for substitution, isolation, work practices, and personnel monitoring discussed for CdTe are applicable to CIS manufacturing as well. In addition, the presence of hydrogen selenide in some CIS fabrication processes requires engineering and administrative controls to safeguard workers and the public against exposure to this highly toxic gas.

Public Health and Environmental Issues

Potential public health issues are related to the use of hydrogen selenide in facilities that use hydrogen selenide as a major feedstock material. Associated hazards can be minimized by using safer alternatives, limiting inventories, and using flow-restricting valves and other safety options discussed in detail elsewhere.[23] Emissions of hydrogen selenide from process tools are controlled with either wet or dry scrubbing. Also, scrubbers that can control accidental releases of this gas are in place in some facilities. Environmental issues are related to the disposal of manufacturing waste and end-of-life modules; these are discussed in the section about PV recycling.

Gallium Arsenide (GaAs) High-Efficiency Solar Cells

Occupational Health and Safety

MOCVD is today's most common process for fabricating III/V PV cells; it employs the highly toxic hydride gases, arsine and phosphine, as feedstocks. Similarly to silane and hydrogen selenide handling, the safe use of these hydrides requires several layers of engineering and administrative controls to safeguard workers and the public against accidental exposure. Such requirements pose financial demands and risks that could create difficulties in scaling up the technology to multi-megawatt levels. One part of the problem is that today's use of the hydrides in MOCVD is highly ineffective. Only about 2–10% are deposited on the PV panels, as a 10–50 times excess of V to III compounds (As to Ga) is required. Metal–organic compounds are used more effectively, with their material utilization ranging from 20 to 50%. In scaling up to 10 MW/year production using MOCVD, the current designs of flat-plate III–V modules will require approximately 23 metric tons of AsH_3, 0.7 tons of PH_3, 7 tons of metal organics, and 1500 tons of hydrogen.[3] These quantities can be effectively delivered only with tube trailers, each carrying 2–3 tons of gases. The potential consequences of a worst-case failure in one of these tube trailers could be catastrophic. On a positive note, however, it is more likely that terrestrial systems will be concentrators, not flat-plates, because the former would be less expensive to manufacture. For example, with a 500 times concentration, the material requirements are 600 times less than those needed for flat plates.[24]

The best way to minimize both the risks associated with certain chemicals and the costs of managing risk is to assess all alternatives during the first steps of developing the technology and designing the facility. These hydrides may be replaced in the future by

---

3 These estimates are based on generic data applicable to the largest current MOCVD reactors (e.g., EMCORE Enterprise E400 and Aixtron AIX3000) and carry some unquantified uncertainty. Production for 24 hours a day by one of these reactors could provide about $100 kW_p$/year, so 100 such reactors will be needed for the 10 MW production basis we are considering herein. Therefore, larger reactors will be needed at this scale, thereby introducing more uncertainty in our estimates.

the use of tertiary butyl arsine (TBAs) and tertiary butyl phosphine (TBP); it appears that there are no intrinsic technical barriers to growing PV-quality GaAs with TBAs and GaAsP or GaInP$_2$ with TBP. Until substitutes are tested and implemented, however, it might be prudent to use arsine and phosphine from reduced-pressure containers, which are commercially available. Research efforts are being made in Europe to replace hydrogen by inert nitrogen. Apparently, there is no inherent reason to prohibit such a substitution. However, since molecular hydrogen decomposes to some extent and atoms participate in the gas-phase chemistry, the PV research community is challenged with learning how to optimize III–V growth conditions with nitrogen.

In summary, the manufacture of PV modules uses some hazardous materials that can present health and safety hazards, if adequate precautions are not taken. Routine conditions in manufacturing facilities should not pose any threats to health and the environment. Hazardous materials could adversely affect occupational health and, in some instances, public health during accidents. Such hazards arise primarily from the toxicity and explosiveness of specific gases. Accidental releases of hazardous gases and vapors are prevented engineering systems, employee training, and safety procedures. As the PV industry continues to vigilantly and systematically approach these issues and mitigation strategies, the risk to the industry, the workers, and the public is minimized.

7.3.3 Recycling Programs

Recycling spent modules may not be seen as an immediate issue with developing solar energy. However, the rapid growth of solar energy eventually will result in problems of waste disposal within 20–30 years as end of life of PV would generate a significant amount of waste (about 100 tons/MW of decommissioned PV modules). Disposal of small quantities of PV modules in landfills should not cause environmental hazards when the modules pass the EPA leaching tests that are designed to simulate release conditions. However, it is widely recognized that in large scales of deployment and decommissioning, recycling end-of-life PV modules would be necessary to prevent risks of environmental pollution and to recover valuable materials. In Europe, the PV industry adopted a proactive approach that served them well during the transition to, a currently required, compliance with Waste Electrical and Electronic Equipment (WEEE) regulations. Germany's Electrical and Electronic Equipment Act (ElectroG) requiring collection and recycling of electrical and electronic equipment (EEE) was extended to PV in mid-2015 and is expected to become a global standard. The United States lacks a national policy and the necessary infrastructure to mandate PV recycling. Environmental regulations can determine the cost and complexity of dealing with end-of-life PV modules. If they were characterized as "hazardous," then special requirements for material handling, disposal, record keeping, and reporting would escalate the cost of decommissioning modules.

Currently, there are well-tested technical solutions (separation and material recovery processes) for c-Si (wafer based) and CdTe PV products, but not for other technologies. The first step in recycling both types of modules is to separate the junction boxes, and for c-Si the aluminum frames. Subsequent steps deal with separating the glass from the solar materials and metals. For c-Si modules thermal treatment burns off the laminates to facilitate the separation processes (called module delamination). The most common way to achieve this is through pyrolysis, heating the module to 450–600°C to decompose the

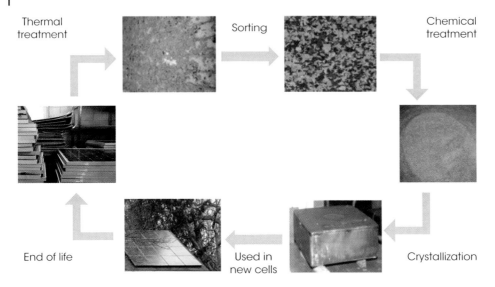

Figure 7.30 Process schematic for recycling of crystalline silicon PV modules.

organic encapsulant. After delamination, the components are manually separated, the glass is sent to a glass recycling facility, and the silicon wafers are processed further, either by polishing and reusing the wafer or by recycling the silicon into a new wafer (Figure 7.30). From the separation steps, copper wire, aluminum frame, glass, silicon, and waste are separated and are sent to recyclers. Plastic is burned off during the thermal treatment and waste goes to a landfill.

First, the unloaded modules transported from the collection sites will be loaded to the automatic conveyor system to enter the recycling process. Then the junction boxes are removed manually. Thermal treatment burns off the laminates to facilitate the separation processes. From the separation steps, copper wire, aluminum frame, glass, and waste are separated. During the next step the solar cells are treated chemically. Surface and diffusion layers are removed subsequently by cleaning steps. Cells and wafer breakage are cleaned by etching techniques. Junction box is processed by an electronic scrap waste treatment company (i.e., collection cost paid by PVTBC). Plastic is burned off after the thermal treatment (i.e., incineration cost paid by PVTBC). Waste goes to a landfill and PVTBC pays the landfill tipping fees. Aluminum can be reused while glass, copper, and silicon can be sold to recycling companies. The thermal process could be improved regarding its throughput, cycle time, and yield. The yield of recovered cells depends largely on the type, design, and state of the modules to be processed. Design-dependent factors that affect the results of the thermal process are type of laminate, crystal type and dimensions of the embedded cells, and the material and dimensions of bonds and soldering.

Recycling of CdTe PV is somewhat more advanced and is employed in all CdTe PV production facilities. It is based on a low-cost, environmentally friendly hydrometal-lurgical technology, which was developed by First Solar and BNL. This technology involves crashing the modules, removing the thin films from the substrate, and recovering the thin-film materials from the solution (Figure 7.31). The modules are cut by

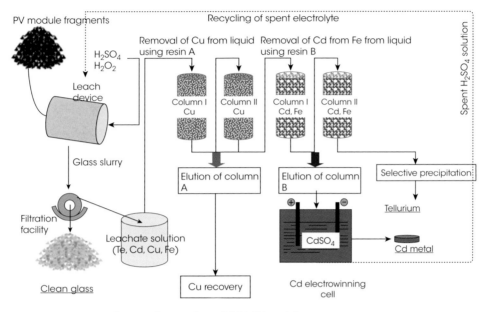

Figure 7.31 Process schematic for recycling of CdTe PV modules.

a shredder and broken in small pieces with a hammer mill. The pieces are then exposed to leaching using a dilute mixture of sulfuric acid and hydrogen peroxide, which extracts the metals (mostly copper) and semiconductor elements (tellurium and cadmium). Ion-exchange column[25] is then used to separate the copper and the cadmium from the solution, resulting in a tellurium-rich solution from where Te is then extracted by selective precipitation. Cadmium is rinsed out the column and is recovered electrolytically.[26] A modified version of this process is used by First Solar. Instead of separating the cadmium and tellurium using ion-exchange columns, First Solar precipitates all of the metals and sends this sludge to a third party for further processing.

As we discussed in the beginning of this chapter, resource availability, affordability, and lowest possible environmental impacts are three major pillars of sustainable PV growth to levels that will enable transition of the current fossil fuel-based electricity to a renewable one. Recycling of spent PV modules addresses all these three dimensions of sustainability (Figure 7.32). Recycling relieves the pressure on material prices as it creates an important secondary resource of materials and it eliminates the potential risks and liabilities associated with waste disposal. End-of-life PV contains materials of high value (e.g., silver, indium, tellurium, and gallium), toxic materials (e.g., cadmium lead, selenium compounds), and a large amount of energy-intensive materials (e.g., glass, copper, steel, aluminum, and solar-grade silicon). For these reasons, we must develop efficient, cost-beneficial recycling processes to aid in creating a PV-recycling economy that is essential to safeguarding and maintaining the SunShot-catalyzed growth of the PV industry. The cost of solar energy is impacted by the price of material inputs such as polysilicon, tellurium, steel, and glass. Their prices are driven by the status of their supply and demand.

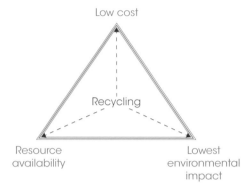

Low cost

Recycling

Resource
availability

Lowest
environmental
impact

Figure 7.32 Recycling strengthens the three major pillars of sustainable large growth of PV (Concept Vasilis Fthenakis).

Recycling helps to avoid shortages of such materials needed for PV production and lowers the cost of PV modules. Currently recycling programs are established for only two types of PV modules: CdTe and c-Si. The first recovers glass and the semiconductor elements for reuse in CdTe synthesis, whereas the second only recovers the aluminum frame and the glass. PV recycling of mature technologies (e.g., c-Si and CdTe) is technically and economically feasible. Accounting for secondary production from recycling and the continuing improvements of module efficiencies and material utilization, a number of studies show that the availability of tellurium in the forthcoming decades is sufficient for a cumulative production at the TW level. In Europe, the PV industry established PV CYCLE, a voluntary program to recycle PV modules (www.pvcycle.org). This type of industry-wide approach to economically manage large-scale recycling should become an essential component of cost reduction roadmaps. Furthermore, we believe that a comparative techno-economic evaluation of existing and proposed PV recycling technologies is necessary; this will aid the commercialization of cost-beneficial recycling and the creation of a PV-recycling economy.

Recycling of defective and end-of-life-PV modules is a necessary component of the sustainability of the large-scale deployment of PVs. There are many studies that urge a proactive systematic approach to resolving issues of the scarcity of materials and to assuring that PV life cycles will not present risks to public health nor the environment in scenarios involving a high penetration of PV in electricity grids. With approximately 16 GWs of PV already installed in the United States through 2014, over a million metric tons of PV module waste would be generated over the next 25 years in this country alone. Meeting the SunShot targets will cause a significant new waste source to the existing waste stream.

The concerns about PV waste are caused by issues associated with electronic waste (e-waste). The Organization for Economic Co-operation and Development (OECD) found e-waste to be one the fastest-growing waste streams in the Unites States. PV waste has the same characteristics as e-waste, that is, it is a combination of valuable materials and toxic ones. It contains materials of high value (e.g., silver, indium, tellurium, and gallium), toxic materials (e.g., cadmium lead, selenium compounds), and a large amount of energy-intensive materials (e.g., glass, copper, aluminum, and solar-grade silicon). We examined the potential of PV in a prospective LCA, focusing on direct costs, resource availability, and environmental impacts, and showed that PV recycling will become increasingly important in resolving cost, resource, and environmental constraints to large scales of sustainable growth. Recycling helps to keep the cost down by creating a significant secondary source and resolves concerns regarding waste disposal (Figure 7.32). A major challenge is setting up cost-effective collection infrastructure; this has been studied by Choi and Fthenakis.[27]

7.4 The Growth of PV is Sustainable and Greatly Needed

Solar energy is a viable alternative to fossil fuel-based energy generation for addressing climate change, meeting growing energy needs, and replacing aging power infrastructure in the United States and many other countries. Solar in a mixture of mostly renewable energy generation can replace the current power infrastructure of the United States. However, solar energy technologies are still evolving, and in most areas their costs are higher than the costs of conventional energy sources. We identified potential roadblocks for solar energy development in the United States in the near future. We show that there is a need for a holistic approach including social and environmental, in addition to economic, considerations, and we discuss policy options for supporting the continuation of PV market growth when the current ITCs expire.

Considerations of social experience and regulatory framework can justify implementing public policy options toward developing solar energy. Different examples of successful policies in the world toward this solar energy goal share a common trait; such policy instruments are implemented through long-term incentive programs. In that sense, US policy makers need to develop the right policy mix for promoting solar energy while sustaining social and political support for it. To ensure sustained growth especially beyond 2017 when ITC program for solar projects is scheduled to expire, investors should be given a variety of incentives for increasing the use of solar energy, such as long-term power contracts, low-cost financing options, incentives for R&D, and credits for displacing CO_2 and toxic emissions in power generation. Part of the full accounting of energy costs lies in the economic determination of externalities so that they are defined, managed, and regulated.

Solar PV should be one of the key components of a transition strategy to a world with stabilized atmospheric CO_2. The United States has incentives to be the leading country in this trend. PV has accomplished great cost reductions, and this trend will continue during the current decade; most importantly, PV does not involve the safety and environmental issues that the other alternative energy resources possess. Besides, investing in solar PV technologies expands the green tech space in the country and eventually helps to eliminate the external costs of fossil fuels. High capital expenditure requirements and safety issues with CCS and nuclear reactors prevent power producers or financial players from investing into such technologies unless revolutionary technological breakthroughs are achieved. Natural gas is considered as an economically viable option and relatively more reliable form of energy. However, natural gas would cut the GHG emissions by less than half compared with coal-fired generation, and it does not offer a long-term solution. Considering the growth in energy needs globally, long-term solutions to decarbonize the energy world should be put forward immediately, not a decade or a couple of decades later.

The readily available potential of solar energy, together with wind and other renewable resources, is far in excess of our energy needs. However, serious investment in solar energy needs to continue to sustain the current trend of cost reductions and grid penetration. Only a sustained policy mix covering all levels of the solar industry will make it the dominant energy source of the 21st century. Governments around the world can and should justify the incentives given to PV by communicating its hidden benefits compared with the external costs of conventional electricity generation.

Self-Assessment Questions

Q7.1 The EBPT of CdTe PV systems in Germany where the average solar irradiation is $1100\,kWh/m^2/year$ is 1.4 years; what would be the EPBT in the Atacama desert of Chile where the latitude tilt solar irradiation is $3000\,kWh/m^2/year$?

Q7.2 What is the energy return on energy investment (EROI) in the above two locations?

Q7.3 How does the land use of PV life cycles compare with that in coal life cycles in the United States?

Q7.4 How does the water use in PV life cycles compare with that in coal life cycles in the United States?

Q7.5 Are emissions of cadmium a problem when CdTe modules encapsulated in glass are engulfed in fires?

Q7.6 What is the source of cadmium and why does sequestering cadmium in environmentally protected products makes sense?

Q7.7 What (if any) are the likely limitations in material availability for TW-scale commercial production of Si, CdTe, and CIGS solar cells?

Q7.8 What type of costs are referred as "external costs" in electricity life cycles and why?

Problems

7.1 If you borrow$10 000 to invest in a PV system, find the amortization costs corresponding to the capital recovery factors for various interest rates shown in the following table.

Term (year)	3%	4%	5%	5.5%	6%
15			0.09634	0.09963	0.10296
20			0.08024	0.08368	0.08718
30			0.06505	0.06881	0.07265

Capital recovery factor {$CRF(i,n)$}. If we take a loan of $\$P$ at interest rate i with a loan term of n years m, then the annual payment would be $A = P \times CRF(i,n)$ where $CRF(i,n) = i(1+i)^n/((1+i)^n - 1)$.

7.2 A 5 kW (DC-rated) PV residential system in New York costs \$2.0/W. If you borrow the money at 4% on a 20-year loan, find the cost of electricity generated if is installed on a south-facing roof with tilt $= L - 15$ and we assume a derated factor of 0.75.

7.3 In the previous example, assume that you are eligible for a federal 10% state tax credit (max \$2000) and a state rebate of \$0.40/W. Also you are in the 30% marginal tax bracket and you can deduct the cost of interest from your tax return. Find the cost of electricity.

7.4 Recycling of CdTe PV modules currently costs about \$2/module, but it could even be profitable and bring about \$0.60/module if the recoveries of Te and glass are optimized. Assuming 120 W modules and lifetime of 30 years, what is the net present value (NPV) of the recycling cost or profit for a 50 MW PV power plant?

7.5 What would be the savings in emissions of CO_2, SO_2, NO_x, and particulates with every kWh of PV electricity displacing grid electricity in your area?

7.6 What would be the water savings with every kWh of PV electricity displacing thermoelectric power generation in your area?

7.7 The Cumulative Energy Demand (CED) for producing multi-crystalline Si PV modules and the associated with roof-top Balance of System (BOS) is about 2700 MJ per m^2 of installation. Assuming 17% efficient modules, calculate the EPBT and the EROI of deploying the system in the south of Germany with insolation (at latitude tilt) of 1300 $kWh/m^2/yr$ and in the south of Greece with insolation of 2000 $kWh/m^2/yr$. (Use a factor of 0.32 to convert primary energy to electricity).

7.8 Assuming the same CED, calculate the EPBT in Las Vegas, Nevada and in Chile's Atacama desert, for one-axis tracking installations (use formulas in chapter 3 to determine the electricity produced by each system).

7.9 What would be the savings in water demand if 100 GW of coal power generation in the US Southwest is displaced with PV? Assume coal power plant capacity of 0.8 and PV 1-axis tracking plant capacity of 0.3.

7.10 Estimate the external environmental costs of electricity derived by coal in the United States by using data presented by well-documented sources, for example by: Paul Epstein et al., "Full cost accounting for the life cycle of coal", Annals of the New York Academy of Sciences, 1219: 73–90, 2011.

Answers to Questions

Q7.1 it will be proportionally lower, thus 0.5 yrs.

Q7.2 For a life expectancy of 30 yrs, EROI would be 20 and 55 correspondingly.

Q7.3   PV uses less land than the coal life cycle in the US because of surface mining.

Q7.4   Thermoelectric power generation uses thousands of times more water than PV.

Q7.5   No, because Cd is sequestered within the molten glass.

Q7.6   It is produced as waste in the production of Zn and if not safely sequestered it would cause a serious disposal management issue.

Q7.7   Ag in Si cells, Te in CdTe, In and Ga in CIGS.

Q7.8   Environmental and social costs that are not accounted for in the price of electricity but are paid by the general public.

References

1 V.M. Fthenakis. metrics for extending thin-film photovoltaics to terawatt levels. *MRS Bulletin*, 37(4), 425–430 (2012).
2 V.M. Fthenakis *et al.* Nitrogen trifluoride emissions from photovoltaics: a life-cycle assessment. *Environmental Science and Technology*, 44(22), 8750–8757 (2010).
3 M. Campbell. The Economics of PV Systems, in A. Reinders *et al.* (eds). *Photovoltaic Solar Energy from Fundamentals to Applications*, John Wiley & Sons, Ltd: Chichester (2017).
4 M. Morjaria *et al.* A grid-friendly plant the role of utility-scale photovoltaic plants in grid stability and reliability. *IEEE Power and Energy Magazine*, 12(3), 87–95 (2014).
5 C. Sener and V. Fthenakis. Energy policy and financing options to achieve solar energy grid penetration targets: accounting for external costs. *Renewable and Sustainable Energy Reviews*, 32, 854–868 (2014).
6 K. Burrows and V. Fthenakis. Glass needs in a growing PV industry. *Solar Energy Materials and Solar Cells*, 132, 455–459 (2015).
7 V.M. Fthenakis and H.C. Kim. Land use and electricity generation: a life cycle analysis. *Renewable and Sustainable Energy Reviews*, 13, 1465–1474 (2009).
8 V.M. Fthenakis and H.C. Kim. Life-cycle of water in U.S. electricity generation. *Renewable and Sustainable Energy Reviews*, 14, 2039–2048 (2010).
9 E. Leccisi *et al.* The energy and environmental performance of ground-mounted photovoltaic systems: a timely update. *Energies*, 9(8), 622 (2016).
10 V. Fthenakis *et al.* Methodology Guidelines on Life Cycle Assessment of Photovoltaic Electricity, 2nd edition, International Energy Agency, Report IEA-PVPS T12-03:2011, November 2011.
11 M. Raugei *et al.* The energy return on energy investment (EROI) of photovoltaics: methodology and comparisons with fossil fuel life cycles. *Energy Policy*, 45, 576–587 (2012).
12 M. Carbajales-Dale *et al.* Energy return on investment (EROI) of solar PV: an attempt at reconciliation. *Proceedings of the IEEE*, 103(7), 995–999 (2015).
13 V. Fthenakis *et al. Life cycle inventories and life cycle assessments of photovoltaic systems, International Energy Agency*, Report IEA-PVPS T12-02:2011. October (2011).

14 V. Fthenakis. Considering the total cost of electricity from sunlight and the alternatives. *Proceedings of the IEEE*, 103(3), 283–286 (2015).

15 V.M. Fthenakis *et al.* Emissions and encapsulation of cadmium in CdTe PV modules during fires. *Progress in Photovoltaics: Research and Applications*, 13, 713–723 (2005).

16 V.M. Fthenakis and H.C. Kim. CdTe photovoltaics: life-cycle environmental profile and comparisons. *Thin Solid Films*, 515, 5961–5963 (2007).

17 V.M. Fthenakis and P.D. Moskowitz. Photovoltaics: environmental, safety and health issues and perspectives. *Progress in Photovoltaics: Research and Applications*, 8, 27–38 (2000).

18 V. Fthenakis (ed.). *Proceedings of the Pb-Free Solder Technology Transfer Workshop*, Vail, CO, Brookhaven National Laboratory Report, October 19, (1999). https://www.bnl.gov/isd/documents/21391.pdf (Accessed on August 21, 2017).

19 E. Ngai and V.M. Fthenakis. *Silane Safety Seminar, 33rd IEEE Photovoltaic Specialists Conference*, May 11–16, 2008, San Diego, CA (2008). https://www.bnl.gov/pv/files/pdf/IEEE_May2008_Silane_Tutorial.pdf (Accessed on August 21, 2017).

20 V.M. Fthenakis. Prevention and control of accidental releases of hazardous materials in PV facilities. *Progress in Photovoltaics: Research and Applications*, 6, 91–98 (1998).

21 V. Fthenakis *et al.* Toxicity of CdTe, CIS and CGS. *Progress in Photovoltaics: Research and Applications*, 7, 489–497 (1999).

22 V.M. Fthenakis and P.D. Moskowitz. Thin-film photovoltaic cells: health and environmental issues in their manufacture, use and disposal. *Progress in Photovoltaics: Research and Applications*, 3(5), 295–306 (1995).

23 V. Fthenakis. Multilayer protection analysis for photovoltaic manufacturing facilities. *Process Safety Progress*, 20(2), 1–8 (2001).

24 V.M. Fthenakis and H.-C. Kim. Life cycle assessment of high-concentration PV systems. *Progress in Photovoltaics: Research and Applications*, 21(3), 379–388 (2013).

25 W. Wang and V.M. Fthenakis. Kinetics study on separation of cadmium from tellurium in acidic solution media using cation exchange resin. *Journal of Hazardous Materials*, B125, 80–88 (2005).

26 V.M. Fthenakis and W. Wang. Extraction and separation of Cd and Te from cadmium telluride photovoltaic manufacturing scrap. *Progress in Photovoltaics: Research and Applications*, 14, 363–371 (2006).

27 (a)J.K. Choi and V.M. Fthenakis. Design and optimization of photovoltaics recycling infrastructure. *Environmental Science and Technology*, 44(22), 8678–8683 (2010); (b)J.K. Choi and V.M. Fthenakis. Economic feasibility of photovoltaic module recycling: survey and model. *Journal of Industrial Ecology*, 14(6), 947–964 (2010); (c)J.K. Choi and V. Fthenakis. Crystalline silicon photovoltaic recycling planning: macro and micro perspectives. *Journal of Cleaner Production*, 66(1), 443–449 (2014).

28 S. Kim and B.E. Dale, Life cycle inventory information of the United States electricity system. *International Journal of Life Cycle Assessment*, 10, 294–310 (2005).

29 V. Fthenakis and H.C. Kim, Greenhouse gas emissions from solar electric and nuclear power: a life cycle study. *Energy Policy*, 35, 2549–2557 (2007).

Index

Electricity from Sunlight: Photovoltaic-Systems Integration and Sustainability, Second Edition.
Vasilis Fthenakis and Paul A Lynn.
© 2018 John Wiley & Sons Ltd. Published 2018 by John Wiley & Sons Ltd.
Companion website: www.wiley.com/go/fthenakis/electricityfromsunlight